Foundations of Quantum Cosmology

AAS Editor in Chief

Ethan Vishniac, Johns Hopkins University, Maryland, USA

About the program:

AAS-IOP Astronomy ebooks is the official book program of the American Astronomical Society (AAS), and aims to share in depth the most fascinating areas of astronomy, astrophysics, solar physics and planetary science. The program includes publications in the following topics:

GALAXIES AND COSMOLOGY

INTERSTELLAR MATTER AND THE LOCAL UNIVERSE

STARS AND STELLAR PHYSICS

EDUCATION, OUTREACH, AND HERITAGE

HIGH-ENERGY PHENOMENA AND FUNDAMENTAL PHYSICS

THE SUN AND THE HELIOSPHERE

THE SOLAR SYSTEM, EXOPLANETS, AND ASTROBIOLOGY

LABORATORY ASTROPHYSICS, INSTRUMENTATION, SOFTWARE, AND DATA

Books in the program range in level from short introductory texts on fast-moving areas, graduate and upper-level undergraduate textbooks, research monographs and practical handbooks.

For a complete list of published and forthcoming titles, please visit iopscience.org/books/aas.

About the American Astronomical Society

The American Astronomical Society (aas.org), established 1899, is the major organization of professional astronomers in North America. The membership (~7,000) also includes physicists, mathematicians, geologists, engineers and others whose research interests lie within the broad spectrum of subjects now comprising the contemporary astronomical sciences. The mission of the Society is to enhance and share humanity's scientific understanding of the universe.

Foundations of Quantum Cosmology

Martin Bojowald
Department of Physics, 104 Davey Lab, University Park, PA 16802, USA

IOP Publishing, Bristol, UK

ISBN   978-0-7503-2460-1 (ebook)
ISBN   978-0-7503-2458-8 (print)
ISBN   978-0-7503-2461-8 (myPrint)
ISBN   978-0-7503-2459-5 (mobi)

DOI   10.1088/2514-3433/ab9c98

Version: 20200901

AAS–IOP Astronomy
ISSN 2514-3433 (online)
ISSN 2515-141X (print)

British Library Cataloguing-in-Publication Data: A catalogue record for this book is available from the British Library.

Published by IOP Publishing, wholly owned by The Institute of Physics, London

IOP Publishing, Temple Circus, Temple Way, Bristol, BS1 6HG, UK

US Office: IOP Publishing, Inc., 190 North Independence Mall West, Suite 601, Philadelphia, PA 19106, USA

Contents

Preamble

Astronomy and cosmology deal with the production, propagation, and detection of various signals in spacetime. According to general relativity, spacetime is not a rigid stage but is curved and expanding, exhibiting its own dynamics subject to physical laws. At a fundamental level, physical laws are quantum. Astronomy and cosmology should therefore be based on a theory of quantum cosmology. However, simple dimensional arguments suggest that any physical quantum-cosmology effects that could not be well approximated by classical general relativity would happen only at extreme scales. As already noticed by Max Planck, it is possible to combine the fundamental constants c (the speed of light), G (Newton's constant of gravity), and $\hbar$ (Planck's constant) and obtain a unique length parameter, the Planck length $\ell_P = \sqrt{\hbar G/c^3} \approx 1.6 \times 10^{-35}$ m. Such tiny distances are currently irrelevant in any lab experiment, let alone cosmology. Similarly, the Planck density $\rho_P = c^5/(\hbar G^2) \approx 5.2 \times 10^{96}$ kg m^{-3} is huge, the equivalent of more than one trillion solar masses contained in a proton-sized region.

At present, and for the foreseeable future, these scales play no role in observationally accessible regimes. Nevertheless, they may have indirect implications, at the very least on a conceptual level. Among the scientific fields, cosmology most directly studies phenomena that were a result of extreme densities in their past. If we use general relativity as a possible theoretical description of cosmology, the energy density of matter in the universe was even infinite at some time in the past, the time of the Big Bang singularity. Subsequent events have shrouded this very early phase from direct observational access because its high temperature ionized matter to a plasma through which electromagnetic waves could not travel freely. Nevertheless, the Big Bang scenario postulates that the visible universe originated in a state of high energy density which, in the absence of any theoretical upper bound in a singular scenario, may well have been as large as or greater than the Planck density. In addition, signals that reach us from the early universe, directly or indirectly, by necessity travel far through spacetime. Even tiny quantum corrections might therefore add up to produce a sizeable effect. Similarly, the current universe is large, and if many small numbers add up coherently over a large volume, a recognizable phenomenon might be produced. Even puzzling late-time ingredients of cosmology, such as dark energy, might therefore benefit by input from quantum cosmology.

Quantum cosmology is expected to be a part of quantum gravity, or some theory that includes a quantum description of general relativity. No such theory is known to date, although several often widely different proposals exist. Because a quantum theory of gravity must address the fundamental question of how spacetime can be described in a quantized way, perhaps subject to quantum jumps and fluctuations, all current proposals are mathematically and conceptually challenging. Moreover, given the extreme scales on which such a theory could be relevant, there are no tests or suggestions from laboratory experiments. Nevertheless, it is important to develop such a theory to a good degree of reliability, even in the absence of current observations. After all, we not only need observations to test a given theory, we also need a good theory to suggest places where we should start looking for relevant observations (or to show us how data from existing observations can be used in new evaluations).

Based on this motivation, various evaluations of proposals of quantum gravity have been performed, also elaborating them into phenomenological scenarios. If one uses qualitative suggestions from proposed theories of quantum gravity, rather than strict derivations, the main mathematical and conceptual challenges can be evaded. In this spirit, like the myths of ancient cosmology, myriad cosmological models have been sprouting like weeds from the fertile soil of primordial matter. There are, therefore, suggestions for observers to start looking for quantum-gravity effects, but the overwhelming flood of often diverging proposals, worked out at different and not always clearly specified levels of rigor, can be paralyzing.

In this book, we will therefore aim to keep close contact with the general foundations of quantum cosmology, maintaining a level of physical rigor that can already be used to rule out some of the more popular proposals of early-universe scenarios. An important consistency condition is given by general covariance, to which an entire chapter is devoted, Chapter 2. The condition of general covariance allows one to classify different consistent modifications of general relativity that may be relevant for quantum cosmology, or for possible descriptions of late-time effects such as dark energy.

It is impossible to discuss quantum cosmology without a certain basis of general relativity. Nevertheless, an attempt has been made in this book to replace, as much as possible, formal considerations in differential geometry by physical reasoning. When necessary, the derivation of key mathematical ingredients of general relativity, such as curvature tensors, is provided in the text. In these and other derivations, the usual convention of omitting summation symbols over index pairs has not been applied, writing all such symbols explicitly. Moreover, all fundamental constants, including the speed of light c and Newton's constant G, are spelled out explicitly and have not been "set equal to one" by a choice of units.

In Chapter 1, we will encounter the main models used in theoretical cosmology. We will derive the Friedmann equation based on general physics principles and the canonical formalism, avoiding at this initial stage the usual differential-geometric concepts of curvature as well as the action principle of general relativity, or Einstein's equation. Important aspects of general covariance, such as reparameterizations of time coordinates or averaging volumes, will play an important role in this derivation. In addition to being independent of more advanced topics of differential geometry, this new derivation has the advantage of showing how tightly constrained the dynamics of cosmological models is by the requirement of general covariance and related physical properties. This insight will be a prelude to the challenges encountered when one tries to modify cosmological equations by including quantum corrections while maintaining the important consistency condition of general covariance.

The same chapter will also show us more involved models, such as anisotropic and spherically symmetric ones, as well as several cosmological applications such as inflation and some basics of black holes. Although the full dynamics of spherically symmetric models will be used in Chapter 1, its derivation is postponed to the second part of Chapter 2. This part will show how the dynamics of spherically symmetric models, as well as those of perturbative inhomogeneity, can be uniquely derived from the principle of covariance. Action principles can again be avoided and replaced by the canonical formalism, although for such field-theory models the latter is more involved than in isotropic models. The first part of Chapter 2 will, however, develop important concepts of curvature and action principles in a more standard form, but without assuming a heavy general-relativity background. In this part, higher-curvature actions of modified gravity will be shown, including a discussion of stability conditions and the general form of Horndeski theories. Chapter 2 can therefore be considered a summary of modified gravity theories consistent with general covariance. By developing crucial ingredients, much can be learned about the possible structures of spacetime.

The rest of the book will include quantum aspects, first presenting the main mathematical properties and physical phenomena of quantum mechanics in Chapter 3. An emphasis will be placed on those ingredients most relevant for quantum cosmology. In particular, the underlying mathematics will be developed based on an algebraic perspective, focusing on properties and implications of non-zero commutators as the main new feature of quantum mechanics, with important physical implications via uncertainty relations.

Quantum cosmology requires a new perspective on quantum states because measurements are done differently in this setting compared with quantum mechanics in, say, atomic physics. In particular, it is not possible to prepare multiple versions of a universe in order to test statistical aspects of quantum mechanics. Moreover, it is unclear what the state of the universe, currently or in its earliest stages, can be assumed to be. Models of quantum gravity rarely provide distinguished states such as ground states because gravity is an inherently unstable phenomenon, as shown by the classical examples of collapse into compact objects or long-term expansion of the whole universe. In quantum cosmology, it is therefore more convenient to attempt derivations of general physical effects based on properties of operators for the main cosmological observables, largely independent of specific states that the universe might be in.

The combination of lessons learned in Chapter 3 from quantum mechanics will be combined with homogeneous cosmological models in Chapter 4. General covariance will again be an important ingredient which restricts possible quantum corrections and also helps when it comes to interpreting their behavior. Different approaches to quantum gravity will be foreshadowed by various representations of quantum mechanics in a cosmological setting. While covariance does play a decisive role in homogeneous cosmological models in spite of the restricted nature of spatial coordinate transformations in this symmetric setting, it will ultimately be seen that homogeneous quantizations cannot be sufficient as a reliable analysis of quantum effects in cosmology and potential observations.

It is therefore necessary to consider at least some inhomogeneous ingredients from full approaches to quantum gravity, which will be exhibited in Chapter 5.

The great variety of different proposals, as well as possible phenomena in each approach, will become clear. Whenever possible, points of contact between different frameworks will be pointed out, but these will mainly be at a technical level or at the level of broader conceptual assumptions. (An example of the latter is the use of imaginary time in various independent calculations.) As in previous chapters, the consistency conditions imposed by general covariance will be emphasized, as well as different degrees to which they have been implemented in the given approaches. The main focus will be on the two currently most successful proposals, causal dynamical triangulations and string theory.

The final chapter, Chapter 6, applies some of the effects suggested by technical constructions in quantum gravity to a setting of inhomogeneous cosmology. The first part describes different versions of "instructive failures" consisting of attempts to amend quantizations of homogeneous cosmological models by coupling them to perturbative inhomogeneity with an aim at studying cosmological structure formation. These proposals, made mainly in the context of loop quantum cosmology, are instructive because they indicate how simple results from the homogeneous setting could play out in a more realistic, inhomogeneous context. However, they are failures because it can be shown that they do not lead to covariant descriptions of spacetime.

These proposals therefore provide useful examples that show how important it is to keep an eye on mathematical consistency conditions even in phenomenological investigations of early-universe cosmology. Some proposals can already be ruled out even before one computes a power spectrum. The remainder of Chapter 6 then continues with a description of consistent spacetime structures that are covariant, but in a generalized form that includes quantum corrections. The final part introduces further consistency conditions of a general conceptual nature, given by infrared renormalization and the trans-Planckian problem, and then presents examples of potential physical scenarios that could explain cosmic inflation or provide alternatives.

This book does not present a single physical prediction of quantum cosmology. Nevertheless, it aspires to be deeply physical by emphasizing the importance of physical rigor in addition to the mathematical rigor that has often been attributed to theories of quantum gravity. The absence of any physical predictions in this book is a consequence precisely of the aimed-for physical rigor: While it is easy to use modified cosmological equations in simple homogeneous models and develop scenarios that might show a potential effect on the cosmic microwave background, it is much more difficult to demonstrate that such a scenario is physically meaningful. To some degree, the physical description depends on the approach used, which is somewhat variable in quantum gravity. Nevertheless, there are general physics principles that any approach should respect, such as general covariance, a meaningful effective field theory at low energy or curvature, and at least some control on approximations such as the assumption of spatial homogeneity in simple cosmological models. By developing these conditions and maintaining an emphasis on them, this book aims to lay the physical foundations of quantum cosmology.

Some of the material in this book is based upon work supported by the National Science Foundation under Grant Nos. PHY-1607414 and PHY-1912168. Any opinions, findings, and conclusions or recommendations expressed in this material are those of the author and do not necessarily reflect the views of the National Science Foundation.

Author biography

Martin Bojowald

Martin Bojowald is a Professor of Physics at The Pennsylvania State University. He is the founder of loop quantum cosmology and of dynamical black-hole models in loop quantum gravity. More recently, his research has turned to a variety of questions in different areas, such as covariance in canonical gravity, semiclassical approximations and effective potentials, as well as non-associative quantum mechanics. These methods have found applications not only in loop quantum gravity but also in other approaches such as causal dynamical triangulations, the no-boundary proposal, and group-field cosmology. His work has earned the First Award in the Gravity Research Foundation Essay Competition (2003), the Xanthopoulos Prize from the International Society on General Relativity and Gravitation (2007), a CAREER Award from the National Science Foundation (2008), a Teaching Award from the Penn State Society of Physics Students (2009), the Faculty Scholar Medal in the Physical Sciences from The Pennsylvania State University (2011), and a Distinguished Professor Lectureship from the Mexican Academy of Sciences (2017). He has been named an Outstanding Reviewer by several scientific journals or organizations, including the American Physical Society (2010), Elsevier journals (several times since 2010), Proceedings of the Royal Society (2017), European Physics Journal (2017), and Universe (2019). He currently serves on the editorial board of SIGMA, Frontiers in Physics (Cosmology) and Physical Review D. His publications include articles in general-audience journals such as Scientific American and Physics Today, as well as the popular-level book Once Before Time.

Foundations of Quantum Cosmology

Martin Bojowald

Chapter 1

The Universe on Large and Small Scales

Physical cosmology is possible because the visible universe appears simple enough for a theoretical description: Observations, mainly by the Sloan Digital Sky Survey (SDSS, Blanton et al. 2017) have confirmed that the matter distribution on large scales, that is, beyond the scale of galaxy clusters, is nearly homogeneous. To a first approximation, its dynamics can therefore be described by a perturbative approach in which the complicated non-linear partial differential equations of general relativity are linearized around a homogeneous background solution. The latter evolves according to a non-linear but ordinary differential equation, the Friedmann equation (Friedmann 1922). Suitable combinations of the fields that appear in a linear perturbation are related to spatial variations of the matter density or spacetime curvature, which can be observed by various means through the effects they have on the propagation of electromagnetic waves.

The attractive nature of the gravitational force explains why the universe is *not* homogeneous on small scales: If we assume some initial matter distribution, overdense regions attract more matter than underdense regions, and therefore grow in mass and eventually collapse through gravitational attraction, becoming even more dense. The behavior of gravity therefore does not provide a simple reason as to why the visible universe should be nearly homogeneous on large scales; it would rather suggest the opposite.

A heuristic argument for large-scale homogeneity is given by models of cosmic inflation (Guth 1981; Linde 1982), which postulate a rapid, accelerated phase of expansion during some brief period in the very early universe. Such a rapid expansion quickly enlarges any tiny region within some initial distribution, to the extent that it becomes larger than the part of the universe that can at present be accessed by observations. If the initial distribution is sufficiently regular, the tiny region out of which the entire visible universe is supposed to have grown does not leave much room for any complicated structure. At the end of inflation, this region would no longer be tiny but still very nearly uniform. During billions of years

doi:10.1088/2514-3433/ab9c98ch1

afterwards, gravity then acts and forms structure, but not beyond the scale of galaxy clusters.

There is no good reason to suppose that the matter distribution before inflation should have been homogeneous on scales larger than the initial tiny seeds of an inflating region. (There may even have been inhomogeneous structure on all scales, including those less than the initial seeds, if the original distribution was fractal, as suggested by Penrose 1990, an observation that may challenge models of inflation.) If we view the universe in time reverse, the early universe is seen in rapid collapse. Different regions filled with matter approach one another, making it easier for gravity to perform its structure-forming role. Such rapid structure formation suggests a complicated initial state beyond the scale of the tiny region enlarged by inflation. The universe between the Big Bang singularity and the onset of inflation may, for all we know, have been in a very complicated shape, too complicated for current theoretical approximations to be able to describe it well.

However, another, surprising reason for homogeneity on small scales in the very early universe has been obtained from mathematical investigations undertaken by Belinskii, Khalatnikov and Lifshitz (BKL; Belinskii et al. 1982): Analyzing Einstein's equation asymptotically close to a spacelike singularity, such as the Big Bang, they found that different points in space seem to decouple from one another, in the sense that the most significant contributions to Einstein's partial differential equations in this regime are associated with time derivatives of the fields, while spatial derivatives are subdominant. While the overall spatial distribution remains highly inhomogeneous and complicated, its time dependence at any given point is equivalent to the evolution of a homogeneous distribution, described by ordinary differential equations. Based on the BKL scenario, there is therefore a reasonable expectation that even the initial state before inflation, or at the Big Bang singularity, may be explainable by simple models at least in terms of its dynamics.

However, the universe at these early times was very dense and hot and curved, infinitely so if we extrapolate Einstein's equations all the way up to the Big Bang singularity they imply. Quantum physics of *anything*, including spacetime and matter, should then be required. In this chapter, we set the stage for such a theory by providing the relevant classical equations of theoretical cosmology, starting from simple homogeneous and isotropic models and moving on to more and more complicated distributions. In this process, we will observe several consistency relations related to the fundamental symmetry of general covariance. In the absence of direct observations of any regime that requires a quantum formulation of cosmology, such consistency conditions are crucial in order to narrow down the otherwise vast set of possible theoretical descriptions.

1.1 Homogeneous and Isotropic Cosmology

Cosmological observations always refer to a finite region around us, bounded by the distance a light ray can travel during the age of the universe. Currently, all messengers used in this context, electromagnetic and gravitational waves, move at the speed of light, and therefore their sources are located on a subset of spacetime swept out by our past light cones during the era of modern cosmology. Since this period is very short compared with the whole age of the universe, the subset may be considered a single past light cone. If we use large-scale homogeneity and isotropy, it is possible to approximate spacetime by a space of fixed topology with a geometry changing in time. (We will later demonstrate this feature in more detail, showing that it does not violate the relativity principle.) The observed past light cone can therefore be projected onto a finite spatial region which determines the farthest distances that have, at any given time, been explored; see Figure 1.1.

A finite cosmological region in an expanding universe has a time-dependent size. It is convenient to express the time dependence of an isotropic universe, which expands uniformly in all directions, through a single function, called the scale factor $a(\tau)$. If the volume of a given spatial region at one time, τ_1, is $V(\tau_1)$, at some other time, τ_2, it is given by

$$V(\tau_2) = \frac{a(\tau_2)^3}{a(\tau_1)^3} V(\tau_1). \tag{1.1}$$

The ratio $V_0 = V(\tau)/a(\tau)^3$ is therefore time-independent. It determines how the volume of a region depends on its extent or shape, independently of the expansion rate. If we measure the volume at any given time, τ, we obtain the value

$$V(\tau) = V_0 a(\tau)^3. \tag{1.2}$$

However, neither V_0 nor $a(\tau)$ can be measured independently; they merely represent convenient quantities for a mathematical description. Their values depend on the choice of units or coordinates made in a geometrical measurement: If we change units such that the previous V_0 is replaced by λV_0 with some $\lambda > 0$, for instance $\lambda = 1 \text{ in}^3/1 \text{ cm}^3$, the actual size of our region should not change. Therefore, for $V(\tau)$ to be independent of λ, $a(\tau)$ should be changed to $a(\tau)/\lambda^{1/3}$ under the rescaling.

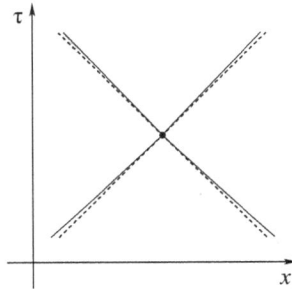

Figure 1.1. Light cone of an observer in an expanding universe. Expanding space drags light rays along with it, deforming the light cone of straight lines in Minkowski spacetime (dashed). The coordinates are x for one spatial dimension and τ for proper time measured by the observer. Cosmological observations show only the behavior of the universe on our past light cone, but not in its interior. The common assumption of approximate homogeneity through the entire spatial region enclosed by the past light cone at any time in the past is therefore an extrapolation.

There should be *some* measurable information contained in the time dependence of $V(\tau)$ or $a(\tau)$ because we can detect the expansion of the universe. One example of a measurable quantity is the ratio $a(\tau_2)/a(\tau_1)$ at two different times: If we replace each $a(\tau)$ with $a(\tau)/\lambda^{1/3}$, the ratio does not change because λ is time-independent. An infinitesimal version of such a ratio is the Hubble parameter

$$H(\tau) = \frac{1}{a(\tau)}\frac{da(\tau)}{d\tau}. \tag{1.3}$$

As a consequence of Einstein's equation specialized to homogeneous and isotropic space, it is related to the energy density of matter, ρ, by the Friedmann equation

$$H^2 = \frac{8\pi G}{3c^2}\rho \tag{1.4}$$

(assuming, for now, that space is not curved).

Non-relativistic matter, for instance a cold chunk of mass M within our region of size V_0, measured in some fixed coordinates and units, has a time-dependent energy density, which decreases owing to dilution in an expanding universe. Because the volume of any region increases in time through a factor of $a(\tau)^3$, the energy density ρ is then equal to a time-independent constant, the energy Mc^2, divided by the volume $V_0 a(\tau)^3$,

$$\rho = \frac{Mc^2}{a^3 V_0}. \tag{1.5}$$

For this kind of matter, we can interpret the Friedmann equation in the form of a Newtonian energy-balance law,

$$-\frac{3}{8\pi}\left(V_0^{1/3}\frac{da}{d\tau}\right)^2 + \frac{GM}{aV_0^{1/3}} = 0, \tag{1.6}$$

to which matter contributes through Newton's potential $GM/(aV_0^{1/3})$ with the distance between two masses replaced by the cosmological distance $aV_0^{1/3} = V^{1/3}$, and a single point mass replaced by the mass M contained in the entire volume at any given time. Moreover, the expansion rate of space, $V_0^{1/3}da/d\tau$, contributes a "kinetic energy" with, rather surprisingly, a negative sign, and the two energies always compensate each other. The latter two features cannot be explained through Newtonian mechanics, but they follow from general relativity and will play important roles in our later discussion of spacetime structure and quantum effects.

The Friedmann equation for non-relativistic matter is solved by

$$a(\tau)V_0^{1/3} = \frac{3}{2}(\sqrt{8\pi GM/3}\; c(\tau - \tau_\infty))^{2/3} \tag{1.7}$$

with an integration constant τ_∞. At $\tau = \tau_\infty$, we have $a(\tau_\infty) = 0$ and therefore $\rho(\tau_\infty) = \infty$ for Equation (1.5). An infinite energy density is not physical, and therefore the solution has a (Big Bang) singularity. However, one may expect that large or infinite densities should be associated with such high temperatures that matter constituents move very fast just by thermal motion, and are no longer non-relativistic. We might even expect that matter gets ionized and forms a plasma, or nuclei disintegrate into a quark–gluon plasma. The energy density of matter at such high temperatures does not follow the simple dilution behavior, Equation (1.5). For instance, a large contribution to the energy density from electromagnetic waves produced by thermal radiation is redshifted in an expanding universe in addition to being diluted, such that their energy density obeys $\rho \propto a^{-4}$. It is then no longer legitimate to extrapolate to early times using Equation (1.6).

Thermodynamics provides us with a way to solve equations for both $a(\tau)$ and $\rho(\tau)$ at the same time, without having to postulate the relationship $\rho(a)$. The energy $E(\tau)$ of a system, identified with the internal energy of thermodynamics, determines the energy density $\rho(\tau) = E(\tau)/V(\tau)$ in a region of volume $V(\tau) = V_0 a(\tau)^3$. During any process, which here is given by the expansion of space acting on the matter it contains, $E(\tau)$ changes according to the first law, $dE = -P dV$, if the pressure is P.

Since our process unfolds in time, the first law implies a differential equation,

$$\frac{dE}{d\tau} = -P\frac{dV}{d\tau},\tag{1.8}$$

which we can rewrite as a differential equation for $\rho(\tau)$ coupled to $a(\tau)$, using $E(\tau) = V_0\rho(\tau)a(\tau)^3$:

$$\frac{d(\rho a^3)}{d\tau} = -P\frac{da^3}{d\tau}\tag{1.9}$$

or

$$\frac{d\rho}{d\tau} + 3H(\rho + P) = 0,\tag{1.10}$$

called the continuity equation.

The Friedmann and continuity equations can be combined to obtain a second-order differential equation for $a(\tau)$, called the Raychaudhuri equation. Taking a derivative on both sides of Equation (1.4), we obtain

$$2H\frac{dH}{d\tau} = \frac{d(H^2)}{d\tau} = \frac{8\pi G}{3c^2}\frac{d\rho}{d\tau} = -\frac{8\pi G}{c^2}H(\rho + P),\tag{1.11}$$

or

$$\frac{dH}{d\tau} = -\frac{4\pi G}{c^2}(\rho + P).\tag{1.12}$$

Since

$$\frac{dH}{d\tau} = \frac{d}{d\tau}\left(\frac{1}{a}\frac{da}{d\tau}\right) = \frac{1}{a}\frac{d^2a}{d\tau^2} - \left(\frac{1}{a}\frac{da}{d\tau}\right)^2,\tag{1.13}$$

we can also write

$$\frac{1}{a}\frac{d^2a}{d\tau^2} = H^2 - \frac{4\pi G}{c^2}(\rho + P) = -\frac{4\pi G}{3c^2}(\rho + 3P). \qquad (1.14)$$

For most matter ingredients, $\rho + 3P \geqslant 0$, such that the expansion of the universe decelerates. We can easily see this for non-relativistic matter, which has pressure $P \approx 0$ negligible compared with its energy density. (The latter, in any relativistic treatment, always contains the huge rest energy density Mc^2/V.) Another example, radiation, has non-negligible pressure, but it is equal to $\frac{1}{3}\rho$ and therefore less than ρ and also positive. This result can be derived from Maxwell's equations, which describe the radiation field, or from the simple argument that radiation in an expanding universe is diluted and redshifted. The latter stretches the wavelength by a factor of a, reducing the frequency or the energy ($\hbar$ times the frequency according to quantum physics) by a factor of $1/a$ in addition to the dilution factor. Radiation therefore has an energy density $\rho(a) \propto a^{-4}$. The continuity equation, slightly rewritten as $a\,d\rho/da + 3(\rho + P) = 0$ then implies $P = \frac{1}{3}\rho$.

Two important problems of modern cosmology are related to the result (1.14):

- Measurements of the expansion history of the universe using type-I super-novae as standard candles indicate that the acceleration $a^{-1}d^2a/d\tau^2$ is currently positive (Riess et al. 1998; Perlmutter et al. 1991). As shown by Equation (1.14), this observation is consistent with general relativity only if the dominant matter ingredient in the current universe (called "dark energy") has strong negative pressure, such that $\rho + 3P < 0$. While negative pressure is not uncommon in laboratory situations, it is difficult to imagine how it could be sustained throughout the universe and be strong enough to overcome the energy density in the combination $\rho + 3P$. The negative pressure required for dark energy can be interpreted as an intrinsic tension of spacetime. As expanding space pulls on this tensed medium, it puts additional energy into it. The energy density of dark energy therefore decreases less in an expanding universe than would be expected from dilution. The behavior required to be consistent with supernovae observations is close to a time-independent energy density of dark energy, $\rho = \Lambda$, which is then called a *cosmological constant*.
- One mathematical possibility to avoid the Big Bang singularity is given by a "bounce" during which the scale factor $a(\tau)$ goes through a local minimum at a positive value (Novello & Bergliaffa 2008). However, at any local minimum of the function $a(\tau)$, the second derivative $d^2a/d\tau^2$ is non-negative, in contrast to what is implied by Equation (1.14) for standard matter. The Big Bang happens at much higher density than the current universe where dark energy seems to be required, making it more conceivable that the inequality $\rho + 3P \geqslant 0$ may be violated. Nevertheless, avoiding the Big Bang singularity by a bounce, or any other means, remains difficult to achieve in self-consistent theoretical models.

As quantum effects always present an opportunity to modify standard properties of matter or spacetime, we will return to these two major problems throughout this book.

1.2 Hamiltonian

With quantization in mind, it is useful to have a Hamiltonian formulation of a given system. So far, we have worked with the configuration variable a, which determines the time-dependent scale of the universe, and its time derivative $da/d\tau$. Neither a nor $da/d\tau$ are directly observable because their values can be changed at will by dividing by a time-independent constant, transforming a to $a/\lambda^{1/3}$. Moreover, the units of a and $da/d\tau$ are unspecified for the same reason, if we allow for dimensionfull λ. In order to have observable quantities with defined units, we should replace a and $da/d\tau$ with $V^{1/3} = V_0^{1/3}a$ and $dV^{1/3}/d\tau = V_0^{1/3}da/d\tau$, respectively.

The length scale $V^{1/3}$ has the usual units of a position variable, and $dV^{1/3}/d\tau$ is its velocity. For a Hamiltonian formulation, we need to replace the velocity with a momentum, but there is no obvious mass parameter associated to the variable $V^{1/3}$. (We should not use the mass parameter M in Equation (1.6) because it refers to the matter content of the universe rather than its extension.) Fortunately, as shown by the form of Newton's potential, we can use Newton's constant, G, in order to turn a length parameter into a mass, also using the speed of light: For the units, this relation implies $[E] = [GM^2/L]$ while we also have $[E] = [Mc^2]$ from special relativity. Eliminating $[E]$, these two equations imply that $[M] = [Lc^2/G]$.

Using $V^{1/3}$ to provide the units of length $[L]$, $V^{1/3}c^2/G$ is a mass parameter, and therefore

$$p_a = -\frac{3c^2}{4\pi G}V^{1/3}\frac{dV^{1/3}}{d\tau} \tag{1.15}$$

can serve as a momentum of $V^{1/3}$. The numerical factor of $3/(4\pi)$ follows conventions related to a Hamiltonian analysis of the action of general relativity restricted to homogeneous and isotropic spacetimes. A negative sign is included in p_a so as to make Hamilton's equations, shown below, consistent with the negative kinetic energy in Equation (1.6). In order to avoid the somewhat inconvenient exponents of $1/3$, we can apply a canonical transformation and work with V and its momentum

$$p_V = -\frac{c^2}{4\pi G}\frac{1}{a}\frac{da}{d\tau} = -\frac{c^2}{4\pi G}H. \tag{1.16}$$

The Hamiltonian of a system is a function of configuration and momentum variables and equals the energy. The Friedmann equation contains the energy density of matter, and can therefore be turned into an energy by multiplying it by the volume. Since the equation relates the matter energy density to the expansion rate, it tells us what the gravitational Hamiltonian should look like. We rewrite Equation (1.4) such that it is a combination of the matter energy $E_{\text{matter}} = V\rho$ and a gravitational contribution, such that all terms taken together are constant in order to ensure energy conservation:

$$E_{\text{matter}} - \frac{3c^2}{8\pi G}VH^2 = 0. \tag{1.17}$$

Since there is no free energy constant for a generic matter content, the only possibility to have a conserved energy is to require a vanishing value. Moreover, the equality of the matter energy density with the Hubble parameter squared, given by the Friedmann equation, implies that matter and gravity enter the energy balance with opposite signs: Gravity behaves like an unstable system with negative kinetic energy, making it possible to have ever-expanding solutions that do not settle down to a stable equilibrium.

In our final step, we rewrite the energy balance in terms of configuration and momentum variables,

$$H(V, p_V) = H_{\text{matter}}(V) - \frac{6\pi G}{c^2} V p_V^2, \tag{1.18}$$

where we have indicated that the matter energy or Hamiltonian H_{matter} usually depends on the volume, but not on the Hubble parameter or p_V. The gravitational contribution is non-standard not only in that it is negative, as already seen, but also because it is a quadratic function of the momentum multiplied with the configuration variable V.

As a simple test of this equation, we can compute Hamilton's equations for evolution with respect to a time coordinate τ, called *proper time*:

$$\frac{dV}{d\tau} = \frac{\partial H}{\partial p_V} = -\frac{12\pi G}{c^2} V p_V \tag{1.19}$$

is equivalent to $H = a^{-1} da/d\tau$ if we use Equation (1.16) as well as $V^{-1} dV/d\tau = 3a^{-1} da/d\tau$. The second Hamilton's equation,

$$\frac{dp_V}{d\tau} = -\frac{\partial H}{\partial V} = -\frac{\partial H_{\text{matter}}}{\partial V} + \frac{6\pi G}{c^2} p_V^2 \tag{1.20}$$

in which the first term is equal to the matter pressure according to thermodynamics, is equivalent to the Raychaudhuri equation.

In the Hamiltonian formulation, therefore, two equations, the Friedmann and Raychaudhuri equations, are implied by a single expression, the Hamiltonian H. As we will see later, this fact, which is deeply related to general covariance, makes it possible to implement quantum effects in a consistent manner: If we find quantum corrections of H, we can derive two mutually consistent equations that amend the classical Friedmann and Raychaudhuri equations by quantum terms. It would be more difficult to modify the Friedmann and Raychaudhuri equations independently and then make sure that the modified versions are mutually consistent.

1.3 Geometry

While homogeneous and isotropic equations used so far are well-justified for a large-scale description of the late universe, gravitational collapse in a dense cosmos implies that anisotropies and inhomogeneities should be an inescapable ingredient in any reliable description of the early universe. Instead of a single function of time,

$a(\tau)$, we then have several functions which depend on time as well as spatial coordinates. Geometry provides efficient means to organize these functions.

1.3.1 The Line Element

If we have a flat space at fixed time, we may choose Cartesian coordinates x^i, labeled by the superscript index $i = 1, 2, 3$, and write the volume V_0 of our finite region $\mathcal{V}$ in the preceding section as $V_0 = \int_{\mathcal{V}} \mathrm{d}^3 x$, with integration ranges depending on the shape and size of $\mathcal{V}$. The choice of coordinates should not matter for observable quantities such as $V(\tau) = V_0 a(\tau)$, which therefore must remain unchanged if we apply a generic coordinate transformation, mapping x^i to new coordinates $x^{i'}(x^i)$.

If this transformation is non-linear, it introduces coordinate-dependent functions in the integral that gives us V_0: The chain rule implies

$$\mathrm{d}x^{i'} = \sum_{j=1}^{3} \frac{\partial x^{i'}}{\partial x^j} \mathrm{d}x^j \tag{1.21}$$

such that, writing $\mathrm{d}^3 x = \mathrm{d}x^1 \mathrm{d}x^2 \mathrm{d}x^3$ as the determinant of a diagonal matrix with components $\mathrm{d}x^i \delta_{ij}$ in terms of the Kronecker delta δ_{ij}, we have

$$\mathrm{d}^3 x' = \det(\mathrm{d}x^{i'} \delta_{i'j'}) = \det\left(\frac{\partial x^{i'}}{\partial x^j}\right) \det(\mathrm{d}x^i \delta_{ij}) = \det\left(\frac{\partial x^{i'}}{\partial x^j}\right) \mathrm{d}^3 x. \tag{1.22}$$

If we transform from Cartesian to polar coordinates, for instance, we obtain the well-known sphere factor $r^2 \sin \vartheta$.

Not only the volume integration but also other geometrical quantities such as distances or areas are transformed when we use new coordinates. It is therefore convenient to define a single object that can describe geometrical properties and is unfazed by all these changes. It is called the *line element* $\mathrm{d}s^2$, and defined as the length of an infinitesimal displacement. In flat space with Cartesian coordinates, the Pythagorean theorem tells us that the line element is given by

$$\mathrm{d}s^2 = \mathrm{d}x^2 + \mathrm{d}y^2 + \mathrm{d}z^2 = \sum_{i=1}^{3} (\mathrm{d}x^i)^2. \tag{1.23}$$

A generic coordinate transformation does not preserve this form. It may introduce cross-terms such as $\mathrm{d}x^i \mathrm{d}x^j$ with $i \neq j$, as well as coordinate-dependent coefficients. However, the line element is always quadratic in the differentials $\mathrm{d}x^i$ because Equation (1.21) is linear in $\mathrm{d}x^i$. This algebraic property is taken as the definition of the line element in general terms, extending the mathematical expression (1.23) to non-Euclidean geometries:

$$\mathrm{d}s^2 = \sum_{i=1}^{3} \sum_{j=1}^{3} g_{ij}(x^k) \mathrm{d}x^i \mathrm{d}x^j. \tag{1.24}$$

The collection of all coefficients g_{ij} in matrix form, which may depend on the position x^k, defines the *metric*. In terms of the metric, we can write the volume of any finite region $\mathcal{V}$ as

$$V = \int_{\mathcal{V}} \sqrt{\det(g_{ij})}\ \mathrm{d}^3x. \tag{1.25}$$

This quantity is independent of our coordinate choice because any factor in d^3x implied by a coordinate transformation (1.22) is canceled out by a corresponding factor in $\sqrt{\det(g_{ij})}$: For Equation (1.24) to be independent of coordinate choices, as required by its definition of an actual (though infinitesimal) length, the chain rule implies that the metric changes according to the *tensor transformation law*

$$g_{i'j'} = \sum_{i=1}^{3} \sum_{j=1}^{3} \frac{\partial x^i}{\partial x^{i'}} \frac{\partial x^j}{\partial x^{j'}} g_{ij}. \tag{1.26}$$

Our previous volume, V_0, was not coordinate invariant, unlike V in Equation (1.25). Only the product $V(\tau) = V_0 a(\tau)^3$ is a measurable quantity independent of coordinate choices. In order to reconcile this result with the geometrical formulation, the scale factor $a(\tau)$ must be part of the metric g_{ij}. If we have Cartesian coordinates on space at any fixed time, the metric, according to Equation (1.23), is given by $g_{ij} = a^2 \delta_{ij}$ where $a > 0$ is spatially constant and the Kronecker delta δ_{ij} represents the unit matrix. If we allow time to change, we can now easily interpret the spatial constant a as a time-dependent function $a(\tau)$, such that $\det(g_{ij}) = a(\tau)^3$ and Equation (1.25) implies $V(\tau) = V_0 a(\tau)^3$. The metric, therefore, contains all the geometrical information relevant for an expanding volume.

The metric of an expanding universe depends on τ as well as spatial coordinates x^i. For a complete treatment, one should allow for differentials $\mathrm{d}\tau$ as well as $\mathrm{d}x^i$ (or rather $c\mathrm{d}\tau$ if all coordinate differentials should have the same units). In order to do so in compact form, we consider $c\tau$ to be the first one of four spacetime coordinates. Since we have already used the label 1 for one of the spatial coordinates, time gets the label zero: $(c\tau, x, y, z) = x^\alpha$, $\alpha = 0, 1, 2, 3$. A general spacetime line element then has the form

$$\mathrm{d}s^2 = \sum_{\alpha=0}^{3} \sum_{\beta=0}^{3} g_{\alpha\beta} \mathrm{d}x^\alpha \mathrm{d}x^\beta \tag{1.27}$$

in which a term $c^2\mathrm{d}\tau^2$ and also cross-terms $c\mathrm{d}\tau\mathrm{d}x^i$ appear. When we change coordinates, the metric $g_{\alpha\beta}$ obeys a spacetime version of the tensor transformation law:

$$g_{\alpha'\beta'} = \sum_{\alpha=0}^{3} \sum_{\beta=0}^{3} \frac{\partial x^\alpha}{\partial x^{\alpha'}} \frac{\partial x^\beta}{\partial x^{\beta'}} g_{\alpha\beta}. \tag{1.28}$$

The components of the spacetime metric $g_{\alpha\beta}$ are, in general, allowed to depend on all four spacetime coordinates. Because the order of factors in the product $dx^\alpha dx^\beta$ does not matter, the metric should obey the relation $g_{\alpha\beta} = g_{\beta\alpha}$ and therefore represent a symmetric matrix: Any matrix M can be written as a unique sum of a symmetric and an antisymmetric one, $M = M_S + M_A$ with $M_S = \frac{1}{2}(M + M^T)$ and $M_A = \frac{1}{2}(M - M^T)$, indicating transposition by a superscript "T." An antisymmetric contribution to $g_{\alpha\beta}$ would cancel out in the sums of Equation (1.27), and therefore not be geometrical because it does not affect the line element. If the metric is supposed to describe the geometry, and only the geometry, it must be symmetric. In Section 5.2.5, we will see how some approaches to quantum gravity, in particular string theory, could be used to motivate "non-geometrical" spacetimes in which an antisymmetric matrix contributes.

In any sufficiently small region, the components $g_{\alpha\beta}$ can be approximated by constants. It is then possible to diagonalize the resulting symmetric matrix, and to absorb all constant coefficients (the eigenvalues) by rescaling coordinates. In this approximation, we obtain the line element of special relativity,

$$ds^2 = -c^2 d\tau^2 + dx^2 + dy^2 + dz^2, \tag{1.29}$$

which can be derived from the principle that the speed of light should take the same value in all coordinate systems, even if they belong to moving observers. In spacetime, we therefore have a metric $g_{\alpha\beta}$ with negative determinant, $\det(g_{\alpha\beta}) < 0$. Moreover, there is always a local coordinate, such as τ, such that the metric restricted to constant τ has positive determinant. These mathematical conditions ensure that there is one and only one time direction in spacetime.

1.3.2 Isotropic Line Elements

If we consider a homogeneous and isotropic universe, the metric cannot depend on spatial coordinates, and it must be such that all references to the spatial index i appear in a rotationally symmetric form. The paradigmatic example of a rotationally symmetric quantity is the radius or its square, $r^2 = x^2 + y^2 + z^2$ in Cartesian coordinates. An example of an isotropic spacetime line element is therefore

$$ds^2 = -N(t)^2 c^2 dt^2 + a(t)^2(dx^2 + dy^2 + dz^2) \tag{1.30}$$

with the scale factor $a(t)$ and a second function, $N(t)$, which is consistent with the symmetry requirements. For a non-degenerate line element that refers to four dimensions, neither a nor N should be zero. To be specific, and without loss of generality, we assume that they are positive.

There is a good reason why we have not yet encountered the second function, $N(t)$: it turns out to be irrelevant for observations. It is always possible to transform the coordinate t to a new one, τ, such that $\tau(t) = \int N(t) dt$. Since $N(t) \neq 0$, the function $\tau(t)$ is monotonic and can be used as a well-defined coordinate transformation everywhere in spacetime, even if we may not be able to determine the integral that defines this function in analytic form. Therefore, $N(t)\mathrm{d}t = \mathrm{d}\tau$, and the line element is

$$\mathrm{d}s^2 = -c^2\mathrm{d}\tau^2 + a(\tau)^2(\mathrm{d}x^2 + \mathrm{d}y^2 + \mathrm{d}z^2). \tag{1.31}$$

The new time coordinate τ (already used before in the example of Minkowski spacetime) is called *proper time*.

It is possible to modify the spatial part of the line element while maintaining the condition of isotropy. In addition to flat space with Cartesian coordinates, there are two further isotropic spaces which have positive curvature (spherical) and negative curvature (hyperboloidal), respectively. Using spatial polar coordinates, the three choices can be parameterized by the sign of a single number k, such that

$$\mathrm{d}s^2 = -c^2\mathrm{d}\tau^2 + a(\tau)^2\left(\frac{\mathrm{d}r^2}{1 - kr^2} + r^2(\mathrm{d}\vartheta^2 + \sin^2\vartheta\,\mathrm{d}\varphi^2)\right). \tag{1.32}$$

For $k = 0$, we have the spatially flat line element in polar coordinates, while $k \neq 0$ leads to two different curved spaces which cannot be obtained from flat space by applying a coordinate transformation. For $k = 1$, the radius r is restricted to the range $0 \leqslant r < 1$. In this case, the coordinate transformation $r = \sin\psi$ from the radius to a third angle, ψ in addition to ϑ and φ, shows the relationship of the spatial part of the line element with a three-dimensional generalization of the two-sphere:

$$\mathrm{d}s^2 = -c^2\mathrm{d}\tau^2 + a(\tau)^2(\mathrm{d}\psi^2 + \sin^2\psi(\mathrm{d}\vartheta^2 + \sin^2\vartheta\,\mathrm{d}\varphi^2)). \tag{1.33}$$

For $k = -1$, a similar transformation exists in which $\sin\psi$ is replaced with the hyperbolic version, $\sinh\psi$.

The line element (1.32) describes a geometry in which space is curved. Curvature is a new geometrical quantity that allows us to distinguish between different spaces or spacetimes independently of their coordinates. Curvature also has physical implications because it affects the motion of light and matter, but unfortunately it is not straightforward to define it without tensor calculus. We therefore postpone a detailed discussion of curvature to the beginning of Chapter 2.

Cosmologically, the main implication of spatial curvature is an additional, potential-like term in the Friedmann equation,

$$H^2 + \frac{kc^2}{a^2} = \frac{8\pi G}{3c^2}\rho. \tag{1.34}$$

Non-zero curvature, $k \neq 0$, therefore does not allow an unchanging vacuum solution with $\rho = 0$ and $H = 0$ like Minkowski spacetime, the standard vacuum of local particle physics. In an expanding universe with increasing $a(\tau)$, the curvature term decreases more slowly than standard matter and is therefore relevant at late times. Positive curvature, $k > 0$, then makes it possible for the Hubble parameter to vanish at some time, when $3kc^4$ equals $8\pi Ga^2\rho$. (Recall that ρ for all standard matter decreases as the universe expands due to dilution, such that any initially large $8\pi Ga^2\rho$ in the early universe will eventually reach the constant $3kc^4$.) At this time, the expansion of the universe may stop and turn into collapse. Positive curvature therefore implies additional gravitational attraction which pulls space together. It could only be overcome by dark energy, which due to its constant undiluted energy density $\rho = \Lambda$ dominates all matter, and even the curvature term, at late times.

Spaces with non-zero curvature $k \neq 0$ are, like flat space, homogeneous and isotropic. It is easy to recognize their spherical symmetry around a center because they are formulated with the same polar coordinates, ϑ and φ, that may be employed in flat space. In particular, we can define a sphere in curved space, centered at $r = 0$, by considering all points with $r = R_0$ constant. At any fixed time such that $a(\tau) = 1$, its area

$$A_0 = \int_0^\pi \int_0^{2\pi} R_0^2 \sin \vartheta d\varphi d\vartheta = 4\pi R_0^2 \qquad (1.35)$$

has the same expression as in flat space, independent of k. However, R_0, introduced as some fixed value of r, is no longer the geometrical radius of the sphere, which is defined as the distance from the center to the sphere. This distance, S_0, should be computed using the line element by integrating $ds = \sqrt{ds^2}$ along a curve in a radial direction, with constant ϑ and φ:

$$S_0 = \int_{r=0}^{r=R_0} ds = \int_0^{R_0} \frac{ds}{dr}\bigg|_{d\vartheta=d\varphi=0} dr = \int_0^{R_0} \frac{dr}{\sqrt{1 - kr^2}}$$

$$= \begin{cases} k^{-1/2} \sin^{-1}(\sqrt{k}\, R_0) & \text{if } k > 0 \\ R_0 & \text{if } k = 0. \\ |k|^{-1/2} \sinh^{-1}(\sqrt{|k|}\, R_0) & \text{if } k < 0 \end{cases} \qquad (1.36)$$

Therefore, $A_0 \neq 4\pi S_0^2$ if $k \neq 0$, and the geometry is not Euclidean.

The scale factor $a(\tau)$ expands or contracts the geometry of curved space, just as it does for flat space. Since we are using non-Cartesian coordinates, we cannot rescale each of them in order to test the scaling behavior. We should rather keep the two angles ϑ and φ unchanged because they do not determine the scale of our geometry. The radial coordinate r, however, is rather arbitrary and can be changed by a scaling transformation, mapping r to $\lambda^{1/3}r$, such that the volume

$$V_0 = \int \frac{r^2}{1 - kr^2} \sin \vartheta d\varphi d\vartheta dr \qquad (1.37)$$

acquires our previous factor, $V_0' = \lambda V_0$. (The integration region here is again a finite region in curved space, which may be all of space for $k > 0$ because the entire space is finite in the spherical case.) In order for the line element to remain unchanged with respect to this transformation, the scale factor should be mapped to $a(\tau)/\lambda^{1/3}$, as before. However, for $k \neq 0$ we also have to transform this parameter, to $k' = k/\lambda^{2/3}$ so as to have an invariant coefficient $1/(1 - kr^2)$ in the line element. The curvature term in the Friedmann equation, kc^2/a^2, is then invariant.

In the curved case, it is convenient to choose the radial coordinate such that $|k| = 1$. (In the spherical model, the maximal radial coordinate is then $r = 1$, referred to as a unit sphere.) Once this choice is made, the scale factor can no longer be rescaled and obtains a more invariant meaning compared with the flat case. Although this choice is not distinguished by geometrical or physical considerations, it is often made. The one-parameter family of isotropic spaces is then reduced to three discrete choices of distinct geometries: $k = 0$, $k = 1$, and $k = -1$.

1.3.3 Anisotropic Models

In addition to introducing spatial curvature, the line element allows us to include anisotropy while, for now, maintaining homogeneity. In the preceding subsection, isotropy restricted the line element to a rather simple form with a single free function. No cross-terms $c\,d\tau\,dx^i$ are allowed in this case because the corresponding metric coefficients, collected in $\sum_{i=1}^{3} M_i c\,d\tau\,dx^i$, would identify a non-zero direction through the vector components M_i, which would not be compatible with isotropy. Moreover, the purely spatial part of the metric must be proportional to the isotropic $dx^2 + dy^2 + dz^2$ or its curved versions, leaving room for just a single function, the scale factor $a(\tau)$.

If we drop the condition of isotropy, cross-term coefficients of the form M_i are allowed, and there are many more choices for the spatial part. In a basic anisotropic line element, we may assume that $M_i = 0$ while the single scale factor $a(\tau)$ is replaced by three independent expansion factors in each of the three Cartesian directions:

$$ds^2 = -c^2 d\tau^2 + (a_1(\tau)^2 d^2x + a_2(\tau)^2 d^2y + a_3(\tau)^2 d^2z). \tag{1.38}$$

These expansion rates obey coupled ordinary differential equations that replace the Friedmann equation. It is convenient to express them in terms of a Hamiltonian for configuration variables α and $\beta_{\pm}$ (called *Misner variables*; Misner 1969) and their momenta p_{α} and $p_{\pm}$, such that

$$\log(L_1 a_1) = \alpha + \beta_+ + \sqrt{3}\beta_- \tag{1.39}$$

$$\log(L_2 a_2) = \alpha + \beta_+ - \sqrt{3}\beta_- \tag{1.40}$$

$$\log(L_3 a_3) = \alpha - 2\beta_+ \tag{1.41}$$

where L_1, L_2 and L_3 are length parameters such that $L_1 L_2 L_3 = V_0$ equals the previous coordinate volume. Since

$$\begin{aligned} \alpha &= \frac{1}{3}(\log(L_1 a_1) + \log(L_2 a_2) + \log(L_3 a_3)) \\ &= \log((V_0 a_1 a_2 a_3)^{1/3}), \end{aligned} \tag{1.42}$$

it plays the role of the logarithmic volume,

$$\alpha = \log\left(V_0^{1/3}a\right) = \log V^{1/3} = \frac{1}{3}\log V, \tag{1.43}$$

as it could also be used in an isotropic model based on the geometric mean a of a_1, a_2 and a_3. The two remaining parameters, $\beta_{\pm}$, represent two independent degrees of freedom for anisotropy. The Hamiltonian is then given by

$$H = \frac{2\pi G}{3c^2}e^{-3\alpha}\left(-p_\alpha^2 + p_+^2 + p_-^2\right) + e^{3\alpha}\rho. \tag{1.44}$$

The two anisotropy momenta $p_{\pm}$ are constant in time because H is independent of the variables $\beta_{\pm}$. Moreover, Hamilton's equation

$$\frac{d\alpha}{d\tau} = \frac{\partial H}{\partial p_\alpha} = -\frac{4\pi G}{3c^2}e^{-3\alpha}p_\alpha \tag{1.45}$$

allows us to rewrite the cosmological energy balance H = 0 as a Friedmann-type equation

$$\begin{aligned}
\left(\frac{1}{a}\frac{da}{d\tau}\right)^2 = \left(\frac{d\alpha}{d\tau}\right)^2 &= \frac{16\pi^2 G^2}{9c^4}\frac{p_\alpha^2}{e^{6\alpha}} \\
&= \frac{8\pi G}{3c^2}\rho + \frac{16\pi^2 G^2}{9c^4}\frac{p_+^2 + p_-^2}{V_0^2 a^6}.
\end{aligned} \tag{1.46}$$

Anisotropies therefore affect the expansion rate in the same way as a matter ingredient whose energy density is proportional to a^{-6} (called a "stiff fluid"). This characteristic behavior of *anisotropic shear* implies that anisotropies are diluted more than non-relativistic matter or radiation in an expanding universe, and should therefore be subdominant at late times. However, they are crucial in the early universe.

Just as it is possible to define homogeneous and isotropic spaces that are not flat, there are several candidates for homogeneous and anisotropic spaces if they are not required to be flat at any fixed time. The possible choices are determined by the Bianchi classification, originally done for three-dimensional Lie algebras, which represent the different types of symmetries that can all be considered homogeneous; see for instance Ellis & MacCallum (1969); MacCallum & Taub (1972). We can use the same configuration variables and momenta for all these models, but their Hamiltonians differ, given for vacuum models by

$$\begin{aligned}
H = {}&\frac{2\pi G}{3c^2}e^{-3\alpha}\left(-p_\alpha^2 + p_+^2 + p_-^2\right) \\
&+ \frac{c^2}{8\pi G}e^{\alpha}\left(n^1 e^{4(\beta_+ + \sqrt{3}\beta_-)} + n^2 e^{4(\beta_+ - \sqrt{3}\beta_-)} + n^3 e^{-8\beta_+}\right. \\
&\left. - 2n^1 n^2 e^{4\beta_+} - 2n^1 n^3 e^{-2(\beta_+ - \sqrt{3}\beta_-)} - 2n^2 n^3 e^{-2(\beta_+ + \sqrt{3}\beta_-)}\right)
\end{aligned} \tag{1.47}$$

Table 1.1. The Different Anisotropic Models According to Their Traditional Classification as Bianchi Types in Class A

Bianchi type	n^1	n^2	n^3
I	0	0	0
II	+1	0	0
VI_0	+1	−1	0
VII_0	+1	+1	0
VIII	+1	+1	−1
IX	+1	+1	+1

Note. The missing numbers III, IV, V as well as versions of VI and VII with the same n^I as listed here but a non-zero value of a new, independent parameter do not result in diagonal spatial metrics (class B).

with constants n^1, n^2 and n^3 that can take values according to Table 1.1. (The Hamiltonian (1.47) describes only those homogeneous models, called "class A," that can be formulated with a diagonal spatial metric determined by three time-dependent scale factors, a_1, a_2, and a_3. Although it is always possible to diagonalize the spatial metric at any given time, there are homogeneous models which evolve such that an initial diagonal spatial metric does not remain diagonal. These models of "class B" are not considered here.)

For the Bianchi-I model with $n^1 = n^2 = n^3 = 0$, the Hamiltonian (1.47) implies the equations of motion

$$\frac{d\alpha}{d\tau} = -\frac{4\pi G}{3c^2}e^{-3\alpha}p_\alpha \tag{1.48}$$

$$\frac{d\beta_+}{d\tau} = \frac{4\pi G}{3c^2}e^{-3\alpha}p_+ \tag{1.49}$$

$$\frac{d\beta_-}{d\tau} = \frac{4\pi G}{3c^2}e^{-3\alpha}p_- \tag{1.50}$$

while p_α and $p_\pm$ are constant. The first equation implies

$$\frac{de^{3\alpha}}{d\tau} = -\frac{4\pi Gp_\alpha}{c^2} \tag{1.51}$$

and is therefore solved by

$$\alpha(\tau) = \frac{1}{3}\log\left(C - \frac{4\pi Gp_\alpha}{c^2}\tau\right) \tag{1.52}$$

with a constant C. We can adjust the zero-point of τ such that $C = 0$. (In an expanding universe, p_α is negative according to Equation (1.51), such that the logarithm is real for positive τ.) Inserting this solution in the equations of motion for $\beta_\pm$ then implies that

$$\beta_\pm(\tau) = \beta_{\pm 0} - \frac{1}{3}\frac{P_\pm}{P_\alpha}\log\left(-\frac{4\pi G P_\alpha}{c^2}\tau\right). \tag{1.53}$$

Therefore,

$$L_1 a_1(\tau) = \exp(\alpha + \beta_+ + \sqrt{3}\beta_-)$$
$$= e^{\beta_{+0} + \sqrt{3}\beta_{-0}}\left(-\frac{4\pi G P_\alpha}{c^2}\tau\right)^{(P_\alpha - P_+ - \sqrt{3}P_-)/(3P_\alpha)} \tag{1.54}$$

$$L_2 a_2(\tau) = \exp(\alpha + \beta_+ - \sqrt{3}\beta_-)$$
$$= e^{\beta_{+0} - \sqrt{3}\beta_{-0}}\left(-\frac{4\pi G P_\alpha}{c^2}\tau\right)^{(P_\alpha - P_+ + \sqrt{3}P_-)/(3P_\alpha)} \tag{1.55}$$

$$L_3 a_3(\tau) = \exp(\alpha - 2\beta_+) = e^{-2\beta_{+0}}\left(-\frac{4\pi G P_\alpha}{c^2}\tau\right)^{(P_\alpha + 2P_+)/(3P_\alpha)}. \tag{1.56}$$

We can absorb any constant factors in the L_i by individually rescaling the spatial coordinates. The anisotropic evolution then follows the *Kasner solutions* (Kasner 1921)

$$a_I(\tau) = \tau^{v_I} \tag{1.57}$$

with constant

$$v_1 = \frac{P_\alpha - P_+ - \sqrt{3}P_-}{3P_\alpha}, \quad v_2 = \frac{P_\alpha - P_+ + \sqrt{3}P_-}{3P_\alpha}, \quad v_3 = \frac{P_\alpha + 2P_+}{3P_\alpha}. \tag{1.58}$$

These exponents satisfy the two relations

$$v_1 + v_2 + v_3 = 1 \tag{1.59}$$

and

$$v_1^2 + v_2^2 + v_3^2 = \frac{P_\alpha^2 + 2P_+^2 + 2P_-^2}{3P_\alpha^2} = 1, \tag{1.60}$$

the last equality following from $p_+^2 + p_-^2 = p_\alpha^2$ because of H = 0 in Equation (1.47) with $n_1 = n_2 = n_3 = 0$.

The two Equations (1.59) and (1.60) determine the allowed triples (v_1, v_2, v_3) as points in space that lie at the intersection of a plane through all three axis units with a unit sphere around the origin; see Figure 1.2. Except for the units themselves, any point in this intersection is such that one of the v_i is negative while the other two are positive. Therefore, while the spatial volume is increasing according to

$$V_0 a_1(\tau)a_2(\tau)a_3(\tau) \propto e^{3\alpha(\tau)} = \tau^{v_1 + v_2 + v_3} = \tau, \tag{1.61}$$

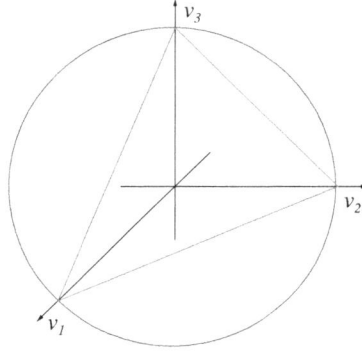

Figure 1.2. The exponents v_i of Kasner solutions (1.57) obey two conditions, (1.59) and (1.60), which imply that the points (v_1, v_2, v_3) lie on the intersection of a plane through the three axis units (red) with a unit circle around the origin (blue). There are no intersection points in the positive octant because the sphere lies "above" the plane in the perspective shown here. Except for the three units, all intersection points are such that one of the v_i is negative while the other two are positive.

one of the three independent spatial dimensions is contracting. This characteristic "Kasner" behavior shows that the dynamics is markedly different from isotropic expansion. (However, matter terms can change this behavior and lead to isotropization at late times; Misner 1968.)

The Bianchi-I model, for which Kasner's solutions are valid, is not the most generic anisotropic model because it requires all three coefficients n^i in Equation (1.47) to be zero. A generic behavior is obtained if all n^i are non-zero, which is the case for the Bianchi VIII and IX models (but, conveniently, none of the class B models). For the Bianchi-IX model, for instance, the Hamiltonian (1.47) contains a complicated anisotropy potential

$$
\begin{aligned}
W(\alpha, \beta_+, \beta_-) = {} & \frac{c^2}{8\pi G} e^\alpha \\
& \times \left(e^{-8\beta_+} - 4e^{-2\beta_+}\cosh(2\sqrt{3}\beta_-) + 2e^{4\beta_+}(\cosh(4\sqrt{3}\beta_-) - 1) \right)
\end{aligned}
\tag{1.62}
$$

in addition to the kinetic term quadratic in momenta. While α merely rescales the potential, the dependence on $\beta_\pm$ confines the anisotropy parameters to a finite region of roughly triangular shape; see Figure 1.3. (The potential is invariant under rotations of (β_+, β_-) by $2\pi/3$.) The typical potential wall is rather steep and of exponential behavior, as can be seen easily in the direction of $\beta_- = 0$ where the dominant behavior of the potential is given by $e^{-8\beta_+}$. While Kasner solutions are approximately realized in regions in which the potential is small, encounters with the walls ("reflections") mix the expansion and contraction behavior of the independent spatial directions (Misner 1969), in a way that implies chaotic dynamics (Damour et al. 2003).

As shown by the a^{-6}-behavior of the anisotropic shear term, seen in Equation (1.46), anisotropies and therefore the generic Bianchi-IX dynamics are relevant in the very early universe, close to the Big Bang singularity. In addition to anisotropies,

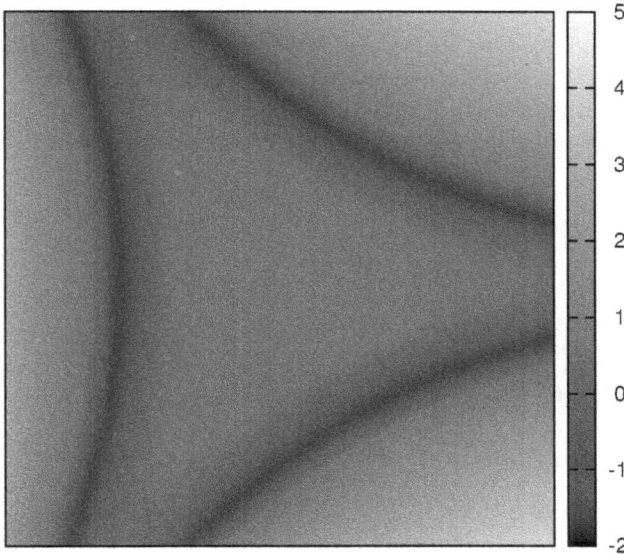

Figure 1.3. Logarithmic plot of the anisotropy potential (1.62) for a Bianchi-IX model, showing $\log|W(\alpha, \beta_+, \beta_-)|$ as a function of β_+ (horizontal) and β_- (vertical) at fixed α. The potential $W(\alpha, \beta_+, \beta_-)$ is negative around $\beta_+ = 0 = \beta_-$ and goes through zero at the black lines, which appear as deep valleys in the logarithmic plot.

inhomogeneity should then be expected because dense matter implies enhanced gravitational collapse. We will return to this model at the end of the present chapter, but first we have to introduce the main features of inhomogeneous spacetimes. Complete inhomogeneity can be described only by solving the rather intractable Einstein equation without any approximations. We will therefore develop inhomogeneous descriptions of spacetime in simpler model systems, which still provide good approximations in certain regimes.

At late times, gravitational collapse happens on small scales but no longer on the scales beyond galaxy clusters, implying large-scale homogeneity. Moreover, the inflation paradigm indicates that even in the rather early universe there was a mechanism at play that homogenized a vast region of the universe, out of which our currently visible part grew. For a considerable time during and after inflation, inhomogeneity was small and can, in an approximation, be described by perturbation theory.

1.4 Perturbative Inhomogeneity

We begin with a homogeneous and isotropic spacetime with line element

$$ds^2 = -c^2 d\tau^2 + a(\tau)^2 (dx^2 + dy^2 + dz^2) \tag{1.63}$$

and introduce small inhomogeneous perturbations of all ten independent components of the metric, given by a symmetric tensor such that $g_{\alpha\beta} = g_{\beta\alpha}$. It is quite convenient to do so after a transformation from proper time τ to *conformal time* η, defined by $c\mathrm{d}\tau/a(\tau) = \mathrm{d}\eta$. The conformal time does not necessarily have time units, depending on the units chosen for a, and it is not invariant under rescalings $x^i \mapsto \lambda x^i$ or changes of units of the spatial coordinates. Nevertheless, for a given scaling, it is a mathematical parameter along curves of evolving metrics.

With this new coordinate, the scale factor can be factored out in the line element

$$\mathrm{d}s^2 = a(\eta)^2\big(- \mathrm{d}\eta^2 + \mathrm{d}x^2 + \mathrm{d}y^2 + \mathrm{d}z^2\big). \tag{1.64}$$

1.4.1 Scalar, Vector, and Tensor Perturbations

We now modify the constant coefficients in the parenthesis of Equation (1.64) by allowing for small inhomogeneity, introducing new spacetime dependent fields ϕ, w_i and h_{ij} such that

$$\mathrm{d}s^2 = a(\eta)^2\left(-(1 + 2\phi)\mathrm{d}\eta^2 + 2 \sum_{i=1}^{3} w_i\mathrm{d}\eta\mathrm{d}x^i + \sum_{i=1}^{3}\sum_{j=1}^{3}(\delta_{ij} + h_{ij})\mathrm{d}x^i\mathrm{d}x^j\right). \tag{1.65}$$

The one scalar field ϕ, three components of the vector field $\vec{w}$, and six independent components of the symmetric spatial matrix $\mathbf{h}$ indeed add up to ten independent functions. The vector and matrix components can be separated further into new scalar and vector fields as follows:

- It is possible to define a vector field starting from a scalar field B, such that $\vec{w} = \nabla B$ is the gradient of B. However, not every vector field is of this form because $\vec{w} = \nabla B$ implies that it has zero curl, $\nabla \times \vec{w} = 0$. In general, any vector field $\vec{w} = \nabla B + \vec{w}_\mathrm{T}$ can be decomposed as the sum of a gradient and a "transverse" vector field $\vec{w}_\mathrm{T}$ such that $\nabla \cdot \vec{w}_\mathrm{T} = 0$. This decomposition is unique because, given any $\vec{w}$, we can first compute B by solving the elliptic partial differential equation $\nabla \cdot \vec{w} = \Delta B$ for B, and defining $\vec{w}_\mathrm{T} = \vec{w} - \nabla B$, which is then guaranteed to be transverse. Using boundary conditions that require all inhomogeneous perturbations to vanish at infinity, the solution B of an elliptic partial differential equation in space is unique for given $\vec{w}$, such that $\vec{w}_\mathrm{T}$ is unique too.
- In a generalization of the previous construction for vectors, we can decompose any symmetric matrix $\mathbf{h}$ as the sum of two scalar, one vector, and one tensor contribution: For the components, we write

$$h_{ij} = -2\delta_{ij}\psi + \left(\frac{\partial^2}{\partial x^i \partial x^j} - \frac{1}{3}\delta_{ij}\Delta\right)E + \left(\frac{\partial v_j^\mathrm{T}}{\partial x^i} + \frac{\partial v_i^\mathrm{T}}{\partial x^j}\right) + h_{ij}^{\mathrm{TT}} \tag{1.66}$$

where $\vec{v}^\mathrm{T}$ is transverse ($\nabla \cdot \vec{v}^\mathrm{T} = 0$, as before) and $\mathbf{h}^{\mathrm{TT}}$ is transverse and traceless (tr $h = \sum_{i=1}^{3} h_{ii}$ is zero, in addition to $\sum_{i=1}^{3}\partial h_{ij}/\partial x^i = 0$). The first scalar, ψ, is derived from the trace such that $-6\psi = \mathrm{tr}\,\mathbf{h}$, which follows from Equation (1.66) using the specific form of coefficients in the E-term as well as the transversality and tracelessness conditions imposed on $\vec{v}^\mathrm{T}$ and $\mathbf{h}^{\mathrm{TT}}$. The traceless part of **h**, derived as $h_{ij}^{\mathrm{T}} = h_{ij} + 2\delta_{ij}\psi$, is not necessarily transverse. Using the required transversality of $\vec{v}^{\mathrm{T}}$ and $\mathbf{h}^{\mathrm{TT}}$ in Equation (1.66), however, we have the equation

$$\sum_{i=1}^{3} \frac{\partial h_{ij}^{\mathrm{T}}}{\partial x^i} = \Delta\left(v_j^{\mathrm{T}} + \frac{2}{3}\frac{\partial E}{\partial x^j}\right) \tag{1.67}$$

which, for a given h_{ij}, can be solved as an elliptic equation for $v_j^{\mathrm{T}} + \frac{2}{3}\partial E/\partial x^j$. Splitting this last vector field into a gradient and a transverse vector field, as done for the vector field $\vec{w}$, determines E and v^{T} independently. Finally, we compute $\mathbf{h}^{\mathrm{TT}}$ by inverting Equation (1.66).

The ten independent perturbations have thus been split into four scalar fields ϕ, ψ, B and E; two transverse vector fields w^{T} and v^{T} each having two independent degrees of freedom; and one transverse-traceless tensor field $\mathbf{h}^{\mathrm{TT}}$ with two independent degrees of freedom. Anticipating a little bit, equations of motion show that the last two degrees of freedom are the only propagating ones, constituting the two polarizations of gravitational waves; see Equation (1.77) below.

1.4.2 Coordinate Dependence

The ten fields that parameterize a perturbed metric are uniquely defined only if we fix our coordinate system. However, we can maintain the perturbative approximation and still perform coordinate transformations that change (η, x, y, z) by adding to each of them a *small* coordinate-dependent function, such as $\eta' = \eta + \xi^0(\eta, x, y, z)$. In addition, we are allowed to perform large reparameterizations of time, which leave the spatial coordinates unchanged, such as transforming from proper time to conformal time or back. Small inhomogeneous transformations and large homogeneous ones are not independent of each other in the sense that their commutator—the difference between performing a large homogeneous transformation and then a small inhomogeneous one and the same transformations applied in reverse order—is not the identity transformation. Such algebraic relationships between transformations will play an important role in our later discussions of covariance, in particular in attempted quantizations. We will return to this question in Chapter 6, but some related subtleties will appear already in the present subsection.

A coordinate transformation is performed by inserting new coordinates everywhere in the perturbed line element, including in the arguments of inhomogeneous fields as well as a in the case of η. We then perform Taylor expansions where necessary, provided the transformation is small, and keep only terms up to linear order in these expansions and their products. For instance, for pure vector modes we are interested only in coordinate transformations by a transverse ξ^i, such that $\eta' = \eta$ and $x^{i'} = x^i + \xi^i$ where $\sum_{i=1}^{3}\partial\xi^i/\partial x^i = 0$. This transformation, inserted in $\mathrm{d}x^i\mathrm{d}x^j$ in Equation (1.65), results in a new decomposition in which the previous $\vec{w}^{\mathrm{T}}$ is replaced by $\vec{w}^{\mathrm{T}} + \mathrm{d}\vec{\xi}/\mathrm{d}\eta$ and $\vec{v}^{\mathrm{T}}$ is replaced by $\vec{v}^{\mathrm{T}} + \vec{\xi}$. The combination

$$\vec{V} = \vec{w}^{\mathrm{T}} - \frac{\mathrm{d}\vec{v}^{\mathrm{T}}}{\mathrm{d}\eta} \tag{1.68}$$

is therefore invariant with respect to linear coordinate transformations, and it is transverse.

Because small coordinate transformations act by adding a displacement vector to the coordinate vector, they do not contain any tensorial contribution. Therefore, the tensor mode $\mathbf{h}^{\mathrm{TT}}$ is automatically invariant with respect to linear coordinate transformations. However, the four scalar fields—ϕ, ψ, B, and E—can be transformed in various ways and require a more detailed analysis in order to find invariant combinations.

To this end, we use the line element

$$\mathrm{d}s^2 = a^2\left(-(1 + 2\phi)\mathrm{d}\eta^2 + 2\sum_{i=1}^{3}\frac{\partial B}{\partial x^i}\mathrm{d}\eta\mathrm{d}x^i\right.$$
$$\left. + \sum_{i=1}^{3}\sum_{j=1}^{3}\left((1 - 2\psi)\delta_{ij} + 2\frac{\partial^2 E}{\partial x^i\partial x^j}\right)\mathrm{d}x^i\mathrm{d}x^j\right) \tag{1.69}$$

in which only the scalar contributions have been maintained. We consider two coordinate transformations of scalar type, one given by the time component ξ^0 and one by the gradient part ξ^i such that $\xi^i = \partial\xi/\partial x^i$:

- If ξ^0 is non-zero but $\xi = 0$, we keep the spatial coordinates x^i but change η to $\eta + \xi^0$. To first order in ξ^0, the differential $\mathrm{d}\eta^2$ then changes to

$$\mathrm{d}\eta^2 + 2\frac{\mathrm{d}\xi^0}{\mathrm{d}\eta}\mathrm{d}\eta^2 + 2\sum_{i=1}^{3}\frac{\partial\xi^0}{\partial x^i}\mathrm{d}\eta\mathrm{d}x^i \tag{1.70}$$

and $a(\eta)^2$ changes to $a(\eta)^2(1 + 2(\xi^0/a)\mathrm{d}a/\mathrm{d}\eta)$ by a Taylor expansion. Therefore, the previous scalar fields are transformed to

$$\phi + \frac{\mathrm{d}\xi^0}{\mathrm{d}\eta} + \frac{1}{a}\frac{\mathrm{d}a}{\mathrm{d}\eta}\xi^0, \quad \psi - \frac{1}{a}\frac{\mathrm{d}a}{\mathrm{d}\eta}\xi^0, \quad B - \xi^0, \quad E. \tag{1.71}$$

Here, we have assumed that $\partial\xi^0/\partial x^i \neq 0$, such that the gradient term containing B does change. If $\partial\xi^0/\partial x^i = 0$, there is no need to change B, while ϕ and ψ do change. There is therefore a subtle difference between spatially dependent ξ^0 and spatially independent, even if both are small.

- If ξ is non-zero but $\xi^0 = 0$, the only term to be transformed is $\sum_{i=1}^{3}\sum_{j=1}^{3}\delta_{ij}\mathrm{d}x^i\mathrm{d}x^j$. It changes to

$$\sum_{i,j=1}^{3}\delta_{ij}\mathrm{d}x^i\mathrm{d}x^j + 2\sum_{i=1}^{3}\frac{\mathrm{d}}{\mathrm{d}\eta}\frac{\partial\xi}{\partial x^i}\mathrm{d}\eta\mathrm{d}x^i + 2\sum_{i=1}^{3}\sum_{j=1}^{3}\frac{\partial^2\xi}{\partial x^i\partial x^j}\mathrm{d}x^i\mathrm{d}x^j. \tag{1.72}$$

Therefore, ϕ and ψ are unchanged while B and E are replaced by $B + \mathrm{d}\xi/\mathrm{d}\eta$ and $E + \xi$, respectively. Therefore, in addition to ϕ and ψ, the combination $B - \mathrm{d}E/\mathrm{d}\eta$ is invariant with respect to ξ-transformations.

Again, if $\partial\xi/\partial x^i = 0$, the ξ-dependent terms in Equation (1.72) drop out, and none of the scalar fields need be changed. However, for spatial transformations this case happens only if $\xi^i = 0$ because we have assumed that $\xi^i = \partial\xi/\partial x^i$ is the gradient of ξ. Therefore, there is no difference between spatially dependent and spatially independent transformations by ξ^i.

The ξ-invariant, $B - \mathrm{d}E/\mathrm{d}\eta$, is changed to $B - \mathrm{d}E/\mathrm{d}\eta - \xi^0$ by spatially dependent ξ^0-transformations. Therefore, we can form two scalar combinations,

$$\Phi := \phi + \frac{1}{a}\frac{\mathrm{d}a}{\mathrm{d}\eta}\left(B - \frac{\mathrm{d}E}{\mathrm{d}\eta}\right) + \frac{\mathrm{d}(B - \mathrm{d}E/\mathrm{d}\eta)}{\mathrm{d}\eta} \quad \text{and}$$

$$\Psi := \psi - \frac{1}{a}\frac{\mathrm{d}a}{\mathrm{d}\eta}\left(B - \frac{\mathrm{d}E}{\mathrm{d}\eta}\right), \tag{1.73}$$

called the Bardeen potentials (Bardeen 1980), which are invariant with respect to all small, spatially dependent coordinate transformations. However, as we saw explicitly, these combinations are *not* invariant with respect to spatially independent coordinate changes ξ^0 in the time direction (which may depend on η).

In cosmology, one often considers scalar matter that provides further scalar fields in addition to ϕ and ψ that come from the metric. Such a scalar field, $\varphi = \bar{\varphi} + \delta\varphi$ expressed as a homogeneous background value $\bar{\varphi}$ plus an inhomogeneous perturbation, changes only under ξ^0-transformations, being replaced by $\delta\varphi + (\mathrm{d}\bar{\varphi}/\mathrm{d}\eta)\xi^0$ according to a Taylor expansion. The combination

$$\Pi = \delta\varphi + \frac{\mathrm{d}\bar{\varphi}}{\mathrm{d}\eta}\left(B - \frac{\mathrm{d}E}{\mathrm{d}\eta}\right) \tag{1.74}$$

is therefore invariant with respect to small, spatially independent coordinate transformations.

If there is a scalar matter field, it may be used instead of $B - \mathrm{d}E/\mathrm{d}\eta$ in Equation (1.73), leading to quantities, called *curvature perturbations*, that are invariant with respect to all small coordinate changes, now including spatially independent ones in the time direction. The two independent curvature perturbations are

$$\mathcal{R}_1 = \psi + \frac{\mathrm{d}a/\mathrm{d}\eta}{a\,\mathrm{d}\bar{\varphi}/\mathrm{d}\eta}\delta\varphi$$

$$\mathcal{R}_2 = \phi - \frac{1}{2}\frac{\mathrm{d}}{\mathrm{d}\eta}\left(\frac{a}{\mathrm{d}a/\mathrm{d}\eta}\right)\psi - \frac{1}{\mathrm{d}\bar{\varphi}/\mathrm{d}\eta}\left(\frac{\mathrm{d}a/\mathrm{d}\eta}{a} - \frac{\mathrm{d}^2\bar{\varphi}/\mathrm{d}\eta^2}{\mathrm{d}\bar{\varphi}/\mathrm{d}\eta}\right)\delta\varphi \tag{1.75}$$

$$+ \frac{1}{2}\frac{a}{\mathrm{d}a/\mathrm{d}\eta}\frac{\mathrm{d}\psi}{\mathrm{d}\eta} - \frac{1}{2\mathrm{d}\bar{\varphi}/\mathrm{d}\eta}\frac{\mathrm{d}\delta\varphi}{\mathrm{d}\eta}.$$

A slight disadvantage is that they may be used only as long as $d\bar{\varphi}/d\eta \neq 0$, but sometimes one can just replace them with $(d\bar{\varphi}/d\eta)\mathcal{R}_1$ and $(d\bar{\varphi}/d\eta)\mathcal{R}_2$ in a regime in which $d\bar{\varphi}/d\eta = 0$ is relevant.

1.4.3 Evolution Equations

A subtle feature of the scalar perturbations is that some of their "invariant combinations" are strictly invariant with respect to small coordinate changes only if the latter are spatially dependent. If they are not invariant with respect to spatially independent time displacements, they obey non-trivial evolution equations, which may be of interest for cosmological structure formation.

However, also those fields that are completely invariant with respect to all small coordinate transformations should obey non-trivial evolution equations, in particular the tensor mode $\mathbf{h}$, which can imply gravitational waves only if it is subject to non-trivial time dependence according to a wave equation. How can this be possible if the modes do not change with respect to small coordinate transformations, including a time displacement $\eta' = \eta + \delta\eta$?

Non-trivial evolution equations exist because a comparison of a mode with its time derivative is sensitive to terms of *second order* in small inhomogeneity, and not just to first order as used in the derivation of invariant combinations. (If fields are required to be invariant under all coordinate changes, they must be spacetime constants; Stewart 1990; and in particular time-independent.)

The time derivative of $\mathbf{h}$ is, after all, defined as $\lim_{\delta\eta \to 0} \mathbf{h}(\eta + \delta\eta)/\delta\eta$. In our previous analysis, we did not consider transformations such as mapping $\sum_{i=1}^{3}\sum_{j=1}^{3} h_{ij}(\eta)dx^i dx^j$ to

$$
\sum_{i=1}^{3}\sum_{j=1}^{3} h_{ij}(\eta + \xi^0)dx^i dx^j
$$

$$
= \sum_{i=1}^{3}\sum_{j=1}^{3}\left(h_{ij}(\eta) + \frac{dh_{ij}(\eta)}{d\eta}\xi^0 + \frac{1}{2}\frac{d^2 h_{ij}(\eta)}{d\eta^2}(\xi^0)^2 + \cdots \right)dx^i dx^j
$$

(1.76)

because both the inhomogeneous perturbation $\mathbf{h}$ and the coordinate transformation ξ^0 are considered small, and we kept only terms linear in small quantities. In the derivative used in an evolution equation with respect to η, by contrast, we divide the expansion by $\delta\eta = \xi^0$. This operation turns the term $(dh_{ij}/d\eta)\xi^0$ in Equation (1.76), which would be considered as second order in our derivation of invariant quantities, into the first-order term, $dh_{ij}/d\eta$. Accordingly, all perturbations, including those for scalar and vector modes, obey non-trivial evolution equations.

Only the tensor mode represents propagating degrees of freedom, subject to the wave equation

$$
-\frac{d^2 \mathbf{h}^{TT}}{d\eta^2} + \Delta\mathbf{h}^{TT} - \frac{1}{a}\frac{da}{d\eta}\frac{d\mathbf{h}^{TT}}{d\eta} = 0,
$$

(1.77)

a hyperbolic partial differential equation that describes gravitational waves. The vector mode is subject to the elliptic equation $\Delta \vec{V} = 0$ and changes in time according to an equation

$$\frac{\mathrm{d}}{\mathrm{d}\eta}\left(\frac{\partial V_j}{\partial x^i} + \frac{\partial V_i}{\partial x^j}\right) = 0 \tag{1.78}$$

which is only of first order in time and does not have wave-like solutions. (It merely implies that $\partial V_j/\partial x^i + \partial V_i/\partial x^j$ does not depend on η for any i and j.)

The scalar modes are subject to several equations, which include the condition that $\Phi = \Psi$ for scalar matter Π, as well as further constraints

$$\frac{\mathrm{d}\Psi}{\mathrm{d}\eta} + \frac{1}{a}\frac{\mathrm{d}a}{\mathrm{d}\eta}\Psi = \frac{4\pi G}{c^2}\frac{\mathrm{d}\bar{\varphi}}{\mathrm{d}\eta}\Pi \tag{1.79}$$

and

$$\Delta\Psi - \frac{3}{a}\frac{\mathrm{d}a}{\mathrm{d}\eta}\left(\frac{\mathrm{d}\Psi}{\mathrm{d}\eta} + \frac{1}{a}\frac{\mathrm{d}a}{\mathrm{d}\eta}\Psi\right) = \frac{4\pi G}{c^2}\delta\rho \tag{1.80}$$

with the density perturbation $\delta\rho$. Moreover, they are subject to a second-order evolution equation

$$\frac{\mathrm{d}^2\Psi}{\mathrm{d}\eta^2} + \frac{3}{a}\frac{\mathrm{d}a}{\mathrm{d}\eta}\frac{\mathrm{d}\Psi}{\mathrm{d}\eta} + \left(\frac{2}{a}\frac{\mathrm{d}^2a}{\mathrm{d}\eta^2} - \left(\frac{1}{a}\frac{\mathrm{d}a}{\mathrm{d}\eta}\right)^2\right)\Psi = 4\pi G\delta P \tag{1.81}$$

with the pressure perturbation δP. (The origin of these equations in general covariance will be explained in Section 2.3.5.) This second-order equation does not have spatial derivatives of Ψ and is therefore not a wave equation. The scalar mode of the metric does not propagate on its own but only reacts to the distribution of scalar matter described by Π (on which $\delta\rho$ and δP depend). The scalar field Π is subject to a wave equation with source terms that depend on the field's self-interaction, described by a potential $W(\bar{\varphi})$:

$$-\frac{\mathrm{d}^2\Pi}{\mathrm{d}\eta^2} + \Delta\Pi - \frac{2}{a}\frac{\mathrm{d}a}{\mathrm{d}\eta}\frac{\mathrm{d}\Pi}{\mathrm{d}\eta} - a^2\frac{\mathrm{d}^2W}{\mathrm{d}\bar{\varphi}^2}\Pi - 2a^2\frac{\mathrm{d}W}{\mathrm{d}\bar{\varphi}}\Psi + 4\frac{\mathrm{d}\bar{\varphi}}{\mathrm{d}\eta}\frac{\mathrm{d}\Psi}{\mathrm{d}\eta} = 0. \tag{1.82}$$

Propagating scalars therefore result only from matter, not from metric perturbations.

The scalar and vector modes have to obey more than one partial differential equation for each field component. Such systems of differential equations cannot be set up arbitrarily, but must be subject to important consistency or integrability conditions. In particular, by looking at the highest-order time derivatives in the equations, we can split them into evolution equations and constraints: The highest-order equations, the second-order Equation (1.81) in the scalar case, determine how the field evolves starting with initial values for Ψ and $\mathrm{d}\Psi/\mathrm{d}\eta$ at some initial η. The remaining equations, (1.79) and (1.80) in the scalar case, restrict the allowed initial values because they depend only on Ψ and $d\Psi/d\eta$ and their spatial derivatives. We therefore have to make sure that any initial values we choose obey these constraint equations. However, Equations (1.79) and (1.80) must hold at all times. Starting with solutions of these two equations as initial values for the evolution Equation (1.81) must therefore result in fields that still obey the constraint equations at any later time. In the given case, one can check by a lengthy calculation that this is indeed the case because time derivatives of Equations (1.79) and (1.80) lead to identities if one uses Equation (1.81) as well as thermodynamics equations obeyed by the matter terms (the latter are inhomogeneous versions of the continuity equation.)

This consistency is not a coincidence but a consequence of general covariance, as we will discuss in more detail in the Chapter 2. Maintaining this consistency will be a major problem once we turn to the questions of how cosmological equations can be modified by quantum effects; see Chapter 6. If we try to modify Equation (1.81), say, we must, at the same time, modify Equations (1.79) and (1.80) in a specific way so as to maintain consistency.

1.4.4 Inflationary Structure Formation

As a classic application of our mode equations, we can introduce the basic features of the inflation scenario. Inflation models begin with an initial state of the matter distribution (described by a scalar field called the inflaton), which is empty except for unavoidable quantum fluctuations of the field. The unstable nature of gravity, in particular in an accelerating universe, is argued (Guth & Pi 1985; Lyth 1985) to turn this distribution into the seeds of structure that can, at a later time, be observed in the cosmic microwave background radiation or in the Galaxy distribution. A complete description requires an understanding of how a quantum state transitions dynamically to an approximate classical distribution (Polarski & Starobinsky 1996; Lesgourgues et al. 1997; Barvinsky et al. 1999; Kiefer et al. 1998, 2007; Burgess et al. 2008), which in practice is often based on decoherence scenarios (Zeh 1970; Giulini et al. 2003; Schlosshauer 2007). However, it remains unclear how the expansion of a homogeneous background spacetime can break translation symmetries, such that a homogeneous vacuum evolves into a structured state (Perez et al. 2006; Mukhopadhyay & Vachaspati 2019).

Inflation rapidly magnifies a small region in a nearly empty universe to a significantly larger scale. It is therefore reasonable to assume that this process can be described by the linear perturbation equations we just introduced. In order to overcome the problem that accelerated expansion requires negative pressure, see Equation (1.14), most inflation models require a scalar field whose potential has a range in which it is nearly constant. For the corresponding values of the potential, the inflation changes only slowly (*slow-roll*) and its momentum and kinetic energy are small.

Perturbations During Exponential Expansion

A homogeneous distribution of a generic scalar field φ with momentum p_φ and potential $W(\varphi)$ has the energy

$$\mathrm{H}_{\text{scalar}} = \frac{c^2}{2} \frac{p_\varphi^2}{V} + VW(\varphi) \tag{1.83}$$

in any spatial region with volume V. (The mass m of a scalar field is determined by the coefficient of the quadratic contribution $\frac{1}{2}m^2c^2\phi^2/\hbar^2$ to the potential $W(\phi)$. Compared with the Hamiltonian in classical mechanics, the kinetic energy has therefore been multiplied by mc^2 in order to remove the mass. See also Section 4.1.1 where the factor of c^2 can be seen to originate in the time component of the metric.) We therefore derive the pressure

$$P_{\text{scalar}} = -\frac{\partial \mathrm{H}_{\text{scalar}}}{\partial V} = \frac{c^2}{2} \frac{p_\varphi^2}{V^2} - W(\varphi). \tag{1.84}$$

If p_ϕ is small, $P_{\text{scalar}} \approx -W(\varphi) \approx -\rho_{\text{scalar}}$ approximately equals the negative energy density, and therefore $\rho_{\text{scalar}} + 3P_{\text{scalar}} < 0$. For this kind of matter state, the continuity Equation (1.10) implies that the energy density is nearly constant in time. This matter is not diluted in an expanding universe, behaving like a cosmological constant or dark energy. The Friedmann equation is then solved by $a(\tau) \propto \exp(\sqrt{8\pi G\rho/3}\,\tau/c)$. In conformal time, $\eta \propto -\exp(-\sqrt{8\pi G\rho/3}\,\tau/c)$, or $a(\eta) \propto \eta^{-1}$.

During slow-roll, derivatives of the flat potential in Equation (1.82) are small and can be ignored. Moreover, since structure has not formed yet, we can ignore $d\Psi/d\eta$ in the last term. Therefore, approximately

$$-\frac{d^2\Pi}{d\eta^2} + \Delta\Pi - \frac{2}{a} \frac{da}{d\eta} \frac{d\Pi}{d\eta} = 0 \tag{1.85}$$

where $a^{-1}da/d\eta = -1/\eta$ during inflation.

This linear partial differential equation can be solved using Fourier transformation in space, expressing the matter field Π as a superposition of plane waves:

$$\Pi(\eta, x, y, z) = \frac{1}{(2\pi)^{3/2}} \int dx dy dz \Pi_{\vec{k}}(\eta) \exp(i\vec{k} \cdot \vec{x}). \tag{1.86}$$

The partial differential Equation (1.85) for Π is then turned into a set of infinitely many (but uncoupled) ordinary differential equations

$$\frac{d^2\Pi_{\vec{k}}}{d\eta^2} + \frac{2}{a} \frac{da}{d\eta} \frac{d\Pi_{\vec{k}}}{d\eta} + |\vec{k}|^2 \Pi_{\vec{k}} = 0 \tag{1.87}$$

for the $\Pi_{\vec{k}}(\eta)$.

The substitution $v_{\vec{k}} = a\Pi_{\vec{k}}$ removes the first-order term: We obtain the *Mukhanov–Sasaki equation*

$$\frac{d^2 v_{\vec{k}}}{d\eta^2} + \left(|\vec{k}|^2 - \frac{1}{a} \frac{d^2 a}{d\eta^2}\right) v_{\vec{k}} = 0. \tag{1.88}$$

During inflation, $a^{-1}\mathrm{d}^2a/\mathrm{d}\eta^2 = 2/\eta^2$. For any given $\vec{k}$, the evolution equation is then solved by

$$v_{\vec{k}}(\eta) = A(\vec{k})e^{-i|\vec{k}|\eta}\left(1 - \frac{i}{|\vec{k}|\eta}\right) + B(\vec{k})e^{i|\vec{k}|\eta}\left(1 + \frac{i}{|\vec{k}|\eta}\right) \tag{1.89}$$

with two free η-independent functions, $A(\vec{k})$ and $B(\vec{k})$, as appropriate for the general solution of a set of second-order ordinary differential equations. (Fourier modes are not required to be real but must be such that $v_{-\vec{k}} = v_{\vec{k}}^*$ to guarantee a real Π. Therefore, $B(\vec{k}) = A(-\vec{k})^*$ and two real functions, given by the real and imaginary parts of $A(\vec{k})$, remain free.)

Since Equation (1.88), for given $\vec{k}$, is of the form of a harmonic-oscillator equation of motion with time-dependent frequency

$$\omega_{\vec{k}} = \sqrt{|\vec{k}|^2 - 2/\eta^2}, \tag{1.90}$$

basic quantum mechanics can be used to deduce properties of a quantized $v_{\vec{k}}$, in particular of its quantum fluctuations that are supposed to determine the initial state; see Chapter 3.

We set up the initial state at the beginning of the inflation phase, where $\tau \to -\infty$ such that $a(\tau) \to 0$, which also corresponds to $\eta \to -\infty$. In this limit, the frequency is time-independent and equals $\omega = |\vec{k}|$. For this frequency, and mass $m = 1$ since no such parameter appears in Equation (1.88), quantum mechanics implies that the quantum fluctuations of a quantized $v_{\vec{k}}$ are given by $\Delta v_{\vec{k}} = \sqrt{\hbar/(2|\vec{k}|)}$ (derived in Equation (3.92)). Time dependence in quantum mechanics always comes by factors of $e^{-i\omega t}$ (implying a positive energy $E = \hbar\omega$), which in the present notation equals $e^{-i|\vec{k}|\eta}$. This function is given by the A-term in our solution (1.89). Therefore, quantum fluctuations suggest the initial conditions $B(\vec{k}) = 0$ and

$$A(\vec{k}) = \sqrt{\hbar/(2|\vec{k}|)}. \tag{1.91}$$

Cosmological observations do not directly measure $v_{\vec{k}}$ or $\Phi_{\vec{k}}$. From the statistical distribution of density variations, $\delta\rho$, one can determine the correlation function $\langle\delta\rho_{\vec{k}}\delta\rho_{\vec{k}'}^*\rangle$. The angular brackets mean that the product of two density variations with wave numbers $\vec{k}$ and $\vec{k}'$, respectively, is computed for all data points and then averaged over the entire ensemble. Using this statistical concept, the *matter power spectrum* is defined as

$$P_\rho(\vec{k}) = \int \mathrm{d}^3k'\left\langle\delta\rho_{\vec{k}}\delta\rho_{\vec{k}'}^*\right\rangle \tag{1.92}$$

and is usually assumed to be of a certain power-law form $P_\rho(\vec{k}) \propto |\vec{k}|^{n_s}$ with the *scalar spectral index* n_s.

The density variation is related to $\Phi_{\vec{k}}$, which represents the relativistic version of Newton's potential, through the Fourier-transformed Poisson equation

$$-|\vec{k}|^2 \Phi_{\vec{k}} = \frac{4\pi G}{c^2} \delta\rho_{\vec{k}}. \tag{1.93}$$

Therefore, the power spectrum

$$P_\Phi(\vec{k}) = \int \mathrm{d}^3 k' \langle \Phi_{\vec{k}} \Phi_{\vec{k}'}^* \rangle \tag{1.94}$$

of the scalar mode is also of a power-law form,

$$P_\Phi(\vec{k}) \propto \frac{P_\rho(\vec{k})}{|\vec{k}|^4} \propto |\vec{k}|^{n_s - 4}. \tag{1.95}$$

If there is no distinguished scale in the universe, we have a *scale-free power spectrum*. Any length scale is then supplied only by the argument of the power spectrum, the wave number $|\vec{k}|$ which has units of $1/[L]$. Counting the k-integrations involved in Fourier transformations and in the definition of the power spectrum, we first note that $\Phi_{\vec{k}}$ has units of $1/[k]^3 = [L]^3$ because it is defined through three-dimensional Fourier transformation. The power spectrum (1.94) contains two such factors, together with another three-dimensional k-integration. Therefore, it has units of $1/[k]^3 = [L]^3$. In a scale-free distribution, all these units must be supplied by $\vec{k}$ itself, and therefore

$$P_\Phi(\vec{k}) \propto |\vec{k}|^{-3}. \tag{1.96}$$

Comparison with Equation (1.95) then implies the spectral index equals $n_s = 1$ for a scale-free spectrum.

In inflation models, we have a solution for $\Pi_{\vec{k}}$ rather than $\Phi_{\vec{k}}$, but their power spectra are the same because the two fields are related by an Equation (1.79) that does not contain spatial derivatives and therefore does not change the scale behavior. The perturbation $\Pi_{\vec{k}} = v_{\vec{k}}/a$, defined just before Equation (1.88), depends on $\vec{k}$ as well as the scale factor a or conformal time η. However, an inspection of Equation (1.87) shows that there is a distinguished time for each mode: During inflation, the coefficient $a^{-1}\mathrm{d}a/\mathrm{d}\eta = \mathrm{d}a/\mathrm{d}\tau = aH$ is increasing. For a given mode $\vec{k}$, once $a^{-1}\mathrm{d}a/\mathrm{d}\eta$ is large enough for $|\vec{k}| \ll a^{-1}\mathrm{d}a/\mathrm{d}\eta$ to hold, the evolution is approximately independent of $\vec{k}$, determined by $\mathrm{d}^2\Pi_{\vec{k}}/\mathrm{d}\eta^2 = -2(a^{-1}\mathrm{d}a/\mathrm{d}\eta)\mathrm{d}\Pi_{\vec{k}}/\mathrm{d}\eta$. This equation has one η-independent solution, and a second solution given by $\Pi_{\vec{k}}^{-1}\mathrm{d}\Pi_{\vec{k}}/\mathrm{d}\eta = -2a^{-1}\mathrm{d}a/\mathrm{d}\eta$, or $\Pi_{\vec{k}}(\eta) = a(\eta)^{-2}$ which decreases and therefore becomes negligible. Long-wavelength modes (small $|\vec{k}|$) therefore do not evolve noticeably. The same mode only starts evolving again long after the end of inflation, when $a^{-1}\mathrm{d}a/\mathrm{d}\eta$ has decreased enough such that $|\vec{k}| \ll a^{-1}\mathrm{d}a/\mathrm{d}\eta$ is no longer fulfilled. At this time, the mode becomes relevant for structure formation, and because it did not evolve since inflation, it still maintains the value it had when $|\vec{k}| = a^{-1}\mathrm{d}a/\mathrm{d}\eta = aH$ was fulfilled the first time. The primordial power spectrum can therefore be computed (as a function only of $|\vec{k}|$) by evaluating our solutions $v_{\vec{k}}(\eta)$ at a time given by $|\vec{k}| = a^{-1}\mathrm{d}a/\mathrm{d}\eta = aH$.

With these ingredients, we compute the power spectrum

$$P_{\Pi}(\vec{k}) \propto |\Pi_{\vec{k}}|^2_{|\vec{k}|=aH} = \left| \frac{v_{\vec{k}}}{a} \right|^2_{|\vec{k}|=aH}. \tag{1.97}$$

For our solution with the specified initial values,

$$\left| \frac{v_{\vec{k}}}{a} \right|^2 = a^{-2}A^2\left(1 + \frac{1}{|\vec{k}|^2\eta^2}\right)$$

$$= \frac{1}{2}\hbar a^{-2}|\vec{k}|^{-1}\left(1 + \frac{a^2H^2}{|\vec{k}|^2}\right) \approx \hbar\frac{H^2}{|\vec{k}|^3} \tag{1.98}$$

using $1/\eta = -\mathrm{d}a/\mathrm{d}\eta = -aH$. Therefore, $P_{\Pi}(\vec{k}) \propto H^2/|\vec{k}|^3$ has, for nearly constant H, the $|\vec{k}|$-dependence (1.96) expected for a scale-free spectrum.

Slow-roll Approximation

If inflation is sourced by a scalar field with non-constant potential $W(\varphi)$, the scale factor is accelerated but not in an exponential manner. Inhomogeneous perturbations then react differently to the expanding background than in exponential expansion, and modifications in the resulting power spectrum can be seen. In particular, the non-constant H in Equation (1.98) contributes to the scale dependence. For slowly changing H, the power spectrum is nearly scale-free, but not exactly scale-free. It is difficult to find exact solutions for the background in such a situation, but deviations from exponential expansion can nevertheless be derived in the *slow-roll approximation*, which expands solutions as a power series in $\mathrm{d}^nW(\varphi)/\mathrm{d}\varphi^n$.

In practice, the first two terms in the slow-roll expansion are sufficient. As dimensionless quantities, they are defined as

$$\epsilon_W = \frac{c^4}{16\pi G}\left(\frac{1}{W}\frac{\mathrm{d}W}{\mathrm{d}\varphi}\right)^2, \quad \eta_W = \frac{c^4}{8\pi G}\frac{1}{W}\frac{\mathrm{d}^2W}{\mathrm{d}\varphi^2}. \tag{1.99}$$

If these parameters are small, the following slow-roll approximation is self-consistent: We assume that the scalar field $\varphi(\tau)$ changes sufficiently slowly compared with the scale factor $a(\tau)$, in the sense that

$$\frac{4\pi G}{c^4}\frac{1}{H^4}\left(\frac{\mathrm{d}^2\varphi}{\mathrm{d}\tau^2}\right)^2 \ll \frac{4\pi G}{c^4}\frac{1}{H^2}\left(\frac{\mathrm{d}\varphi}{\mathrm{d}\tau}\right)^2 \ll 1. \tag{1.100}$$

We can then approximate the Friedmann equation

$$H^2 = \frac{8\pi G}{3c^2}\left(\frac{1}{2c^2}\left(\frac{d\varphi}{d\tau}\right)^2 + W(\varphi)\right), \tag{1.101}$$

with an energy density $\rho = H_{\text{scalar}}/V$ derived from the Hamiltonian (1.83), by

$$H^2 \approx \frac{8\pi G}{3c^2}W(\varphi), \tag{1.102}$$

and the continuity equation

$$\frac{d^2\varphi}{d\tau^2} + 3H\frac{d\varphi}{d\tau} + c^2\frac{dW}{d\varphi} = 0 \tag{1.103}$$

by

$$3H\frac{d\varphi}{d\tau} \approx -c^2\frac{dW}{d\varphi}. \tag{1.104}$$

Combinations of these equations show how self-consistency of the slow-roll approximation is related to smallness of the slow-roll parameters. Using first Equation (1.104) and then Equation (1.102), we have

$$\frac{4\pi G}{c^4}\frac{1}{H^2}\left(\frac{d\varphi}{d\tau}\right)^2 \approx \frac{4\pi G}{9}\frac{1}{H^4}\left(\frac{dW}{d\varphi}\right)^2 \approx \frac{c^4}{16\pi G}\left(\frac{1}{W}\frac{dW}{d\varphi}\right)^2 \tag{1.105}$$
$$= \epsilon_W \ll 1.$$

Moreover, if the first inequality in Equation (1.100) is fulfilled, using Equation (1.102) several times we have

$$\frac{d(3H\,d\varphi/d\tau)}{d\tau} = 3\frac{dH}{d\tau}\frac{d\varphi}{d\tau} + 3H\frac{d^2\varphi}{d\tau^2} \approx 3\frac{dH}{d\tau}\frac{d\varphi}{d\tau}$$
$$\approx \frac{1}{c}\frac{d\sqrt{24\pi GW(\varphi)}}{d\varphi}\left(\frac{d\varphi}{d\tau}\right)^2 \tag{1.106}$$
$$= \sqrt{\frac{6\pi G}{c^2}}\frac{1}{\sqrt{W}}\frac{dW}{d\varphi}\left(\frac{d\varphi}{d\tau}\right)^2 \approx \frac{4\pi G}{c^2}\frac{1}{H}\frac{dW}{d\varphi}\left(\frac{d\varphi}{d\tau}\right)^2.$$

The same expression, using Equation (1.104) twice, should equal

$$\frac{d(3H\,d\varphi/d\tau)}{d\tau} \approx -c^2\frac{d}{d\tau}\frac{dW}{d\varphi} = -c^2\frac{d^2W}{d\varphi^2}\frac{d\varphi}{d\tau} \approx \frac{c^4}{3H}\frac{d^2W}{d\varphi^2}\frac{dW}{d\varphi}. \tag{1.107}$$

Combining these equations, the second inequality in Equation (1.100) requires

$$\frac{4\pi G}{c^4}\frac{1}{H^2}\left(\frac{d\varphi}{d\tau}\right)^2 \approx \frac{c^2}{3}\frac{1}{H^2}\frac{d^2W}{d\varphi^2} \approx \frac{c^4}{8\pi G}\frac{1}{W}\frac{d^2W}{d\varphi^2} = \eta_W \ll 1. \tag{1.108}$$

While a vanishing ϵ_W implies constant non-zero $W(\varphi)$ (a cosmological constant) and therefore eternal exponential inflation, a small but non-zero ϵ_W ensures that inflation lasts long enough, even while it will eventually end when the scalar field settles in at a local minimum of its non-constant potential. The duration of inflation is conveniently expressed in terms of the *number of e-folds*, defined as the natural logarithm of the ratio of the scale factors at two times, given by the end of inflation, τ_{fin}, and some other relevant time during inflation, τ_*, at which modes are currently still observable:

$$N_* = \ln \frac{a(\tau_{\text{fin}})}{a(\tau_*)}. \tag{1.109}$$

This number can be obtained by integrating the Hubble parameter, and then simplified in a slow-roll approximation, using Equations (1.102) and (1.104):

$$N_* = \ln a(\tau_{\text{fin}}) - \ln a(\tau_*) = \int_{\tau_*}^{\tau_{\text{fin}}} \frac{da}{a}$$

$$= \int_{\tau_*}^{\tau_{\text{fin}}} H(\tau)d\tau \approx \int_{\varphi_*}^{\varphi_{\text{fin}}} \sqrt{\frac{8\pi GW}{3c^2}} \frac{3Hd\varphi}{-c^2 dW/d\varphi} \tag{1.110}$$

$$= \int_{\varphi_{\text{fin}}}^{\varphi_*} \frac{8\pi G}{c^4} \frac{W}{dW/d\varphi} d\varphi = \int_{\varphi_{\text{fin}}}^{\varphi_*} \sqrt{\frac{4\pi G}{c^4 \epsilon_W}} d\varphi.$$

For instance, a power-law potential $W(\phi) = \lambda\varphi^n$ has a slow-roll parameter

$$\epsilon_n(\varphi) = \frac{c^4 n^2}{16\pi G} \frac{1}{\varphi^2}. \tag{1.111}$$

The number of *e*-folds is therefore related to φ_* by

$$N_* \approx \frac{8\pi G}{nc^4}\left(\varphi_*^2 - \varphi_{\text{fin}}^2\right) \approx \frac{8\pi G}{nc^4}\varphi_*^2 \tag{1.112}$$

where we ignore φ_{fin} because, by definition of the end of inflation, it is near the minimum of the power-law potential and therefore much smaller than φ_*. The slow-roll parameter at φ_* can then be expressed as a function of N_*:

$$\epsilon_n(\varphi_*) \approx \frac{n}{2N_*}. \tag{1.113}$$

Using perturbation equations on a background with corrections implied by the slow-roll approximation, the procedure described previously for exponential inflation (Mukhanov et al. 1992) leads to a spectral index which differs from one according to

$$n_s = 1 + 2\eta_W - 6\epsilon_W. \tag{1.114}$$

Moreover, the tensor-to-scalar ratio, given by the ratio of the amplitude of tensor perturbations by the amplitude of scalar perturbations, is determined by

$$r = 16\epsilon_W. \tag{1.115}$$

This quantity should therefore be small in any inflation model that can be described by a slow-roll approximation, but the factor of 16 is rather large. Some models can (and do) therefore get in conflict with observations of this value (Ade & Planck Collaboration 2014). Current upper bounds on r require

$$r < 0.11 \tag{1.116}$$

at 95% confidence level, while

$$n_s = 0.9649 \pm 0.0042 \tag{1.117}$$

at 68% confidence level.

For power-law inflation, using Equation (1.113), we obtain

$$r \approx \frac{8n}{N_*} = 0.13n \tag{1.118}$$

for 60 e-folds, $N_* = 60$. Power-law inflation is therefore ruled out by observations of the tensor-to-scalar ratio. In Section 2.2.4 we will see a potential that is allowed by observations.

1.5 Spherically Symmetric Models

Perturbative inhomogeneity is useful in an analysis of the early universe, at least within a region of spacetime that can be described by the inflation paradigm. If we would like to understand the initial state before inflation, however, we do not have any good reason to rely on near homogeneity, and non-perturbative inhomogeneity is essential. A complete treatment, solving Einstein's equation without any assumptions about the matter distribution, is not possible using analytical solutions, and numerical methods are not very useful if we do not know specific properties of initial values. However, there are several models which are neither exactly nor approximately homogeneous and can still be solved. An important example is given by spherically symmetric models, which reveal properties of black holes and may also be used for cosmological purposes.

1.5.1 Line Element and Equations of Motion

A spherically symmetric line element can depend only on time and a radial coordinate r, but not on the two polar angles ϑ and φ. It must then be of the form

$$ds^2 = -c^2 N^2 dt^2 + L^2(dr + Mcdt)^2 + S^2(d\vartheta^2 + \sin^2\vartheta d\varphi^2) \tag{1.119}$$

with four functions N, L, M and S depending on t and r. The function N also appeared in isotropic and homogeneous cosmological models, where it depended only on t and could be eliminated by introducing a new time coordinate τ such that $d\tau = N(t)dt$; see Section 1.3.2. Here, however, $N(t, r)$ in general depends also on r, such that it cannot unambiguously be integrated only over t. We will keep it as an explicit term in the equations of motion.

In Section 2.3.3 we will derive the Hamiltonian that determines equations of motion for spherically symmetric systems. We will then also see that only the two functions L and S have momenta, p_L and p_S, respectively, while N and M do not. This property is related to the fact that the N in isotropic models could be completely eliminated by introducing a new time coordinate. It therefore does not represent a physical degree of freedom and should not have a momentum.

While we cannot eliminate $N(t, r)$ here, in general, certain isotropic models are a subset of spacetimes described by Equation (1.119), for instance when $M = 0$ and $L = a(t)/\sqrt{1 - kr^2}$ while $S = a(t)r$, see Equation (1.32). If N had an independent momentum for the general (1.119), it would remain independent after specializing to Equation (1.32), which would not be consistent with our results from Section 2.3.3. (One might think that a momentum of N could exist in spherically symmetric models, provided it contained a factor of $L - a(t)/\sqrt{1 - kr^2}$ or $S - a(t)r$ such that it would vanish in an isotropic subsystem. However, the dependence on L or S would mean that it would *not* be independent of the momenta of L or S, respectively.)

The equations of motion, derived from Hamilton's equations in a system which does not have non-zero momenta for N and M, are of second order in time only for L and S, while N and M may appear together with spatial derivatives up to second order, as well as time derivatives up to first order. These equations, in the vacuum case, are

$$
-\frac{1}{N}\frac{\partial}{\partial t}\left(\frac{1}{N}\frac{\partial(SL)}{\partial t}\right) + \frac{1}{N}\frac{\partial}{\partial t}\left(\frac{1}{N}\frac{\partial(MSL)}{\partial r}\right)
$$
$$
+ \frac{1}{N}\frac{\partial}{\partial r}\left(\frac{M}{N}\left(\frac{\partial(SL)}{\partial t} - \frac{\partial(MSL)}{\partial r}\right)\right)
$$
$$
+ \frac{1}{N^2}\left(\frac{\partial L}{\partial t} - \frac{\partial(ML)}{\partial r}\right)\left(\frac{\partial S}{\partial t} - M\frac{\partial S}{\partial r}\right) \tag{1.120}
$$
$$
= -\frac{1}{L}\frac{\partial^2 S}{\partial r^2} - \frac{1}{LN}\frac{\partial N}{\partial r}\frac{\partial S}{\partial r} + \frac{1}{L^2 N}\frac{\partial L}{\partial r}\frac{\partial(NS)}{\partial r} - \frac{S}{LN}\frac{\partial^2 N}{\partial r^2}
$$

and

$$
-\frac{1}{N}\frac{\partial}{\partial t}\left(\frac{S}{N}\left(\frac{\partial S}{\partial t} - M\frac{\partial S}{\partial r}\right)\right) + \frac{M}{N}\frac{\partial}{\partial r}\left(\frac{S}{N}\left(\frac{\partial S}{\partial t} - M\frac{\partial S}{\partial r}\right)\right)
$$
$$
+ \frac{1}{2N}\left(\frac{\partial S}{\partial t} - M\frac{\partial S}{\partial r}\right)^2 \tag{1.121}
$$
$$
= -\frac{1}{NL^2}\frac{\partial S}{\partial r}\frac{\partial(NS)}{\partial r} + \frac{1}{2L^2}\left(\frac{\partial S}{\partial r}\right)^2 + \frac{1}{2}.
$$

The second-order terms in these partial differential equations can be recognized as squares of the derivative operator

$$\hat{t}_S = \frac{1}{N}\left(\frac{\partial}{\partial t} - M\frac{\partial}{\partial r}\right) \tag{1.122}$$

acting on S, and

$$\hat{t}_L = \frac{1}{N}\left(\frac{\partial}{\partial t} - \frac{\partial}{\partial r}M\right) \tag{1.123}$$

acting on L or SL. (There is a slight difference in the derivative operators for S and L, respectively, because L, unlike S, has a density weight. This geometrical property is explained in more detail in Section 2.3.4; see Equations (2.178) and (2.181).)

Therefore, N and M determine the direction and the rate at which time evolves in these equation for S and L, with respect to the coordinates t and r that have initially been assumed without any reference to the specific dynamics. For this reason, N and M do not appear with second-order time derivatives and do not have independent momenta because they merely determine a direction but are not dynamical. Alluding to this role, N is called the *lapse function* and M the *shift vector*. (In a model with full spatial inhomogeneity, there would be three components of a vector M^i in a contribution $dx^i + M^i dt$ to the line element, generalizing the radial term in Equation (1.119). In a spherically symmetric spacetime, $\vec{M}$ must be radial and has just one non-zero component, M.)

At first sight, Equations (1.120) and (1.121) do not seem to be covariant because some of the second-order time derivatives are not accompanied by corresponding second-order space derivatives. Nevertheless, these equations *are* covariant—provided the functions S and L together with their first-order derivatives are subject to the constraints

$$-\frac{S}{N^2}\left(\frac{\partial L}{\partial t} - \frac{\partial(ML)}{\partial r}\right)\left(\frac{\partial S}{\partial t} - M\frac{\partial S}{\partial r}\right) - \frac{L}{2N^2}\left(\frac{\partial S}{\partial t} - M\frac{\partial S}{\partial r}\right)^2$$
$$+ \frac{S}{L}\frac{\partial^2 S}{\partial r^2} - \frac{S}{L^2}\frac{\partial S}{\partial r}\frac{\partial L}{\partial r} + \frac{1}{2L}\left(\frac{\partial S}{\partial r}\right)^2 - \frac{L}{2} = 0 \tag{1.124}$$

$$-\frac{1}{N}\frac{\partial S}{\partial r}\left(\frac{\partial(SL)}{\partial t} - \frac{\partial(MSL)}{\partial r}\right) + L\frac{\partial}{\partial r}\left(\frac{S}{N}\left(\frac{\partial S}{\partial t} - M\frac{\partial S}{\partial r}\right)\right) = 0. \tag{1.125}$$

Moreover, these constraints are preserved by the evolution equations, and therefore need be imposed only on initial values. We will demonstrate all these statements about covariance and the direction of evolution in Section 2.3, once we have developed the formalism of canonical gravity which is most useful for this purpose. In Section 2.3.4 we will derive the equations of motion and constraints of spherically symmetric general relativity.

1.5.2 Lemaître–Tolman–Bondi Models

As an example, we may assume that coordinates can be chosen such that $N = 1$ and $M = 0$, in which case the time coordinate t behaves like proper time. In a cosmological context, the line element (1.119) with this choice is referred to as a Lemaître–Tolman–Bondi (LTB) model (Lemaître 1933; Tolman 1934; Bondi 1947); see Enqvist (2008) for a recent review of cosmological applications. The two remaining metric components, L and S, are then subject to the differential equations

$$-\frac{\partial^2(SL)}{\partial t^2} + \frac{1}{L}\frac{\partial^2 S}{\partial r^2} + \frac{\partial L}{\partial t}\frac{\partial S}{\partial t} - \frac{1}{L^2}\frac{\partial L}{\partial r}\frac{\partial S}{\partial r} = 0 \tag{1.126}$$

from Equation (1.120),

$$-S\frac{\partial^2 S}{\partial t^2} - \frac{1}{2}\left(\frac{\partial S}{\partial t}\right)^2 + \frac{1}{2L^2}\left(\frac{\partial S}{\partial r}\right)^2 - \frac{1}{2} = 0 \tag{1.127}$$

from Equation (1.121), and

$$-S\frac{\partial L}{\partial t}\frac{\partial S}{\partial t} - \frac{1}{2}L\left(\frac{\partial S}{\partial t}\right)^2 + \frac{S}{L}\frac{\partial^2 S}{\partial r^2} - \frac{S}{L^2}\frac{\partial S}{\partial r}\frac{\partial L}{\partial r} + \frac{1}{2L}\left(\frac{\partial S}{\partial r}\right)^2 - \frac{L}{2} = 0 \tag{1.128}$$

$$-\frac{\partial S}{\partial r}\frac{\partial L}{\partial t} + L\frac{\partial^2 S}{\partial t\partial r} = 0 \tag{1.129}$$

from the constraints.

Substituting $T = \partial S/\partial r$ in the last equation, it takes the form

$$\frac{1}{L}\frac{\partial L}{\partial t} = \frac{1}{T}\frac{\partial T}{\partial t} \tag{1.130}$$

which has the general solution

$$L(t, r) = f(r)T(t, r) = f(r)\frac{\partial S}{\partial r} \tag{1.131}$$

with a free function $f(r)$. This function depends on the choice of the radial coordinate because a transformation $r \mapsto \tilde{r}$ should map L in Equation (1.119) to $(d\tilde{r}/dr)L$. Such a free function is therefore required in the general solution to guarantee coordinate invariance, but it does not have direct physical significance in terms of observable quantities.

Since we know L (for given f), Equation (1.127) is turned into a differential equation only for S,

$$\begin{aligned}
0 &= -S\frac{\partial^2 S}{\partial t^2} - \frac{1}{2}\left(\frac{\partial S}{\partial t}\right)^2 + \frac{1/f(r)^2 - 1}{2} \\
&= -\frac{1}{2\partial S/\partial t}\left(\frac{\partial}{\partial t}\left(S\left(\frac{\partial S}{\partial t}\right)^2\right) - \left(\frac{1}{f(r)^2} - 1\right)\frac{\partial S}{\partial t}\right).
\end{aligned} \tag{1.132}$$

Setting the last parenthesis equal to zero, we integrate once and obtain

$$S\left(\frac{\partial S}{\partial t}\right)^2 = -k(r)S + g(r) \tag{1.133}$$

with a second free function, $g(r)$, where we defined $k(r) = 1 - 1/f(r)^2$. The first equation of motion, (1.126), now takes the form

$$-f\frac{\partial}{\partial r}\left(S\frac{\partial^2 S}{\partial t^2} + \frac{1}{2}\left(\frac{\partial S}{\partial t}\right)^2 + \frac{k(r)}{2}\right) = 0 \tag{1.134}$$

and is identically satisfied if Equation (1.133) holds.

We can rewrite Equation (1.133) as

$$\left(\frac{1}{S}\frac{\partial S}{\partial t}\right)^2 + \frac{k(r)}{S^2} = \frac{g(r)}{S^3} \tag{1.135}$$

which at any fixed position r is a Friedmann-like equation for the position-dependent scale factor $S(t, r)$. The spatial curvature parameter is replaced by the function $k(r)$, and we have dust-like matter with position-dependent energy density $(3c^2/(8\pi G))g(r)/S(t, r)^3$. In strictly homogeneous models, the function $k(r) = 0$ or $k(r) = \pm r^2$, depending on the topology, is fixed by making a choice of radial coordinate such that $S(t, r) = a(t)r$ in Equation (1.119). The time-dependent function $a(t)$ is then interpreted as the scale factor.

At this stage, the energy density is determined by a free function, $g(r)$. However, we have not implemented all equations yet. Inserting Equation (1.131) in (1.128), we obtain

$$\begin{aligned}
0 = & -Sf(r)\frac{\partial^2 S}{\partial t \partial r}\frac{\partial S}{\partial t} - \frac{1}{2}f(r)\frac{\partial S}{\partial r}\left(\frac{\partial S}{\partial t}\right)^2 \\
& + \frac{S}{f(r)\partial S/\partial r}\frac{\partial^2 S}{\partial r^2} - \frac{S}{f(r)^2 \partial S/\partial r}\left(\frac{\partial f(r)}{\partial r}\frac{\partial S}{\partial r} + f(r)\frac{\partial^2 S}{\partial r^2}\right) \\
& + \frac{1}{2f(r)}\frac{\partial S}{\partial r} - \frac{f(r)}{2}\frac{\partial S}{\partial r} \\
= & -\frac{1}{2}f(r)\frac{\partial}{\partial r}\left(S\left(\frac{\partial S}{\partial t}\right)^2 + k(r)S\right) = -\frac{1}{2}f(r)\frac{\partial g}{\partial r},
\end{aligned} \tag{1.136}$$

using Equation (1.135) in the last step.

Therefore, $g(r)$ is constant, and Equation (1.135), for suitable $k(r)$, is nothing but the Friedmann equation with dust matter. However, the equations we solved assumed vacuum, without a dust energy added to the Hamiltonian, which corresponds to Equation (1.128). The constant g must, in fact, be zero if we impose a suitable boundary condition at some point, r_0, where there is no matter. Since $g(r_0) = 0$ and g is constant, g must then vanish everywhere.

We can generalize the equations to dust matter by adding G/c^2 times the local dust energy, $4\pi L S^2 \rho(t, r)$ with $L = f(r)\partial S/\partial r$, to the Hamiltonian on the left-hand side of Equation (1.128). (In order to obtain the local dust energy in a small shell between radii r and $r + dr$, we multiply the density ρ with (i) the area $4\pi S(t, r)^2$ of a sphere of constant r and (ii) the radial scale factor $L(t, r)$ so as to turn dr into the actual distance Ldr measured by the line element.) Adding the same term to the right-hand side of Equation (1.136), which was derived from Equation (1.128), we obtain the equation

$$\frac{1}{2}f(r)\frac{\partial g}{\partial r} = \frac{4\pi G}{c^2}f(r)S^2\frac{\partial S}{\partial r} \tag{1.137}$$

with solution

$$g(t, r) = \frac{8\pi G}{c^2}\int_0^r \rho(t, \tilde{r})S^2\frac{\partial S}{\partial \tilde{r}}d\tilde{r} \tag{1.138}$$

using a flat-space volume integration in the final step. The value $g(t, r)$ at fixed t and r is therefore proportional to the total dust energy in a volume of radius r. For spatially constant $\rho(t) = Mc^2/(V_0 a(t)^3)$ and $S(t, r) = a(t)r$ as in Friedmann models, we obtain

$$\frac{g(t, r)}{S(t, r)^3} = \frac{8\pi GM}{V_0 a(t)^3 r^3}\int_0^r \tilde{r}^2 d\tilde{r} = \frac{8\pi G}{3c^2}\rho(t) \tag{1.139}$$

which is the expected contribution to the Friedmann Equation (1.135) in the case of strict homogeneity.

1.5.3 Canonical Formulation and Static Solutions

Here, we will use only one further ingredient from the canonical formalism, which is the form of momenta of S and L, related to their first-order derivatives:

$$p_S = -\frac{1}{N}\left(\frac{\partial(SL)}{\partial t} - \frac{\partial(MSL)}{\partial r}\right), \quad p_L = -\frac{S}{N}\left(\frac{\partial S}{\partial t} - M\frac{\partial S}{\partial r}\right). \tag{1.140}$$

Notice again the appearance of derivative operators (1.122) and (1.123). These momenta are derived in Section 2.3.4.

The canonical constraints and evolution equations take on more compact forms:

$$-\frac{p_S p_L}{S} + \frac{Lp_L^2}{2S^2} + \frac{S}{L}\frac{\partial^2 S}{\partial r^2} - \frac{S}{L^2}\frac{\partial S}{\partial r}\frac{\partial L}{\partial r} + \frac{1}{2L}\left(\frac{\partial S}{\partial r}\right)^2 + \frac{1}{4}LSV(S) = 0 \tag{1.141}$$

$$p_S\frac{\partial S}{\partial r} - L\frac{\partial p_L}{\partial r} = 0 \tag{1.142}$$

for the two constraints, and

$$
\frac{1}{N}\frac{\partial p_S}{\partial t} = -\frac{p_S p_L}{S^2} + \frac{L p_L^2}{S^3} - \frac{1}{L}\frac{\partial^2 S}{\partial r^2} - \frac{1}{LN}\frac{\partial N}{\partial r}\frac{\partial S}{\partial r}
$$
$$
+ \frac{1}{L^2 N}\frac{\partial L}{\partial r}\frac{\partial(NS)}{\partial r} - \frac{S}{LN}\frac{\partial^2 N}{\partial r^2} \tag{1.143}
$$
$$
- \frac{L}{4}\frac{d(SV(S))}{dS} + \frac{\partial(Mp_S)}{\partial r}
$$

$$
\frac{1}{N}\frac{\partial p_L}{\partial t} = -\frac{p_L^2}{2S^2} - \frac{S}{NL^2}\frac{\partial S}{\partial r}\frac{\partial N}{\partial r} - \frac{1}{2L^2}\left(\frac{\partial S}{\partial r}\right)^2
$$
$$
- \frac{1}{4}SV(S) + M\frac{\partial p_L}{\partial r} \tag{1.144}
$$

for the evolution equations.

In addition to introducing momenta, we have generalized the equations by introducing a potential $V(S)$, such that $V(S) = -2/S$ for spherically symmetric gravity. This potential classifies all local, covariant theories in two-dimensional spacetime (t, r) with second-order equations of motion. These generalized versions of spherically symmetric gravity are called *two-dimensional dilaton gravity models* (Strobl 1999). (See a parenthetical remark after Equation (2.111) in the next chapter for a general definition of dilaton gravity.) Other examples in this class of models include spherically symmetric restrictions of gravity in $D > 3$ spacetime dimensions, in which case $V(S) = -(D - 2)(D - 3)S^{-2/(D-2)}$. Any modification of the equations and constraints that preserves the momenta and second-order nature and is local, but cannot be written as a choice of some suitable $V(S)$, violates general covariance. This statement will play an important role in Chapter 6.

Postponing for now a full proof of the mutual consistency of Equations (1.141)–(1.144), we demonstrate that they do have non-trivial joint solutions. We specialize our equations to a case of *static* geometries, such that all functions depend only on r but not on t, and we have $M = 0$. Moreover, we explore their relationship with covariance by stipulating that we should be able to choose our radial coordinate, r, freely by applying any monotonic function f, mapping r to $\tilde{r} = f(r)$. In particular, because S is assumed to depend only on r in the static case, we can simply use the function $f(r) = S(r)$. (The function $S(r)$ should be monotonic because the area $4\pi S(r)^2$ of spheres with radius r is expected to increase with r.) Transforming to the new coordinate $\tilde{r}$ and also using $M = 0$, the line element (1.119) takes the form

$$
ds^2 = -c^2 N^2 dt^2 + \tilde{L}^2 d\tilde{r}^2 + \tilde{r}^2(d\vartheta^2 + \sin^2\vartheta d\varphi^2) \tag{1.145}
$$

with $\tilde{L}(\tilde{r}(r)) = L(dS/dr)^{-1}$. Dropping the tilde for the sake of convenience, we may therefore assume that we start with a line element (1.119) with only two free functions, $N(r)$ and $L(r)$, while $S(r) = r$.

Given the staticity conditions, the momenta p_S and p_L vanish, simplifying the equations of motion and constraints. In particular, the second constraint, (1.142), is identically satisfied, while the first constraint, (1.141) with $S = r$, takes the form

$$\frac{2r}{L^3}\frac{dL}{dr} - \frac{1}{L^2} - \frac{1}{2}rV(r) = 0. \tag{1.146}$$

Rewriting this equation for the function $K = 1/L^2$ instead of L, such that

$$\frac{d(rK)}{dr} + \frac{1}{2}rV(r) = 0, \tag{1.147}$$

we obtain $rK(r) = -\frac{1}{2}\int rV(r)dr$, or

$$L(r) = \frac{1}{\sqrt{r^{-1}\left(\ell - \frac{1}{2}\int_0^r zV(z)dz\right)}} \tag{1.148}$$

with a constant of integration introducing the length parameter ℓ. For spherically symmetric general relativity in four spacetime dimensions, such that $V(r) = -2/r$, we have

$$L(r) = \frac{1}{\sqrt{1 + \ell/r}}. \tag{1.149}$$

We have now solved both constraints. The remaining metric component, N, is determined by the condition that the evolution equations, (1.144) and (1.143), must be consistent with having zero momenta at all times: $p_S = 0$ and $p_L = 0$ must imply $\partial p_S/\partial t = 0$ and $\partial p_L/\partial t = 0$. Still using $V(r) = -2/r$, we obtain two additional equations,

$$\frac{dN}{dr} + \frac{1 - L^2}{2r}N = 0 \tag{1.150}$$

and

$$r^2\frac{d^2N}{dr} + r\frac{dN}{dr}\left(1 - \frac{r}{L}\frac{dL}{dr}\right) - \frac{r}{L}\frac{dL}{dr}N = 0. \tag{1.151}$$

Equation (1.151) is identically satisfied if we take a derivative by r of Equation (1.150) and insert the result as well as Equation (1.150) itself in Equation (1.151) in order to eliminate all derivatives of N from the latter equation. Any solution of Equation (1.150) is therefore a joint solution of Equations (1.150) and (1.151).

For spherically symmetric general relativity, using Equation (1.149), we rewrite Equation (1.150) as

$$\frac{d\log N}{dr} = -\frac{\ell}{2r^2(1 + \ell/r)} = \frac{1}{2}\frac{d\log(1 + \ell/r)}{dr} \tag{1.152}$$

solved by

$$N(r) = C\sqrt{1 + \ell/r} \qquad (1.153)$$

with an integration constant C that can be set equal to $C = 1$ upon rescaling the time coordinate. There is thus only one free constant, ℓ, that cannot be absorbed in redefined coordinates. With the usual convention $\ell = -2Gm/c^2$ with a new constant m, our solution implies the *Schwarzschild line element*

$$\mathrm{d}s^2 = -\left(1 - \frac{2Gm}{c^2 r}\right)c^2 \mathrm{d}t^2 + \frac{\mathrm{d}r^2}{1 - 2Gm/(c^2 r)} + r^2(\mathrm{d}\vartheta^2 + \sin^2\vartheta \mathrm{d}\varphi^2). \qquad (1.154)$$

This line element describes the geometry of spacetime in the vacuum region around a spherical object of mass m. The ratio

$$R_S = \frac{2Gm}{c^2} = -\ell \qquad (1.155)$$

is its *Schwarzschild radius*.

1.5.4 Non-rotating Black Holes

If a spherical object of mass m extends to a radius greater than the corresponding Schwarzschild radius, $R > R_S$, the vacuum solution (1.154) can only be used for $r > R > R_S$. In this region, all coefficients of Equation (1.154) are regular. However, a more compact object of the same mass may have a matter region restricted to some radius $R < R_S$. The vacuum solution should then apply even in an intermediate region where $r > R$ but $r < R_S$. In this region, $1 - R_S/r < 0$, such that the coefficient of $c^2 \mathrm{d}t^2$ in Equation (1.154) is positive, while the coefficient of $\mathrm{d}r^2$ is negative. The coordinate r/c therefore behaves like time in the line element, while ct behaves like a spatial coordinate. We should not put too much weight on the fact that we denoted t by t because we thought it was a time coordinate, which is after all just a label and justified by our intuitive understanding of near-Minkowski spacetimes only as long as $R_S/r \ll 1$, far from the compact central object. When we consider a region with $r < R_S$, we should correct this mistaken understanding by switching the role of time and radius, using a coordinate transformation such that $T = r/c$ is time and $x = ct$ a spatial coordinate. The Schwarzschild line element is then rewritten as

$$\mathrm{d}s^2 = -\frac{c^2 \mathrm{d}T^2}{2Gm/(c^3 T) - 1} + \left(\frac{2Gm}{c^3 T} - 1\right)\mathrm{d}x^2 + c^2 T^2(\mathrm{d}\vartheta^2 + \sin^2\vartheta \mathrm{d}\varphi^2). \qquad (1.156)$$

This new line element is no longer static because its coefficients depend on T. (It is nevertheless consistent with our previous solutions, derived from staticity, because the conditions we used can be rephrased as using independence of one of the coordinates other than ϑ and φ. Formally, the geometry is static in the direction of an imaginary time, ix.) The previous symmetry which gave us staticity remains, but it is now a spatial symmetry because the coefficients are independent of x. The solution (1.156) therefore describes a time-dependent *spatially homogeneous* spacetime, just as we discussed in cosmological models. In addition to spatial homogeneity, given by translations of x as well as rotations of any sphere at constant x, the symmetry includes an additional rotational axis because the rotation group, SO(3), is three-dimensional. It turns out that this additional symmetry cannot be obtained from any model in the Bianchi classification discussed in Section 1.3.3. It is the only homogeneous geometry of this kind, called the Kantowski–Sachs model (Kantowski & Sachs 1966).

For $r < R_S$, the dynamical nature of the Kantowski–Sachs model indicates that spacetime itself is no longer stable. This region of spacetime cannot be considered collapsing matter in an otherwise stable spacetime background because it is part of the vacuum solution, where no matter resides. The solution describes the interior of a non-rotating black hole, which is delimited not by any surface of a material object but rather by the subset of spacetime defined by $r = R_S$. For this value, the line element (1.154) is not well-defined, but it can be extended to a well-defined geometry by introducing another coordinate transformation.

In order to understand the transition region, we look at radii r close to R_S, writing $r = R_S + \delta r$ with $\delta r \ll R_S$. The coefficients in the line element can therefore be simplified by a Taylor expansion, writing

$$1 - \frac{R_S}{r} = 1 - \frac{1}{1 + \delta r / R_S} = \frac{\delta r}{R_S} + O((\delta r / R_S)^2). \tag{1.157}$$

Before we evaluate the resulting line element, we restrict it to two dimensions, ignoring the angles ϑ and φ, and multiply it with $\delta r / R_S$. Defining $X = \delta r / R_S$, we obtain the line element

$$\mathrm{d}s^2 = -X^2 c^2 \mathrm{d}t^2 + \mathrm{d}X^2. \tag{1.158}$$

Multiplication with X rescales the geometry, but it preserves the behavior of light rays, defined by $\mathrm{d}s^2 = 0$. The new line element therefore describes the same light cones and causal structure as the previous one. (Since we have ignored the angles, we now consider only radial light rays.)

Light rays in a spacetime described by Equation (1.158) are lightlike worldlines with coordinates t and X such that

$$c\frac{\mathrm{d}t}{\mathrm{d}X} = \pm\frac{1}{X} \tag{1.159}$$

solved by

$$ct(X) = D \pm \log|X| \tag{1.160}$$

with constant D.

We can parameterize these one-dimensional curves by a real number λ, such that $X = X(\lambda)$ and $ct = ct(\lambda) = D \pm \log|X(\lambda)|$. Any monotonic reparameterization of λ may be applied without changing the nature of solutions as light rays, solving $\mathrm{d}s^2 = 0$. Increasing λ by a certain amount therefore has no physical meaning, and λ does not describe the length or duration of a stretch of a light ray, which in fact has no length because the only invariant distance measure, ds^2, vanishes on it. It is nevertheless possible to introduce a distinguished parameterization, defined by the condition that it be obtained as a limiting case of timelike or spacelike geodesics approaching a lightlike direction. This *affine parameterization* by λ is therefore defined by the condition that $t(\lambda)$ and $X(\lambda)$ solve not only the first-order differential Equation (1.159) implied by $ds^2 = 0$, but also the second-order equations that result from requiring $\int \sqrt{ds^2}$ to be an extremum among all possible worldlines between two given events.

Writing

$$\int \sqrt{ds^2} = \int \sqrt{-X^2 c^2 \left(\frac{dt}{d\sigma}\right)^2 + \left(\frac{dX}{d\sigma}\right)^2} \, d\sigma = \int L[t(\sigma), X(\sigma)] d\sigma \qquad (1.161)$$

with the Lagrangian

$$L[t(\sigma), X(\sigma)] = \sqrt{-X^2 c^2 \left(\frac{dt}{d\sigma}\right)^2 + \left(\frac{dX}{d\sigma}\right)^2} \qquad (1.162)$$

in terms of some generic curve parameter σ, we can use the Euler–Lagrange equations

$$-\frac{d}{d\sigma} \frac{\partial L}{\partial (dt/d\sigma)} + \frac{\partial L}{\partial t} = 0, \quad -\frac{d}{d\sigma} \frac{\partial L}{\partial (dX/d\sigma)} + \frac{\partial L}{\partial X} = 0 \qquad (1.163)$$

in order to determine extremal worldlines. For our specific L, we obtain

$$c^2 \frac{d}{d\sigma}\left(\frac{X^2}{L} \frac{dt}{d\sigma}\right) = 0 \qquad (1.164)$$

and

$$-\frac{d}{d\sigma}\left(\frac{1}{L} \frac{dX}{d\sigma}\right) - \frac{c^2 X}{L}\left(\frac{dt}{d\sigma}\right)^2 = 0. \qquad (1.165)$$

If we divide these equations by L and define the affine parameter λ by $d\lambda = L d\sigma$, we obtain the geodesic equations

$$\frac{d}{d\lambda}\left(X^2 \frac{dt}{d\lambda}\right) = 0 \qquad (1.166)$$

$$\frac{d^2 X}{d\lambda^2} + X\left(c \frac{dt}{d\lambda}\right)^2 = 0. \qquad (1.167)$$

The first geodesic equation implies that $dt/d\lambda \propto 1/X(\lambda)^2$, which can be solved only if we know $X(\lambda)$. Without solving for $t(\lambda)$, we can insert $dt/d\lambda \propto 1/X^2$ in the second equation, which is then a second-order partial differential equation only for X. However, it is easier to use the light-ray condition (1.159), which implies

$$c\frac{\mathrm{d}t}{\mathrm{d}\lambda} = \pm\frac{1}{X}\frac{\mathrm{d}X}{\mathrm{d}\lambda},$$ (1.168)

in the second geodesic equation:

$$\frac{\mathrm{d}^2 X}{\mathrm{d}\lambda^2} + \frac{1}{X}\left(\frac{\mathrm{d}X}{\mathrm{d}\lambda}\right)^2 = 0.$$ (1.169)

This equation is still of second order, but it can easily be integrated after rewriting

$$\frac{\mathrm{d}^2 X}{\mathrm{d}\lambda^2} + \frac{1}{X}\left(\frac{\mathrm{d}X}{\mathrm{d}\lambda}\right)^2 = \frac{\mathrm{d}}{\mathrm{d}\lambda}\left(X\frac{\mathrm{d}X}{\mathrm{d}\lambda}\right) = 0,$$ (1.170)

such that $X\mathrm{d}X/\mathrm{d}\lambda = A$ is some λ-independent constant, A, and therefore

$$X(\lambda) = \sqrt{2A(\lambda - \lambda_0)}.$$ (1.171)

The second integration constant, λ_0, can be set equal to zero upon shifting the zero-point of λ.

1.5.5 Horizons

With our solutions for light rays in affine parameterization, we can begin to analyze the causal structure of the region around $r = R_S$, or around $X = 0$ in Equation (1.158). The causal structure is determined by light rays, which are given by the solutions we just obtained. In order to probe whether our spacetime has a meaningful extension to the subset where $r = R_S$, we can aim light rays toward this worldline in spacetime, starting from values of r slightly greater or slightly smaller than the Schwarzschild radius. In our two-dimensional spacetime, we use the same kind of light rays but aimed at $X = 0$.

It is then convenient to discard the original coordinates, t and X, in favor of coordinates that directly describe the light rays. The first such coordinate is the parameter D in Equation (1.160). Since it is constant on any given light ray, varying it like a coordinate gives us a continuous family of light rays. The second coordinate is the affine parameter λ, which is a useful measure of how much we have progressed on a given light ray, determined by a fixed D. We therefore transform coordinates from (t, X) to

$$D = ct + \log X \quad \text{and} \quad \lambda = X^2.$$ (1.172)

(It is sufficient to consider only left-moving light rays at positive X, thus the choice of a positive sign in D. The constant $2A$ in Equation (1.171) is irrelevant in this coordinate transformation and is therefore ignored in the second equation.)

Implementing this coordinate transformation in the line element (1.158), we obtain

$$\mathrm{d}s^2 = -\lambda\mathrm{d}D^2 + \mathrm{d}D\mathrm{d}\lambda.$$ (1.173)

This line element no longer defines a diagonal metric because there is a cross-term $\mathrm{d}D\mathrm{d}\lambda$. Nevertheless, it has a great advantage compared with the original line element because it is not singular at $\lambda = 0$: While the coefficient of $\mathrm{d}D^2$ vanishes at $\lambda = 0$, the determinant of the metric,

$$\det g = \det \begin{pmatrix} -\lambda & \dfrac{1}{2} \\ \dfrac{1}{2} & 0 \end{pmatrix} = -\frac{1}{4}, \tag{1.174}$$

is non-zero everywhere. Therefore, we can extend spacetime beyond $\lambda = 0$ by allowing for all real λ, including $\lambda < 0$, and not just the positive λ implied by the second equation of (1.172).

A spacetime diagram visualizes the structure of spacetime according to our new extension. But which coordinate, D or λ or some function of them, should play the role of time? The line element (1.173), taken on its own, does not uniquely answer this question because it is not of the form of a slightly modified Minkowski line element. We have to be more formal and define what it means for a coordinate, $\tilde{T}$, to be considered a time coordinate. If we choose a constant value of $\tilde{T}$, what is left in the line element should be pure space. And pure, timeless space is characterized by a line element which is always positive, provided there is any separation between point. Neither D nor λ fulfill this condition everywhere because $\mathrm{d}s^2|_{D=\mathrm{const}} = 0$ is always zero, while $\mathrm{d}s^2|_{\lambda=\mathrm{const}} = -\lambda \mathrm{d}D^2$ is negative for $\lambda > 0$. The coordinate λ therefore behaves across $X = 0$ much like the original r in Equation (1.154) across $r = R_{\mathrm{S}}$.

While neither D nor λ are time coordinates around $\lambda = 0$, the simple linear combination $\tilde{T} = D - \lambda$ fulfills this purpose: $\mathrm{d}s^2|_{\tilde{T} = \mathrm{const}} = (1 - \lambda)\mathrm{d}\lambda^2 > 0$ for $\lambda < 1$, which does not cover all of spacetime but a large-enough region around $\lambda = 0$. We do not need to rewrite the line element using this new time coordinate (together with λ). It is sufficient to express the equations for light rays, determined by

$$0 = \mathrm{d}s^2 = (-\lambda \mathrm{d}D + \mathrm{d}\lambda)\mathrm{d}D, \tag{1.175}$$

as functions $\tilde{T}(\lambda)$. In terms of D and λ, there are two families of solutions, given by $\mathrm{d}D = 0$ and thus $D(\lambda) = \mathrm{const}$, as well as $\mathrm{d}D = \lambda^{-1}\mathrm{d}\lambda$ and thus $D(\lambda) = \log|\lambda| + \mathrm{const}$. Moreover, there is a solitary solution, given by $\lambda = 0$.

In terms of $\tilde{T}$ and λ, the two families of light rays are given by $\tilde{T}(\lambda) = \mathrm{const} - \lambda$ and $\tilde{T}(\lambda) = \log|\lambda| - \lambda + \mathrm{const}$, respectively, while $\lambda = 0$ remains $\lambda = 0$. The first set of solutions looks just like light rays in Minkowski spacetime, while the second family is not represented by straight lines but rather by curves asymptoting toward the solitary solution $\lambda = 0$ toward $\tilde{T} \to -\infty$. For $\lambda < 0$, both families of light rays have negative slope. Any light ray emitted in this region therefore can only move to the left; see Figure 1.4. It cannot cross $\lambda = 0$ and reach positive values. Light is captured in the region of negative λ, just as in a black hole. This region is bounded by the *horizon* at $\lambda = 0$, which is not a material surface but, as seen by our solutions, one of the light rays.

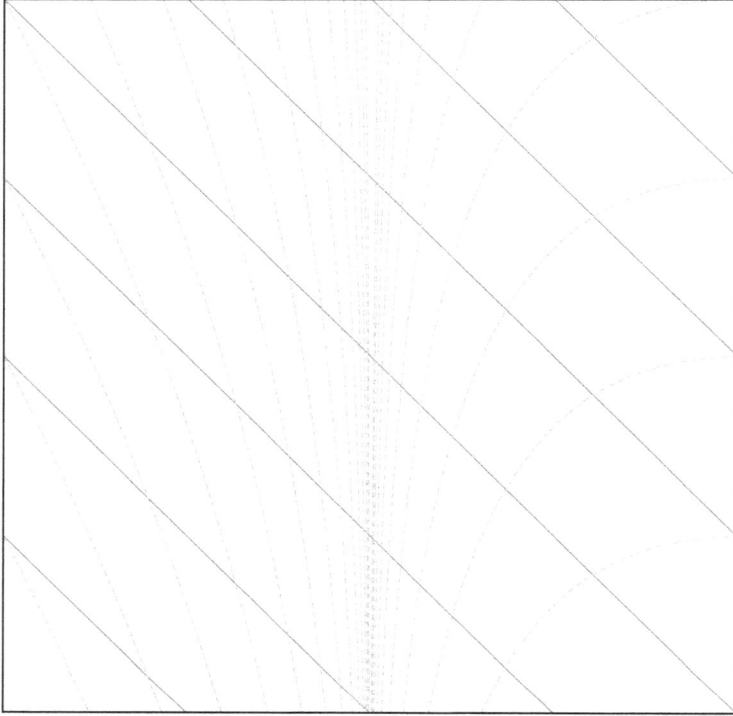

Figure 1.4. Spacetime diagram of the extended line element (1.173). The causal structure is visualized by families of light rays, shown in red for $\tilde{T}(\lambda) = \text{const} - \lambda$ and in green for $\tilde{T}(\lambda) = \log|\lambda| - \lambda + \text{const}$. All light rays, and therefore the entire future light cones, point to the left for negative λ, in the region on the left of the central horizon at $\lambda = 0$.

1.6 Full Inhomogeneity

Without any assumptions about symmetries or perturbations, Einstein's equations are complicated and not much is known about their general solutions. However, mathematically they seem to simplify close to a spacelike singularity such as the Big Bang, even though the physics, such as the thermodynamic behavior of matter, is less clear in this regime. The simplified, near-singular form of spacetime is explored in the Belinskii–Khalatnikov–Lifshitz (BKL) scenario (Belinskii et al. 1982).

We can eliminate uncertainties about matter by working with vacuum solutions, focusing on properties of near-singular geometry. The behavior can be illustrated in spherically symmetric models, for which we have homogeneous solutions as well as evolution equations for non-perturbative inhomogeneity. The homogeneous solution cannot be of the Bianchi types from Section 1.3.3 because spherical symmetry implies rotational invariance around any given point. In terms of the anisotropic scale factors a_i, therefore, two of them must be equal, $a_2 = a_3$.

We have obtained a solution of this form from the Schwarzschild line element for $r < 2Gm/c^2$ (within the Schwarzschild radius) where the coefficients $(1 - 2Gm/(c^2 r))^{\pm 1}$ are negative. Therefore, the role of time is played by $T = r/c$, while $x = ct$ behaves like a radial coordinate. Simply substituting these new names for our old coordinates, we obtain the Kantowski–Sachs line element

$$ds^2 = -\frac{c^2 dT^2}{T_S/T - 1} + (T_S/T - 1)dx^2 + c^2T^2(d\vartheta^2 + \sin^2\vartheta d\varphi^2), \qquad (1.176)$$

here using the "Schwarzschild time"

$$T_S = \frac{R_S}{c} = \frac{2Gm}{c^3}. \qquad (1.177)$$

This line element is homogeneous because all metric components are independent of x, but unlike the Schwarzschild solution for $r > R_S$ it is not static.

We transform to proper time τ by integrating

$$d\tau = \frac{dT}{\sqrt{T_S/T - 1}} = -d\left(T\sqrt{\frac{T_S}{T} - 1} + \frac{1}{2}T_S \arctan \frac{\frac{1}{2}T_S/T - 1}{\sqrt{T_S/T - 1}} \right). \qquad (1.178)$$

For $T \ll T_S$ (close to the singularity), there is a simple relation between T and proper time τ, obtained by integrating

$$d\tau \approx \sqrt{\frac{T}{T_S}}\, dT = \frac{2}{3}\frac{dT^{3/2}}{\sqrt{T_S}}. \qquad (1.179)$$

Therefore, $T = T_S^{1/3}(3\tau/2)^{2/3}$ and the line element is turned into

$$ds^2 \approx - c^2 d\tau^2 + T_S^{2/3}(3\tau/2)^{-2/3}dx^2$$
$$+ c^2 T_S^{2/3}(3\tau/2)^{4/3}(d\vartheta^2 + \sin^2\vartheta d\varphi^2). \qquad (1.180)$$

This solution is of Kasner form with exponents $v_1 = -1/3$ and $v_2 = v_3 = 2/3$. In terms of the functions L and S of a general spherically symmetric line element, we have

$$L(\tau) = T_S^{1/3}(3\tau/2)^{-1/3}, \quad S(\tau) = cT_S^{1/3}(3\tau/2)^{2/3} \qquad (1.181)$$

while $N = 1$ and $M = 0$.

We can now test the behavior of inhomogeneity by choosing functions $L(\tau, x) \propto \tau^{v_1(x)}$ and $S(\tau, x) \propto \tau^{v_2(x)}$ with position-dependent Kasner exponents, still obeying the inequalities $v_1 < 0$ and $v_2 > 0$. The time dependence of $v_1(x)$ can be studied by inserting these specific L and S in the equations of motion generated by the spherically symmetric Hamiltonian. Close to $\tau = 1$, the x-dependence of L and S is weak and we have a nearly homogeneous initial solution. Close to the singularity at $\tau = 0$, however, the position dependence of $v_i(x)$ is significant. The weak position dependence at an initial time is therefore magnified as we evolve toward the singularity, which generically is expected to be very inhomogeneous.

However, it turns out that the dynamics even of such an inhomogeneous geometry near $\tau = 0$ is simple. In the equations of motion, coefficients of time derivatives diverge faster than coefficients of space derivatives. For $L(\tau, x) \propto \tau^{v_1(x)}$, we compute

$$\frac{\partial L}{\partial \tau} \propto \frac{L}{\tau}, \quad \frac{\partial^2 L}{\partial \tau^2} \propto \frac{L}{\tau^2} \tag{1.182}$$

and

$$\frac{\partial L}{\partial r} \propto \frac{\mathrm{d}v_1}{\mathrm{d}r} L \log \tau, \quad \frac{\partial^2 L}{\partial r^2} \propto \left(\frac{\mathrm{d}^2 v_1}{\mathrm{d}r^2} + \left(\frac{\mathrm{d}v_1}{\mathrm{d}r} \right)^2 \log \tau \right) L \log \tau, \tag{1.183}$$

and similarly for S. Since $\lim_{\tau \to 0}(\tau \log \tau) = 0$, $1/\tau$ diverges faster than $\log \tau$. One can now go through all terms in the evolution equations and conclude that those with time derivatives diverge as quickly as $1/(\tau^2 L) \propto 1/\tau^{2-v_1}$, while those with spatial derivatives diverge at most like $(\log \tau)^2 S/\tau \propto (\log \tau)^2 /\tau^{1-v_2}$. Since $v_1 < 0$ and $v_2 > 0$, spatial derivatives are subdominant in the equations of motion. Ignoring them is therefore self-consistent.

Time derivatives dominate, suggesting an approximation of the complicated inhomogeneous dynamics by a set of ordinary differential equations for each point in space. The metric at each point evolves like the metric in a single homogeneous model, but the initial values at any other point are in general different. Close to a spacelike singularity, inhomogeneity can be expressed through position-dependent initial values, while the equations of motion are the same at each point. This Belinskii–Khalatnikov–Lifshitz (BKL) scenario indicates that it may not be hopeless to study the approach to the Big Bang singularity with analytical approximations. The scenario has successfully passed several numerical and various analytical tests (Garfinkle 2004; Berger 2002; Uggla et al. 2003; Rendall 2005; Andersson & Rendall 2001; Lim et al. 2009; Rendall & Weaver 2001; Saotome et al. 2010).

Invoking the BKL scenario, we can use our homogeneous models for a description of evolution toward a spacelike singularity. However, we cannot use the full homogeneous space of such a model, but should select a small region with tiny coordinate volume V_0. As we approach the singularity, this volume should be made smaller and smaller in order to maintain the approximation. In this description of the local geometry, V_0 serves as an infrared cutoff, bounding from above the distance scales that can be modeled. In classical equations, the value of V_0 does not show up, which is why the local classical dynamics is indistinguishable from that of a full homogeneous model. However, close to a singularity we expect quantum effects to become important. In quantum physics, it is well known that predictions depend on the scale set by an infrared cutoff. For instance, the Casimir effect implies a measurable force as a consequence of the quantum structure of the vacuum between two metal plates. The plates set boundary conditions, functioning as an infrared cutoff for possible quantum fluctuations of fields between them. Changing the distance between the plates, a process that may be described by infrared renormalization, is analogous to changing V_0 in a quantum version of the BKL scenario. We therefore expect quantum effects to depend on the progress of V_0, which traces the enhanced structure formation in a collapsing universe. We will confirm this property explicitly in our discussions of Chapters 4 and 6.

In a cosmological application, however, we must face another question: How do we implement quantum corrections and the V_0-dependence they imply without violating covariance of the classical equations? In the next chapter, we continue with the classical discussion but delve more deeply into the structure of spacetime implied by general covariance.

References

Ade, P. A. R., & Planck Collaboration, 2014, A&A, 571, A22

Andersson, L., & Rendall, A. D. 2001, CMaPh, 218, 479

Bardeen, J. M. 1980, PhRvD, 22, 1882

Barvinsky, A. O., Kamenshchik, A. Y., Kiefer, C., & Mishakov, I. V. 1999, NuPhB, 551, 374

Belinskii, V. A., Khalatnikov, I. M., & Lifschitz, E. M. 1982, AdPhys, 13, 639

Berger, B. K. 2002, LRR, 5, 1

Blanton, M. R., Bershady, M. A., Abolfathi, B., et al. 2017, AJ, 154, 28

Bondi, H. 1947, MNRAS, 107, 410

Burgess, C. P., Holman, R., & Hoover, D. 2008, PhRvD, 77, 063534

Damour, T., Henneaux, M., & Nicolai, H. 2003, CQGra, 20, R145

Ellis, G. F. R., & MacCallum, M. A. H. 1969, CMaPh, 12, 108

Enqvist, K. 2008, GReGr, 40, 451

Friedmann, A. 1922, ZPhy, 10, 377

Garfinkle, D. 2004, PhRvL, 93, 161101

Giulini, D., Kiefer, C., Joos, E., et al. 2003, Decoherence and the Appearance of a Classical World in Quantum Theory (Berlin: Springer)

Guth, A. H. 1981, PhRvD, 23, 347

Guth, A. H., & Pi, S. Y. 1985, PhRvD, 32, 1899

Kantowski, R., & Sachs, R. K. 1966, JMaPh, 7, 443

Kasner, E. 1921, AmJM, 43, 217

Kiefer, C., Lohmar, I., Polarski, D., & Starobinsky, A. A. 2007, CQGra, 24, 1699

Kiefer, C., Polarski, D., & Starobinsky, A. A. 1998, IJMPD, 7, 455

Lemaître, G. 1933, ASSB, 53, 51

Lesgourgues, J., Polarski, D., & Starobinsky, A. A. 1997, NuPhB, 497, 479

Lim, W. C., Andersson, L., Garfinkle, D., & Pretorius, F. 2009, PhRvD, 79, 123526

Linde, A. D. 1982, PhLB, 108, 389

Lyth, D. H. 1985, PhRvD, 31, 1792

MacCallum, M. A. H., & Taub, A. H. 1972, CMaPh, 25, 173

Misner, C. W. 1968, ApJ, 151, 431

Misner, C. W. 1969, PhRv, 186, 1319

Misner, C. W. 1969, PhRvL, 22, 1071

Mukhanov, V. F., Feldman, H. A., & Brandenberger, R. H. 1992, PhR, 215, 203

Mukhopadhyay, M., & Vachaspati, T. 2019, arXiv:1907.03762

Novello, M., & Bergliaffa, S. E. P. 2008, PhR, 463, 127

Penrose, R. 1990, in Proc. of the 14th Texas Symp. on Relativistic Astrophysics, ed. E. Fenves (New York: The New York Academy of Sciences)

Perez, A., Sahlmann, H., & Sudarsky, D. 2006, CQGra, 23, 2317

Perlmutter, S., et al. 1991, ApJ, 517, 565

Polarski, D., & Starobinsky, A. A. 1996, CQGra, 13, 377

Rendall, A. D. 2005, in 100 Years of Relativity—Space-Time Structure: Einstein and Beyond, ed. A. Ashtekar (Singapore: World Scientific)

Rendall, A. D., & Weaver, M. 2001, CQGra, 18, 2959

Riess, A. G., Filippenko, A. V., Challis, P., et al. 1998, AJ, 116, 1009

Saotome, R., Akhoury, R., & Garfinkle, D. 2010, CQGra, 27, 165019

Schlosshauer, M. 2007, Decoherence and the Quantum-to-classical Transition (Berlin: Springer)

Stewart, J. M. 1990, CQGra, 7, 1169

Strobl, T. 1999, arXiv:hep-th/0011240

Tolman, R. C. 1934, PNAS, 20, 169

Uggla, C., van Elst, H., Wainwright, J., & Ellis, G. F. R. 2003, PhRvD, 68, 103502

Zeh, H. D. 1970, FoPh, 1, 69
Reprinted in ed. J. A. Wheeler, & W. H. Zurek 1983, Quantum theory and measurement (Princeton: Princeton University Press)

Foundations of Quantum Cosmology

Martin Bojowald

Chapter 2

Covariance

The field equations of gravitational systems are subject to consistency conditions which ensure that all constraints required for initial values are preserved in time. In Lemaître–Tolman–Bondi models, for instance, we found that general solutions of the two Equations (1.128) and (1.129), which are of first order in time derivatives and therefore constrain the allowed initial values for second-order evolution Equations (1.126) and (1.127), are maintained at all times by the evolution equations.

A related property is the fact that general solutions are available for any choice of coordinates. For instance, in the same models, solving one of the constraints led to an equation, (1.131), which, inserted in the line element (1.119), implies a free time-independent function $f(r)^2$ multiplying the radial differential dr^2. A solution of this form therefore remains available after an arbitrary coordinate transformation $r \mapsto \tilde{r}(r)$ of the radial coordinate, such that $d\tilde{r} = (d\tilde{r}/dr)dr$. Choosing a solution with $f(r) = d\tilde{r}/dr$ then corresponds to the coordinate change. (Time-dependent coordinate transformations would have been possible in our general solutions had we not set $M = 0$ in our transition from Equation (1.119) to Lemaître–Tolman–Bondi form.)

It is the condition of general covariance that implies these consistency relations not only for classical systems based on general relativity but also for modified gravity theories such as higher-curvature actions. However, making sure that quantum corrections do not violate general covariance often presents challenging problems. An understanding of different consistent implementations of covariance is therefore important.

2.1 Lagrangian and Hamiltonian Formulations

This chapter introduces different perspectives on covariance. Although covariance is a unique physical condition, its mathematical treatment is formally split into Lagrangian and Hamiltonian versions, describing two rather different but equally

doi:10.1088/2514-3433/ab9c98ch2

important formulations of the same geometrical properties of spacetime. These two formulations are equivalent to each other, but they differ quite significantly in their technical details.

The Lagrangian viewpoint is based on spacetime tensors and works by constructing coordinate-independent functions of the metric and its partial derivatives that can be used as action principles. Geometrically, this approach makes use of several properties of Riemannian curvature. General covariance is represented by invariance with respect to arbitrary coordinate transformations on spacetime, or spacetime diffeomorphisms in an infinitesimal version.

The Hamiltonian approach, by contrast, seems to violate covariance in its initial setup, foregoing spacetime tensors and instead formulating spacetime dynamics as an evolving *spatial metric*, defined on spatial slices by setting a time coordinate equal to constant values. The spatial metric and its time derivative replace positions and velocities, respectively, of Hamiltonian mechanics. Geometrically, this approach implements covariance by requiring that the metric evolution does not depend on the choice of spatial slices, or on the choice of a time coordinate that defines these slices. The underlying symmetry is given by deformations of spatial hypersurfaces within spacetime. Although the mathematics of these transformations is rather different from spacetime diffeomorphisms, physical effects of these invariance conditions are equivalent. As we will see, hypersurface deformations are a non-linear generalization of the better-known transformations of axes in spacetime diagrams of special relativity, subject to Lorentz boosts.

2.1.1 Equivalence

When it comes to quantization, the Hamiltonian approach is often considered more fundamental, and indeed it may give rise to more general spacetime structures than can be incorporated in an action principle.

From classical mechanics, we are used to the fact that the Lagrangian and Hamiltonian viewpoints are equivalent. In a Lagrangian formulation, given a function $L(q, \dot{q})$ of position and velocity components, q and $\dot{q} = dq/dt$, we derive equations of motion by computing extrema of the action functional

$$S[q(t), \dot{q}(t)] = \int L(q(t), \dot{q}(t)) dt. \tag{2.1}$$

The result is determined by Euler–Lagrange equations

$$-\frac{d}{dt}\frac{\partial L}{\partial \dot{q}} + \frac{\partial L}{\partial q} = 0. \tag{2.2}$$

Since L usually depends only on q and $\dot{q}$, but not on $\ddot{q}$ or higher derivatives, the equations of motion are second-order in time derivatives.

The Hamiltonian is obtained through a Legendre transformation of L, first defining momentum components

$$p = \frac{\partial L}{\partial \dot{q}} \tag{2.3}$$

and then computing the Hamiltonian

$$H(q, p) = p\dot{q}(p) - L(q, \dot{q}(p)) \tag{2.4}$$

where $\dot{q}(p)$ is defined as a function of p, inverting Equation (2.3). Equations of motion are then given by Hamilton's equations

$$\frac{dq}{dt} = \frac{\partial H}{\partial p}, \quad \frac{dp}{dt} = -\frac{\partial H}{\partial q}. \tag{2.5}$$

These are first-order differential equations, but we can rewrite them as a single second-order equation by computing $\ddot{q}$ as an additional time derivative of $\dot{q}$ in the first equation of (2.5), which by the chain rule applied to the right-hand side of the same equation then depends on $\dot{p}$ (assuming that H, as usual is not linear in p). This derivative is equal to the function $-\partial H/\partial q$ which, in general, may depend on both q and p. If the first equation in (2.5) can be inverted for $p(\dot{q})$, $\ddot{q}$ can be written as a function of q and $\dot{q}$, which is our second-order equation of motion. (This invertibility condition is guaranteed for systems usually encountered in classical mechanics, where L depends quadratically on $\dot{q}$, and therefore H depends quadratically on p such that $\dot{q} = \partial H/\partial p$ is linear in p.)

Using Equations (2.3)–(2.5), Equation (2.2) becomes the identity

$$-\frac{dp}{dt} + \frac{\partial L}{\partial q} = \frac{\partial H}{\partial q} + \frac{\partial L}{\partial q} = 0. \tag{2.6}$$

Therefore, given Equations (2.3) and (2.4), the two equations in (2.5) imply the same dynamics as Equation (2.2). This equivalence does not depend on the specific form of Lagrangians and Hamiltonians and is therefore realized broadly.

2.1.2 Constrained Theories Subject to Gauge Symmetries

However, when applied to spacetime theories, the equivalence of Lagrangian and Hamiltonian formulations rests on additional assumptions. To establish the first assumption, it is important to distinguish between gauge symmetries and non-gauge symmetries. A *gauge symmetry* relates different mathematical solutions of a theory that describe exactly the same physics and therefore express a redundance in the mathematical formulation. A *symmetry* that is not gauge relates mathematical solutions that are similar but physically distinct, for instance because they correspond to different choices of initial values. In the example of gravity, a spherical central mass implies a symmetry because any elliptical orbit of a planet can be rotated and remains a valid solution. The initial position and velocity at the same initial time, evaluated for the original orbit and the rotated orbit, are different, such that the two solution describe physically distinct situations. If we transform a solution obtained in spherical coordinates to Cartesian coordinates, however, we do not change the physics even though we have two distinct mathematical descriptions.

Such a transformation, which is part of general covariance, constitutes a gauge symmetry.

Gauge symmetries such as general covariance imply that some of the degrees of freedom used in an initial setup of the theory are redundant, for instance some components $g_{\alpha\beta}$ of the spacetime metric. Since general covariance requires that we can freely transform coordinates, a changing value of a single component $g_{\alpha\beta}$ may have two different origins: It could change dynamically because spacetime is evolving, or it could just have been adjusted to a new coordinate system introduced to simplify a calculation. Therefore, not all the time derivatives $\dot{g}_{\alpha\beta}$ can be independent physical degrees of freedom that appear in the Lagrangian. The derivation of momenta via Equation (2.3) then leads to constraints on the momenta: The number of independent canonical momenta, initially written as some tensor $p^{\alpha\beta}$, is smaller than the number of time derivatives of configuration variables, $g_{\alpha\beta}$. The analog of the first equation in (2.5) then might not be invertible uniquely to express the momenta as functions of time derivatives of the metric, and the arguments between Equations (2.5) and (2.6) no longer apply in the given form.

It is possible to establish an equivalence between Lagrangian and Hamiltonian formulations even in the presence of gauge symmetries. One can describe the reduced number of momentum components, compared with time derivatives of configuration variables, by starting with an auxiliary set $p^{\alpha\beta}$ of as many momentum components as there are configuration variables, that is, ten independent components in a symmetric tensor $g_{\alpha\beta}$. In a second step, this auxiliary theory is then reduced to the correct number of independent momentum components by imposing constraints (Bergmann 1949; Dirac 1950; Anderson & Bergmann 1951; Dirac 1969, 1958a, 1958b; Arnowitt et al. 1962) on the components of $p^{\alpha\beta}$ beyond the symmetry condition $p^{\alpha\beta} = p^{\beta\alpha}$ that is also required for $g_{\alpha\beta}$. As we will see in more detail later, only the spatial components g_{ij} of the metric, determined in some coordinate system that defines the spatial hypersurfaces of the Hamiltonian formulation, have independent momenta, p^{ij}, while the time–time and time–space components g_{00} and g_{0i} do not have momenta (Dirac 1958a; Arnowitt et al. 1962).

This result is intuitive if we think of a spacetime geometry as a succession of spatial geometries which depend on time. The components g_{ij} then indeed describe the evolving spatial geometry, while g_{00} and g_{0i} merely tell us how a choice of coordinates determines how we should evolve from one spatial geometry to another one in a given time step. As in the example of spherically symmetric gravity in Section 1.5, these components determine the direction of evolution, Equations (1.122) and (1.123), but do not evolve themselves. Since this argument refers to coordinates, it must be put into a suitable mathematical formulation that ensures overall covariance. We will develop the corresponding framework of canonical gravity in Section 2.3; for more details, see Bojowald (2010).

2.1.3 Spacetime Theories

The second crucial feature of spacetime theories is that they are field theories in which q and p are replaced by position-dependent fields, such as $\phi(x)$ and $p_\phi(x)$. All realistic matter theories are field theories because matter distributions are usually position-dependent. A general formalism for fields in a Lagrangian or Hamiltonian perspective is therefore readily available. Hamilton's equations and Euler–Lagrange equations are formulated for fields by replacing partial derivatives such as $\partial/\partial q$ with *functional derivatives*, such as $\delta/\delta\phi(x)$. The latter are defined in an infinite-dimensional space because a function $\phi(x)$ amounts to infinitely many independent values if x is varied. There are therefore subtle mathematical issues in a complete formulation, but in practice functional derivatives work much like partial derivatives. (We will see more detailed examples.)

Lagrangian and Hamiltonian formulations based on functional derivatives for fields respect the equivalence statement. But crucially, when we formulate an action principle in the Lagrangian viewpoint of a spacetime theory, the time integration in Equation (2.1) is replaced by a spacetime integration, $\mathrm{d}t\mathrm{d}^3x$ if we can use Cartesian coordinates, or more generally $\sqrt{-\det g}\,\mathrm{d}t\mathrm{d}^3x$. The availability of such coordinates, at least locally in a small region of curved spacetime, depends on the structure of spacetime. If we use this integration measure, we implicitly assume that the structure of time, as it appears in the definition of the action on which Euler–Lagrange equations are based, remains classical: It is still described by a metric tensor in Riemannian geometry.

This very assumption may be violated in quantum cosmology, for instance if spacetime acquires a discrete structure. Even an effective, large-scale description that does not directly use a discrete space may retain a trace of the underlying discreteness, reflected in a modified integration. While some versions of such a modification can be formulated at the level of an action, for instance using non-commutative geometry, the Hamiltonian approach remains more general because, while it still uses spatial integrations, it does not require an extension to spacetime integrations. Therefore the equivalence between Hamiltonian and Lagrangian formulations need not hold anymore when quantum spacetime effects are considered. New models of quantum cosmology may then be presented by Hamiltonian or canonical gravity.

2.2 Action Principles

The equations of isotropic cosmology can be derived from an action principle which identifies them with stationary points of the function

$$S^{\mathrm{iso}}[a(t),\,N(t),\,\phi(t)] = -\frac{3c^2V_0}{8\pi G}\int \mathrm{d}t\left(\frac{a\dot{a}^2}{N} - kc^2aN\right)$$
$$+ S^{\mathrm{iso}}_{\mathrm{matter}}[a(t),\,N(t),\,\phi(t)], \tag{2.7}$$

depending on the scale factor $a(t)$, the lapse function $N(t)$, and some matter field $\phi(t)$ whose dynamics is determined by the matter action $S^{\mathrm{iso}}_{\mathrm{matter}}$. This action corresponds to the Lagrangian

$$L^{\text{iso}}[a, N] = -\frac{3c^2 V_0}{8\pi G}\left(\frac{a\dot{a}^2}{N} - kc^2 aN\right) + L^{\text{iso}}_{\text{matter}}[a, N, \phi].\tag{2.8}$$

Since $\dot{N}$ does not appear in the action, it is one of the metric components whose momenta are constrained to vanish in a Hamiltonian formulation:

$$p_N = \frac{\partial L^{\text{iso}}}{\partial \dot{N}} = 0.\tag{2.9}$$

In a Lagrangian formulation, the usual Euler–Lagrange equation for this variable,

$$-\frac{\text{d}}{\text{d}t}\frac{\partial L^{\text{iso}}}{\partial \dot{N}} + \frac{\partial L^{\text{iso}}}{\partial N} = 0,\tag{2.10}$$

simplifies to $\partial L^{\text{iso}}/\partial N = 0$. This equation is equivalent to the Friedmann equation if we identify the matter energy density with

$$\rho = \frac{1}{V_0 a^3}\frac{\partial L^{\text{iso}}_{\text{matter}}}{\partial N}.\tag{2.11}$$

The Euler–Lagrange equation for a can be seen to be equivalent to the Raychaudhuri equation, with pressure

$$P = \frac{1}{3 V_0 a^2}\frac{\partial L^{\text{iso}}_{\text{matter}}}{\partial a}.\tag{2.12}$$

To obtain this equation, we assume the usual condition that the matter Lagrangian does not depend on $\dot{a}$.

While the terms in Equation (2.7) or (2.8) may look almost random at first sight, much of them is determined by the condition of general covariance or coordinate independence.

2.2.1 Coordinate Independence

The first contribution to Equation (2.7), proportional to $V_0 \int \text{d}t a\dot{a}^2/N$, is not arbitrary but restricted by the condition of coordinate independence. We do not often measure the action directly, but in quantum thermodynamics it represents the phase-space volume in multiples of the constant of nature $\hbar$. Therefore, the action is measurable in principle, and therefore must be coordinate independent.

The action is always an integral over time, which requires the appearance of the time coordinate via $\text{d}t$. If there were spatial dependence, the action would also integrate fields over the three spatial coordinates. In a homogeneous model, these integrations are replaced by the single volume factor V_0, measuring the finite spatial region we choose to represent the geometry. Neither $\text{d}t$ nor V_0 are coordinate independent. The specific way in which they fail to be coordinate independent guides the possible dependence of the observable S on other coordinate-dependent quantities, given for isotropic spacetimes by a, $\dot{a}$, and N.

We usually expect that an action of a classical theory depends on "velocities" in a quadratic manner, here referring to the expansion velocity $\dot{a}$. The dot is a derivative by the same t that appears in dt, but $\dot{a}^2 dt = (da)^2/dt$ depends on the choice of the time coordinate. However, the line element tells us that the combination $N dt$ is independent of this choice. Therefore, we should divide the quadratic term in the action by N in order to have dt appear only in the combination $N dt$, as it indeed happens in Equation (2.7). In addition, we need a factor of a in this term (without a time derivative) because $V_0 \dot{a}^2$ would not be invariant with respect to rescaling the spatial coordinates, while $V_0 a \dot{a}^2$ is invariant. Similarly, the curvature parameter k scales like a^2 because kr^2 in the curved-space line element (1.32) must be coordinate independent. Therefore, the combination $V_0 ka$ is invariant, and a factor of N is needed to obtain the invariant $N dt$. Up to numerical factors and sign choices, all terms in Equation (2.7) are therefore determined by general covariance.

For more general spacetimes, it would be tedious to go through such considerations in order to determine invariant actions. Fortunately, geometry helps us because curvature, just like the "kinetic" terms in an action, is a quantity that depends on second-order derivatives or squares of first-order derivatives of the metric. (This property mimics the well-known relationship between the second-order derivative of a function $f(x)$ and the curvature of its graph.)

In spacetime, curvature has several independent components, corresponding to the different ways in which curves moving in independent dimensions can bend: In general, there are 20 independent curvature components in four-dimensional spacetime. Fortunately, a smaller number of combinations of curvature components that are independent of coordinate choices are relevant for the formulation of action principles. Applied to isotropic models, there is only one such function of up to second-order derivatives of the scale factor, given by the *Ricci scalar*

$$R = 6\left(\frac{\ddot{a}}{N^2 a c^2} + \frac{\dot{a}^2}{N^2 a^2 c^2} - \frac{\dot{a}\dot{N}}{a N^3 c^2} + \frac{k}{a^2} \right). \tag{2.13}$$

Like most curvature parameters, the Ricci scalar is defined here such that it has units of $1/[L]^2$. We obtain Equation (2.7) from the Ricci scalar if we multiply the latter with the integration measure

$$\sqrt{-\det g} = r^2 \sin(\vartheta) N(t) \frac{a(t)^3}{\sqrt{1 - kr^2}}, \tag{2.14}$$

as well as the common factor of $c^4/(16\pi G)$ which provides the units of an action, or energy times time, and then integrate by parts to eliminate the second-order derivative of a.

Tensor Transformation
In general, the derivation of coordinate independent action principles for metric theories, or of curvature quantities such as Equation (2.13), starts with the tensor transformation law (1.28): We know that the metric transforms under a coordinate change by

$$g_{\alpha'\beta'} = \sum_{\alpha=0}^{3} \sum_{\beta=0}^{3} \frac{\partial x^\alpha}{\partial x^{\alpha'}} \frac{\partial x^\beta}{\partial x^{\beta'}} g_{\alpha\beta}. \tag{2.15}$$

A field theory for the metric should depend on its partial derivatives of at least first order. In general, however, partial derivatives of tensors do not obey the tensor transformation law.

While a partial derivative of a function, such as a scalar matter field ϕ, does transform according to the 1-index tensor transformation law

$$\frac{\partial \phi}{\partial x^{\alpha'}} = \sum_{\alpha=0}^{3} \frac{\partial x^\alpha}{\partial x^{\alpha'}} \frac{\partial \phi}{\partial x^\alpha} \tag{2.16}$$

thanks to the chain rule, a partial derivative of the metric (or some other tensor) does *not* obey the required law. For instance, the partial derivative $\partial g_{\alpha'\beta'}/\partial x^{\gamma'}$ can be transformed by inserting the tensor transformation law of the metric as well as the chain rule (2.16):

$$\frac{\partial g_{\alpha'\beta'}}{\partial x^{\gamma'}} = \sum_{\alpha=0}^{3} \sum_{\beta=0}^{3} \sum_{\gamma=0}^{3} \frac{\partial x^\gamma}{\partial x^{\gamma'}} \frac{\partial}{\partial x^\gamma} \left(\frac{\partial x^\alpha}{\partial x^{\alpha'}} \frac{\partial x^\beta}{\partial x^{\beta'}} g_{\alpha\beta} \right)$$

$$= \sum_{\alpha=0}^{3} \sum_{\beta=0}^{3} \sum_{\gamma=0}^{3} \frac{\partial x^\alpha}{\partial x^{\alpha'}} \frac{\partial x^\beta}{\partial x^{\beta'}} \frac{\partial x^\gamma}{\partial x^{\gamma'}} \frac{\partial g_{\alpha\beta}}{\partial x^\gamma} \tag{2.17}$$

$$+ \sum_{\alpha=0}^{3} \sum_{\beta=0}^{3} \sum_{\gamma=0}^{3} \left(\frac{\partial x^\alpha}{\partial x^{\alpha'}} \frac{\partial^2 x^\beta}{\partial x^{\beta'}\partial x^{\gamma'}} + \frac{\partial^2 x^\alpha}{\partial x^{\alpha'}\partial x^{\gamma'}} \frac{\partial x^\beta}{\partial x^{\beta'}} \right) g_{\alpha\beta}. \tag{2.18}$$

While line (2.17) would be the correct tensor transformation law for a three-index tensor, mimicking Equation (1.28) by multiplying with a factor of $\partial x^\alpha/\partial x^{\alpha'}$ or $\partial x^{\alpha'}/\partial x^\alpha$, respectively, for each subscript or superscript index on the tensor, line (2.18) is not of this form: It contains second-order derivatives of the coordinate transformation which are in general non-zero. Tensor derivatives and invariant actions built from them therefore require suitable combinations of partial derivatives of the metric in which unwanted terms such as Equation (2.18) cancel out.

Christoffel Symbol

For several different purposes it is useful to introduce certain linear combinations of partial derivatives of the metric. A common expression appears not only in curvature tensors but also when one derives partial differential equations for geodesics in curved spacetime, that is, curves of minimum length in spacelike directions or of maximum length (duration) in timelike directions. Focusing on the timelike case, the duration of a worldline $x^\alpha(\sigma)$—the position x^α as a function of some real number σ taking values in a suitable range—is given by the proper-time difference

$$\Delta\tau_{AB} = \frac{1}{c}\int_A^B \sqrt{-\mathrm{d}s^2} = \frac{1}{c}\int_A^B \sqrt{-\sum_{\alpha=0}^3 \sum_{\beta=0}^3 g_{\alpha\beta}\mathrm{d}x^\alpha \mathrm{d}x^\beta}$$

$$= \frac{1}{c}\int_{\sigma_A}^{\sigma_B} \sqrt{-\sum_{\alpha=0}^3 \sum_{\beta=0}^3 g_{\alpha\beta}\frac{\mathrm{d}x^\alpha}{\mathrm{d}\sigma}\frac{\mathrm{d}x^\beta}{\mathrm{d}\sigma}}\ \mathrm{d}\sigma$$

(2.19)

between two events, A and B.

Extrema of $\Delta\tau_{AB}$, considering all timelike worldlines connecting two fixed events A and B as illustrated in Figure 2.1, are determined by Euler–Lagrange equations of (2.19), viewed as an action principle for $x^\alpha(\sigma)$ with velocity components $\mathrm{d}x^\beta(\sigma)/\mathrm{d}\sigma$: $x^\alpha(\sigma)$ must be such that

$$-\frac{\mathrm{d}}{\mathrm{d}\sigma}\frac{\partial L}{\partial(\mathrm{d}x^\alpha/\mathrm{d}\sigma)} + \frac{\partial L}{\partial x^\alpha} = 0$$

(2.20)

for all $\alpha = 0, 1, 2, 3$, using the Lagrangian

$$L(x^\alpha, \mathrm{d}x^\delta/\mathrm{d}\sigma) = \sqrt{-\sum_{\beta=0}^3 \sum_{\gamma=0}^3 g_{\beta\gamma}(x^\epsilon)\frac{\mathrm{d}x^\beta}{\mathrm{d}\sigma}\frac{\mathrm{d}x^\gamma}{\mathrm{d}\sigma}}\ .$$

(2.21)

Position and velocity components in spacetime are independent, a condition which is expressed by the basic partial derivatives

$$\frac{\partial x^\alpha}{\partial x^\beta} = \delta^\alpha_\beta, \quad \frac{\partial(\mathrm{d}x^\alpha/\mathrm{d}\sigma)}{\partial(\mathrm{d}x^\beta/\mathrm{d}\sigma)} = \delta^\alpha_\beta$$

(2.22)

with the Kronecker delta δ^α_β. Therefore,

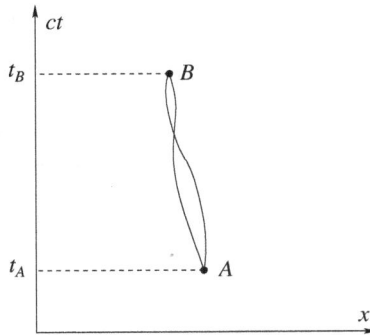

Figure 2.1. Timelike worldlines connecting two events, A and B.

$$\frac{\partial L}{\partial(dx^\alpha/d\sigma)} = -\frac{1}{2L} \sum_{\beta=0}^{3} \sum_{\gamma=0}^{3} g_{\beta\gamma}\left(\delta_\alpha^\beta \frac{dx^\gamma}{d\sigma} + \frac{dx^\beta}{d\sigma}\delta_\alpha^\gamma\right)$$

$$= -\frac{1}{L} \sum_{\beta=0}^{3} g_{\alpha\beta}\frac{dx^\beta}{d\sigma} \tag{2.23}$$

after renaming some indices. The Euler–Lagrange equations read

$$\frac{d}{d\sigma}\left(\frac{1}{L} \sum_{\beta=0}^{3} g_{\alpha\beta}\frac{dx^\beta}{d\sigma}\right) - \frac{1}{2L} \sum_{\beta=0}^{3} \sum_{\gamma=0}^{3} \frac{\partial g_{\beta\gamma}}{\partial x^\alpha} \frac{dx^\beta}{d\sigma}\frac{dx^\gamma}{d\sigma} = 0. \tag{2.24}$$

The combination $L d\sigma = d\tau$, which appears a few times in this equation, equals the proper-time differential according to the definition (2.21) of L. Therefore, the Euler–Lagrange equations can be written as

$$L\left(\frac{d}{d\tau} \sum_{\beta=0}^{3}\left(g_{\alpha\beta}\frac{dx^\beta}{d\tau}\right) - \frac{1}{2}\sum_{\beta=0}^{3}\sum_{\gamma=0}^{3}\frac{\partial g_{\beta\gamma}}{\partial x^\alpha}\frac{dx^\beta}{d\tau}\frac{dx^\gamma}{d\tau}\right)$$

$$= L\left(\sum_{\beta=0}^{3} g_{\alpha\beta}\frac{d^2x^\beta}{d\tau^2} + \sum_{\beta=0}^{3}\sum_{\gamma=0}^{3}\frac{\partial g_{\alpha\beta}}{\partial x^\gamma}\frac{dx^\gamma}{d\tau}\frac{dx^\beta}{d\tau} - \frac{1}{2}\sum_{\beta=0}^{3}\sum_{\gamma=0}^{3}\frac{\partial g_{\beta\gamma}}{\partial x^\alpha}\frac{dx^\beta}{d\tau}\frac{dx^\gamma}{d\tau}\right) = 0.$$

If $L \neq 0$, we rewrite the last equation as

$$\sum_{\beta=0}^{3} g_{\alpha\beta}\frac{d^2x^\beta}{d\tau^2} = -\frac{1}{2}\sum_{\beta=0}^{3}\sum_{\gamma=0}^{3}\left(\frac{\partial g_{\alpha\beta}}{\partial x^\gamma} + \frac{\partial g_{\alpha\gamma}}{\partial x^\beta} - \frac{\partial g_{\beta\gamma}}{\partial x^\alpha}\right)\frac{dx^\beta}{d\tau}\frac{dx^\gamma}{d\tau} \tag{2.25}$$

where we symmetrized the middle term in Equation (2.25) with respect to β and γ because the product $(dx^\beta/d\tau)(dx^\gamma/d\tau)$ at the end has the same symmetry. With this symmetrization, it is possible to extract the coefficients of $(dx^\beta/d\tau)(dx^\gamma/d\tau)$ uniquely. After multiplication with the inverse matrix of $g_{\alpha\beta}$ on both sides, the coefficients are given by

$$\Gamma^\alpha_{\beta\gamma} = \frac{1}{2}\sum_{\delta=0}^{3} g^{\alpha\delta}\left(\frac{\partial g_{\gamma\delta}}{\partial x^\beta} + \frac{\partial g_{\beta\delta}}{\partial x^\gamma} - \frac{\partial g_{\beta\gamma}}{\partial x^\delta}\right). \tag{2.26}$$

In terms of indices, it is common to denote the inverse of the metric simply by raising the two subscripts on $g_{\alpha\beta}$ to superscripts, $(g^{-1})^{\alpha\beta} = g^{\alpha\beta}$. Because the metric is required to be invertible, g and g^{-1} contain the same information about spacetime geometry, and it is not necessary to change the symbol in any way other than by moving its indices.

Covariant Derivative

Taking the inverse of the matrix $g_{\alpha\beta}$ is indicated in Equation (2.26) by using superscripts, $g^{\alpha\beta}$. This convention is consistent with the tensor transformation law because

$$g^{\alpha'\beta'} = \sum_{\alpha=0}^{3} \sum_{\beta=0}^{3} \frac{\partial x^{\alpha'}}{\partial x^{\alpha}} \frac{\partial x^{\beta'}}{\partial x^{\beta}} g^{\alpha\beta} \tag{2.27}$$

is then implied by inverting all matrices in the tensor transformation of $g_{\alpha\beta}$. For tensors other than $g_{\alpha\beta}$, we cannot be sure that they are invertible. Turning their subscripts into superscripts, if necessary, is then accomplished by defining

$$T^{\alpha\cdots}_{\cdots} = \sum_{\beta=0}^{3} g^{\alpha\beta} T^{\cdots}_{\beta\cdots} \tag{2.28}$$

where any number of superscript or subscript indices can be included without changing them on the two sides. This definition ensures that $T^{\alpha\cdots}_{\cdots}$ obeys the correct tensor transformation law provided $T^{\cdots}_{\beta\cdots}$ does so.

Similarly, a superscript can be turned into a subscript via

$$T^{\cdots}_{\alpha\cdots} = \sum_{\beta=0}^{3} g_{\alpha\beta} T^{\beta\cdots}_{\cdots}. \tag{2.29}$$

This definition works also for $g^{\alpha\beta}$ itself. But since we already defined the meaning of both $g_{\alpha\beta}$ and $g^{\alpha\beta}$, we obtain a new result,

$$g^{\gamma}_{\alpha} = \sum_{\beta=0}^{3} g_{\alpha\beta} g^{\beta\gamma} = \delta^{\gamma}_{\alpha}, \tag{2.30}$$

which identifies the mixed-index metric g^{γ}_{α} with the Kronecker delta.

The coefficients (2.26), taken together, are called the *Christoffel symbol*. Like individual partial derivatives of the metric, they do not transform according the tensor transformation law, and therefore we are not allowed to call them a tensor. (They are "just" a symbol.) Nevertheless, they are convenient for several purposes. In particular, the *geodesic equation* is obtained from Equation (2.25) in the compact form

$$\frac{\mathrm{d}^2 x^{\alpha}}{\mathrm{d}\tau^2} = -\sum_{\beta=0}^{3} \sum_{\gamma=0}^{3} \Gamma^{\alpha}_{\beta\gamma} \frac{\mathrm{d}x^{\beta}}{\mathrm{d}\tau} \frac{\mathrm{d}x^{\gamma}}{\mathrm{d}\tau}. \tag{2.31}$$

On the left-hand side, we have the time derivative of the relativistic velocity, or the relativistic acceleration, which should transform according to the tensor transformation law.

Moreover, the time derivative along a given worldline is, in general, equal to the scalar product of the tangent vector of the worldline, $dx^\beta/d\tau$, with the gradient, $\partial/\partial x^\beta$:

$$\frac{d}{d\tau} = \sum_{\beta=0}^{3} \frac{dx^\beta}{d\tau} \frac{\partial}{\partial x^\beta}. \tag{2.32}$$

Since the right-hand side of the geodesic equation appears in the form of such a scalar product and must be true for any $dx^\beta/d\tau$, we can combine both sides, eliminate a $\sum_{\beta=0}^{3} dx^\beta/d\tau$ in each term, and write

$$\frac{\partial}{\partial x^\beta} \frac{dx^\alpha}{d\tau} + \sum_{\gamma=0}^{3} \Gamma^\alpha_{\beta\gamma} \frac{dx^\gamma}{d\tau} = 0. \tag{2.33}$$

Since the right-hand side is always zero for a geodesic, in any coordinate system, the whole equation must be covariant and therefore obeys the tensor transformation law, even though $\partial v^\alpha/\partial x^\beta$ and $\sum_{\gamma=0}^{3} \Gamma^\alpha_{\beta\gamma} v^\gamma$ individually do not do so for any vector v^α. The left-hand side of Equation (2.33) therefore defines the *covariant derivative*

$$\nabla_\beta v^\alpha = \frac{\partial v^\alpha}{\partial x^\beta} + \sum_{\gamma=0}^{3} \Gamma^\alpha_{\beta\gamma} v^\gamma \tag{2.34}$$

which can also be applied to tensors with more than one index provided each index receives a Christoffel term (as well as a negative sign if it is a subscript). For instance,

$$\begin{aligned}
\nabla_\alpha g_{\beta\gamma} &= \frac{\partial g_{\beta\gamma}}{\partial x^\alpha} - \sum_{\delta=0}^{3} \Gamma^\delta_{\alpha\beta} g_{\delta\gamma} - \sum_{\delta=0}^{3} \Gamma^\delta_{\alpha\gamma} g_{\beta\delta} \\
&= \frac{\partial g_{\beta\gamma}}{\partial x^\alpha} - \frac{1}{2}\left(\frac{\partial g_{\gamma\alpha}}{\partial x^\beta} + \frac{\partial g_{\gamma\beta}}{\partial x^\alpha} - \frac{\partial g_{\alpha\beta}}{\partial x^\gamma} \right) - \frac{1}{2}\left(\frac{\partial g_{\gamma\beta}}{\partial x^\alpha} + \frac{\partial g_{\alpha\beta}}{\partial x^\gamma} - \frac{\partial g_{\alpha\gamma}}{\partial x^\beta} \right) \\
&= 0
\end{aligned} \tag{2.35}$$

using Equation (2.26) and the inverse relationship of $g_{\alpha\beta}$ and $g^{\alpha\beta}$, as well as symmetry of $g_{\alpha\beta}$. With respect to the covariant derivative, the metric is therefore always constant, in any coordinate system, even if its components depend on the position or time.

Transformation of Partial Derivatives
The Christoffel symbol is not a tensor because it does not obey the tensor transformation law. Instead, its transformation can be derived by inserting the tensor transformation of the metric:

$$\Gamma^{\alpha}_{\beta\gamma} = \frac{1}{2} \sum_{\alpha'=0}^{3} \sum_{\beta'=0}^{3} \sum_{\gamma'=0}^{3} \sum_{\delta'=0}^{3} \sum_{\epsilon=0}^{3} \sum_{\delta=0}^{3} \frac{\partial x^{\alpha}}{\partial x^{\alpha'}} \frac{\partial x^{\delta}}{\partial x^{\delta'}} g^{\alpha'\delta'}$$

$$\times \left(\frac{\partial}{\partial x^{\gamma}} \left(\frac{\partial x^{\epsilon'}}{\partial x^{\delta}} \frac{\partial x^{\beta'}}{\partial x^{\beta}} g_{\epsilon'\beta'} \right) + \frac{\partial}{\partial x^{\beta}} \left(\frac{\partial x^{\epsilon'}}{\partial x^{\delta}} \frac{\partial x^{\gamma'}}{\partial x^{\gamma}} g_{\epsilon'\gamma'} \right) - \frac{\partial}{\partial x^{\delta}} \left(\frac{\partial x^{\beta'}}{\partial x^{\beta}} \frac{\partial x^{\gamma'}}{\partial x^{\gamma}} g_{\beta'\gamma'} \right) \right)$$

$$= \sum_{\alpha'=0}^{3} \sum_{\beta'=0}^{3} \sum_{\gamma'=0}^{3} \frac{\partial x^{\alpha}}{\partial x^{\alpha'}} \frac{\partial x^{\beta'}}{\partial x^{\beta}} \frac{\partial x^{\gamma'}}{\partial x^{\gamma}} \Gamma^{\alpha'}_{\beta'\gamma'}$$

$$+ \frac{1}{2} \sum_{\alpha'=0}^{3} \sum_{\beta'=0}^{3} \sum_{\gamma'=0}^{3} \sum_{\delta'=0}^{3} \sum_{\epsilon=0}^{3} \sum_{\delta=0}^{3} \frac{\partial x^{\alpha}}{\partial x^{\alpha'}} \frac{\partial x^{\delta}}{\partial x^{\delta'}} g^{\alpha'\delta'}$$

$$\times \left(\frac{\partial}{\partial x^{\gamma}} \left(\frac{\partial x^{\epsilon'}}{\partial x^{\delta}} \frac{\partial x^{\beta'}}{\partial x^{\beta}} \right) g_{\epsilon'\beta'} + \frac{\partial}{\partial x^{\beta}} \left(\frac{\partial x^{\epsilon'}}{\partial x^{\delta}} \frac{\partial x^{\gamma'}}{\partial x^{\gamma}} \right) g_{\epsilon'\gamma'} - \frac{\partial}{\partial x^{\delta}} \left(\frac{\partial x^{\beta'}}{\partial x^{\beta}} \frac{\partial x^{\gamma'}}{\partial x^{\gamma}} \right) g_{\beta'\gamma'} \right).$$

Taking partial derivatives only of the metric components produces the correct term for a tensor transformation law, but second-order derivatives of the coordinate transformation do not completely cancel out, just as in Equation (2.18).

Fortunately, there are at least some cancellations after renaming suitable indices, which can be used to simplify the extra term

$$\sum_{\beta'=0}^{3} \sum_{\gamma'=0}^{3} \sum_{\epsilon'=0}^{3} \left(\frac{\partial}{\partial x^{\gamma}} \left(\frac{\partial x^{\epsilon'}}{\partial x^{\delta}} \frac{\partial x^{\beta'}}{\partial x^{\beta}} \right) g_{\epsilon'\beta'} + \frac{\partial}{\partial x^{\beta}} \left(\frac{\partial x^{\epsilon'}}{\partial x^{\delta}} \frac{\partial x^{\gamma'}}{\partial x^{\gamma}} \right) g_{\epsilon'\gamma'} - \frac{\partial}{\partial x^{\delta}} \left(\frac{\partial x^{\beta'}}{\partial x^{\beta}} \frac{\partial x^{\gamma'}}{\partial x^{\gamma}} \right) g_{\beta'\gamma'} \right)$$

$$= \sum_{\beta'=0}^{3} \sum_{\gamma'=0}^{3} \sum_{\epsilon'=0}^{3} \left(\frac{\partial}{\partial x^{\gamma}} \left(\frac{\partial x^{\epsilon'}}{\partial x^{\delta}} \frac{\partial x^{\beta'}}{\partial x^{\beta}} \right) g_{\epsilon'\beta'} + \frac{\partial}{\partial x^{\beta}} \left(\frac{\partial x^{\epsilon'}}{\partial x^{\delta}} \frac{\partial x^{\beta'}}{\partial x^{\gamma}} \right) g_{\epsilon'\beta'} - \frac{\partial}{\partial x^{\delta}} \left(\frac{\partial x^{\beta'}}{\partial x^{\beta}} \frac{\partial x^{\epsilon'}}{\partial x^{\gamma}} \right) g_{\beta'\epsilon'} \right)$$

$$= 2 \sum_{\beta'=0}^{3} \sum_{\epsilon'=0}^{3} g_{\epsilon'\beta'} \frac{\partial x^{\epsilon'}}{\partial x^{\delta}} \frac{\partial^2 x^{\beta'}}{\partial x^{\gamma}} \partial x^{\beta}.$$

(We also used the fact that partial derivatives by coordinates in the same system commute with one another.) Therefore,

$$\Gamma^{\alpha}_{\beta\gamma} = \sum_{\alpha'=0}^{3} \sum_{\beta'=0}^{3} \sum_{\gamma'=0}^{3} \frac{\partial x^{\alpha}}{\partial x^{\alpha'}} \frac{\partial x^{\beta'}}{\partial x^{\beta}} \frac{\partial x^{\gamma'}}{\partial x^{\gamma}} \Gamma^{\alpha'}_{\beta'\gamma'} + \sum_{\beta'=0}^{3} \frac{\partial x^{\alpha}}{\partial x^{\beta'}} \frac{\partial^2 x^{\beta'}}{\partial x^{\beta} \partial x^{\gamma}}. \tag{2.36}$$

As a test calculation, we can demonstrate explicitly that the covariant derivative ∇_{β} is indeed covariant, that is, $\nabla_{\beta} v^{\alpha} = \partial_{\beta} v^{\alpha} + \sum_{\gamma=0}^{3} \Gamma^{\alpha}_{\beta\gamma} v^{\gamma}$ transforms like a tensor for any vector v^{γ}. From Equation (2.36) together with $v^{\gamma'} = \sum_{\gamma=0}^{3} (\partial x^{\gamma'} / \partial x^{\gamma}) v^{\gamma}$, we obtain

$$\sum_{\gamma=0}^{3} \Gamma^{\alpha}_{\beta\gamma} v^{\gamma} = \sum_{\alpha'=0}^{3} \sum_{\beta'=0}^{3} \sum_{\gamma'=0}^{3} \frac{\partial x^{\alpha}}{\partial x^{\alpha'}} \frac{\partial x^{\beta'}}{\partial x^{\beta}} \Gamma^{\alpha'}_{\beta'\gamma'} v^{\gamma'}$$
$$+ \sum_{\gamma=0}^{3} \sum_{\beta'=0}^{3} \sum_{\gamma'=0}^{3} \frac{\partial x^{\alpha}}{\partial x^{\beta'}} \frac{\partial x^{\gamma}}{\partial x^{\gamma'}} \frac{\partial^2 x^{\beta'}}{\partial x^{\beta} \partial x^{\gamma}} v^{\gamma'}. \tag{2.37}$$

The partial derivative transforms according to

$$\frac{\partial v^{\alpha'}}{\partial x^{\beta'}} = \sum_{\alpha=0}^{3} \sum_{\beta=0}^{3} \frac{\partial x^{\beta}}{\partial x^{\beta'}} \frac{\partial}{\partial x^{\beta}} \left(\frac{\partial x^{\alpha'}}{\partial x^{\alpha}} v^{\alpha} \right)$$
$$= \sum_{\alpha=0}^{3} \sum_{\beta=0}^{3} \left(\frac{\partial x^{\beta}}{\partial x^{\beta'}} \frac{\partial x^{\alpha'}}{\partial x^{\alpha}} \frac{\partial v^{\alpha}}{\partial x^{\beta}} + \sum_{\gamma'=0}^{3} \frac{\partial x^{\beta}}{\partial x^{\beta'}} \frac{\partial^2 x^{\alpha'}}{\partial x^{\beta} \partial x^{\alpha}} \frac{\partial x^{\alpha}}{\partial x^{\gamma'}} v^{\gamma'} \right). \tag{2.38}$$

Multiplying with inverse matrices of coordinate transformations, we derive

$$\frac{\partial v^{\alpha}}{\partial x^{\beta}} = \sum_{\alpha'=0}^{3} \sum_{\beta'=0}^{3} \frac{\partial x^{\beta'}}{\partial x^{\beta}} \frac{\partial x^{\alpha}}{\partial x^{\alpha'}} \frac{\partial v^{\alpha'}}{\partial x^{\beta'}} - \sum_{\alpha'=0}^{3} \sum_{\gamma'=0}^{3} \sum_{\gamma=0}^{3} \frac{\partial x^{\alpha}}{\partial x^{\alpha'}} \frac{\partial^2 x^{\alpha'}}{\partial x^{\beta} \partial x^{\gamma}} \frac{\partial x^{\gamma}}{\partial x^{\gamma'}} v^{\gamma'}. \tag{2.39}$$

After renaming α' as β' in the last term, it is evident that the extra terms cancel out in the sum of Equations (2.39) and (2.37), which defines the covariant derivative.

Curvature Tensors
A second possibility to obtain covariant objects from the Christoffel symbol is given by using suitable combinations of the symbol with its partial derivatives. We begin with the long expression

$$\frac{\partial \Gamma^{\alpha}_{\beta\gamma}}{\partial x^{\delta}} = \sum_{\alpha'=0}^{3} \sum_{\beta'=0}^{3} \sum_{\gamma'=0}^{3} \sum_{\delta'=0}^{3} \frac{\partial x^{\alpha}}{\partial x^{\alpha'}} \frac{\partial x^{\beta'}}{\partial x^{\beta}} \frac{\partial x^{\gamma'}}{\partial x^{\gamma}} \frac{\partial x^{\delta'}}{\partial x^{\delta}} \frac{\partial \Gamma^{\alpha'}_{\beta'\gamma'}}{\partial x^{\delta'}}$$
$$+ \sum_{\alpha'=0}^{3} \sum_{\beta'=0}^{3} \sum_{\gamma'=0}^{3} \left(\sum_{\delta'=0}^{3} \frac{\partial^2 x^{\alpha}}{\partial x^{\alpha'} \partial x^{\delta'}} \frac{\partial x^{\beta'}}{\partial x^{\beta}} \frac{\partial x^{\gamma'}}{\partial x^{\gamma}} \frac{\partial x^{\delta'}}{\partial x^{\delta}} \right.$$
$$\left. + \frac{\partial x^{\alpha}}{\partial x^{\alpha'}} \frac{\partial^2 x^{\beta'}}{\partial x^{\beta} \partial x^{\delta}} \frac{\partial x^{\gamma'}}{\partial x^{\gamma}} + \frac{\partial x^{\alpha}}{\partial x^{\alpha'}} \frac{\partial x^{\beta'}}{\partial x^{\beta}} \frac{\partial^2 x^{\gamma'}}{\partial x^{\gamma} \partial x^{\delta}} \right) \Gamma^{\alpha'}_{\beta'\gamma'} \tag{2.40}$$
$$+ \sum_{\beta'=0}^{3} \sum_{\delta'=0}^{3} \frac{\partial^2 x^{\alpha}}{\partial x^{\delta'} \partial x^{\beta'}} \frac{\partial x^{\delta'}}{\partial x^{\delta}} \frac{\partial^2 x^{\beta'}}{\partial x^{\beta} \partial x^{\gamma}} + \sum_{\beta'=0}^{3} \frac{\partial x^{\alpha}}{\partial x^{\beta'}} \frac{\partial^3 x^{\beta'}}{\partial x^{\beta} \partial x^{\gamma} \partial x^{\delta}}$$

transforming the first-order partial derivatives of the Christoffel symbol.

The last term, the only one that contains a third-order partial derivative, is symmetric in β and δ. It therefore cancels out in the antisymmetric combination

$$\frac{\partial \Gamma^{\alpha}_{\delta\gamma}}{\partial x^{\beta}} - \frac{\partial \Gamma^{\alpha}_{\beta\gamma}}{\partial x^{\delta}} = \sum_{\alpha'=0}^{3} \sum_{\beta'=0}^{3} \sum_{\gamma'=0}^{3} \sum_{\delta'=0}^{3} \frac{\partial x^{\alpha}}{\partial x^{\alpha'}} \frac{\partial x^{\beta'}}{\partial x^{\beta}} \frac{\partial x^{\gamma'}}{\partial x^{\gamma}} \frac{\partial x^{\delta'}}{\partial x^{\delta}} \left(\frac{\partial \Gamma^{\alpha'}_{\delta'\gamma'}}{\partial x^{\beta'}} - \frac{\partial \Gamma^{\alpha'}_{\beta'\gamma'}}{\partial x^{\delta'}} \right)$$

$$+ \sum_{\alpha'=0}^{3} \sum_{\beta'=0}^{3} \sum_{\gamma'=0}^{3} \sum_{\delta'=0}^{3} \frac{\partial^2 x^{\alpha}}{\partial x^{\alpha'} \partial x^{\delta'}} \frac{\partial x^{\gamma'}}{\partial x^{\gamma}} \left(\frac{\partial x^{\beta'}}{\partial x^{\delta}} \frac{\partial x^{\delta'}}{\partial x^{\beta}} - \frac{\partial x^{\beta'}}{\partial x^{\beta}} \frac{\partial x^{\delta'}}{\partial x^{\delta}} \right) \Gamma^{\alpha'}_{\beta'\gamma'}$$

$$+ \sum_{\alpha'=0}^{3} \sum_{\beta'=0}^{3} \sum_{\gamma'=0}^{3} \frac{\partial x^{\alpha}}{\partial x^{\alpha'}} \left(\frac{\partial x^{\beta'}}{\partial x^{\delta}} \frac{\partial^2 x^{\gamma'}}{\partial x^{\gamma} \partial x^{\beta}} - \frac{\partial x^{\beta'}}{\partial x^{\beta}} \frac{\partial^2 x^{\gamma'}}{\partial x^{\gamma} \partial x^{\delta}} \right) \Gamma^{\alpha'}_{\beta'\gamma'} \tag{2.41}$$

$$+ \sum_{\beta'=0}^{3} \sum_{\delta'=0}^{3} \frac{\partial^2 x^{\alpha}}{\partial x^{\delta'} \partial x^{\beta'}} \left(\frac{\partial x^{\delta'}}{\partial x^{\beta}} \frac{\partial^2 x^{\beta'}}{\partial x^{\delta} \partial x^{\gamma}} - \frac{\partial x^{\delta'}}{\partial x^{\delta}} \frac{\partial^2 x^{\beta'}}{\partial x^{\beta} \partial x^{\gamma}} \right).$$

Products of second-order partial derivatives of the coordinate transformation, as seen exclusively in the last line, can also be obtained from products of Christoffel symbols. The product of two Christoffel symbols has six indices, but if two of them are summed over, four remain and can match the index structure of Equation (2.41). In particular, the combination

$$\sum_{\epsilon=0}^{3} \Gamma^{\alpha}_{\beta\epsilon} \Gamma^{\epsilon}_{\delta\gamma} = \sum_{\alpha'=0}^{3} \sum_{\beta'=0}^{3} \sum_{\gamma'=0}^{3} \sum_{\delta'=0}^{3} \sum_{\epsilon'=0}^{3} \frac{\partial x^{\alpha}}{\partial x^{\alpha'}} \frac{\partial x^{\beta'}}{\partial x^{\beta}} \frac{\partial x^{\gamma'}}{\partial x^{\gamma}} \frac{\partial x^{\delta'}}{\partial x^{\delta}} \Gamma^{\alpha'}_{\beta'\epsilon'} \Gamma^{\epsilon'}_{\delta'\gamma'}$$

$$+ \sum_{\alpha'=0}^{3} \sum_{\beta'=0}^{3} \sum_{\epsilon'=0}^{3} \frac{\partial x^{\alpha}}{\partial x^{\alpha'}} \frac{\partial x^{\beta'}}{\partial x^{\beta}} \frac{\partial^2 x^{\epsilon'}}{\partial x^{\gamma} \partial x^{\delta}} \Gamma^{\alpha'}_{\beta'\epsilon'}$$

$$+ \sum_{\beta'=0}^{3} \sum_{\gamma'=0}^{3} \sum_{\delta'=0}^{3} \sum_{\epsilon'=0}^{3} \sum_{\epsilon=0}^{3} \frac{\partial x^{\alpha}}{\partial x^{\beta'}} \frac{\partial^2 x^{\beta'}}{\partial x^{\beta} \partial x^{\epsilon}} \frac{\partial x^{\epsilon}}{\partial x^{\epsilon'}} \frac{\partial x^{\delta'}}{\partial x^{\delta}} \frac{\partial x^{\gamma'}}{\partial x^{\gamma}} \Gamma^{\epsilon'}_{\delta'\gamma'} \tag{2.42}$$

$$+ \sum_{\beta'=0}^{3} \sum_{\epsilon'=0}^{3} \sum_{\epsilon=0}^{3} \frac{\partial x^{\alpha}}{\partial x^{\beta'}} \frac{\partial^2 x^{\beta'}}{\partial x^{\beta} \partial x^{\epsilon}} \frac{\partial x^{\epsilon}}{\partial x^{\epsilon'}} \frac{\partial^2 x^{\epsilon'}}{\partial x^{\delta} \partial x^{\gamma}}$$

removes half the extra terms from Equation (2.41), while the remaining half is removed by subtracting Equation (2.42) with β and δ exchanged. To see this, the identity

$$\sum_{\beta'=0}^{3} \frac{\partial x^{\alpha}}{\partial x^{\beta'}} \frac{\partial^2 x^{\beta'}}{\partial x^{\beta} \partial x^{\epsilon}} = - \sum_{\beta'=0}^{3} \sum_{\epsilon'=0}^{3} \frac{\partial x^{\beta'}}{\partial x^{\beta}} \frac{\partial^2 x^{\alpha}}{\partial x^{\beta'} \partial x^{\epsilon'}} \frac{\partial x^{\epsilon'}}{\partial x^{\epsilon}} \tag{2.43}$$

is useful, which follows after taking a partial derivative of

$$\sum_{\beta'=0}^{3} \frac{\partial x^{\alpha}}{\partial x^{\beta'}} \frac{\partial x^{\beta'}}{\partial x^{\beta}} = \delta^{\alpha}_{\beta} \tag{2.44}$$

by x^{ϵ}.

Combining all four terms, we obtain a four-index tensor, the *Riemann curvature tensor*

$$R_{\beta\gamma\delta}{}^{\alpha} = \frac{\partial \Gamma^{\alpha}_{\delta\gamma}}{\partial x^{\beta}} - \frac{\partial \Gamma^{\alpha}_{\beta\gamma}}{\partial x^{\delta}} + \sum_{\epsilon=0}^{3} \Gamma^{\alpha}_{\beta\epsilon}\Gamma^{\epsilon}_{\delta\gamma} - \sum_{\epsilon=0}^{3} \Gamma^{\alpha}_{\delta\epsilon}\Gamma^{\epsilon}_{\beta\gamma}. \tag{2.45}$$

It is antisymmetric in β and δ because we had to antisymmetrize in order to remove third-order partial derivatives of the coordinate transformation. Other symmetries follow from the detailed expression in terms of metric derivatives. Taking these into account, the Riemann tensor has $d^2(d^2 - 1)/12$ independent components in d dimensions, or 20 components in four-dimensional spacetime.

Relevant information about curvature is often contained in tensors with fewer independent components, obtained from the Riemann tensor by summing over indices. The *Ricci tensor* is defined by identifying α and β in Equation (2.45) and summing over them:

$$R_{\gamma\delta} = \sum_{\alpha=0}^{3} \left(\frac{\partial \Gamma^{\alpha}_{\delta\gamma}}{\partial x^{\alpha}} - \frac{\partial \Gamma^{\alpha}_{\alpha\gamma}}{\partial x^{\delta}} \right) + \sum_{\alpha=0}^{3} \sum_{\epsilon=0}^{3} \left(\Gamma^{\alpha}_{\alpha\epsilon}\Gamma^{\epsilon}_{\delta\gamma} - \Gamma^{\alpha}_{\delta\epsilon}\Gamma^{\epsilon}_{\alpha\gamma} \right). \tag{2.46}$$

The Ricci tensor is symmetric and therefore has ten independent components. The remaining two indices can be summed over after raising one of them using the inverse metric, $g^{\alpha\beta}$. The result is the *Ricci scalar*

$$R = \sum_{\gamma=0}^{3} \sum_{\delta=0}^{3} g^{\gamma\delta} R_{\gamma\delta}, \tag{2.47}$$

which is equal to the trace of the matrix given by $R_{\gamma\delta}$.

Any other identifications of indices of the Riemann tensor would result in an object that vanishes identically, owing to antisymmetries in the indices. The unique scalar depending on second-order metric derivatives is therefore given by R. It appears in the *Einstein–Hilbert action* of general relativity

$$S[g_{\alpha\beta}, \phi] = \frac{c^4}{16\pi G} \int dt d^3x \sqrt{-\det g} (R - 2\Lambda) + S_{\text{matter}}[g_{\alpha\beta}, \phi] \tag{2.48}$$

which, at second order of derivatives and for a given matter action, is uniquely determined by general covariance up to the choice of two constants, Newton's constant G and the cosmological constant Λ.

Example: Spatially Flat Friedmann–Robertson–Walker Models
The symmetries of a line element such as

$$ds^2 = -c^2 d\tau^2 + a(\tau)^2 (dx^2 + dy^2 + dz^2) \tag{2.49}$$

greatly simplify the derivation of curvature tensors. Starting with the Christoffel symbol, the diagonal nature of the metric in this case implies that non-zero contributions to Equation (2.26) result only from terms with $\alpha = \delta$, collapsing the sum to a single term. Moreover, the only non-zero partial derivatives of the metric are of the form $\partial g_{ii}/\partial \tau$ where $i = 1, 2, 3$. Therefore, $\Gamma^\alpha_{\beta\gamma}$ is non-zero only if one index is τ and the two others are spatial and equal to each other. Isotropy then implies that the same result is obtained for all three choices of x^i, and there are only two different non-zero components:

$$\Gamma^\tau_{x^i x^i} = \frac{1}{2} g^{\tau\tau} \left(-\frac{\partial g_{x^i x^i}}{\partial \tau} \right) = \frac{a}{c^2} \frac{da}{d\tau} \tag{2.50}$$

$$\Gamma^{x^i}_{\tau x^i} = \Gamma^{x^i}_{x^i \tau} = \frac{1}{2} g^{x^i x^i} \frac{\partial g_{x^i x^i}}{\partial \tau} = \frac{1}{a} \frac{da}{d\tau}. \tag{2.51}$$

Isotropy further implies that the Ricci tensor is diagonal with identical $R_{x^i x^i}$ for $i = 1, 2, 3$. The two independent components are

$$
\begin{aligned}
R_{\tau\tau} &= -\left(\frac{\partial \Gamma^x_{x\tau}}{\partial \tau} + \frac{\partial \Gamma^y_{y\tau}}{\partial \tau} + \frac{\partial \Gamma^z_{z\tau}}{\partial \tau} \right) - \left((\Gamma^x_{x\tau})^2 + (\Gamma^y_{y\tau})^2 + (\Gamma^z_{z\tau})^2 \right) \\
&= -3\frac{d}{d\tau}\left(\frac{1}{a}\frac{da}{d\tau} \right) - 3\left(\frac{1}{a}\frac{da}{d\tau} \right)^2 = -\frac{3}{a}\frac{d^2 a}{d\tau^2}
\end{aligned}
\tag{2.52}
$$

and

$$
\begin{aligned}
R_{xx} &= \frac{\partial \Gamma^\tau_{xx}}{\partial \tau} + \Gamma^\tau_{xx}\left(\Gamma^x_{\tau x} + \Gamma^y_{\tau y} + \Gamma^z_{\tau z} \right) - 2\Gamma^x_{\tau x}\Gamma^\tau_{xx} \\
&= \frac{1}{c^2}\left(\frac{d}{d\tau}\left(a\frac{da}{d\tau} \right) + \left(\frac{da}{d\tau} \right)^2 \right) = \frac{a}{c^2}\frac{d^2 a}{d\tau^2} + \frac{2}{c^2}\left(\frac{da}{d\tau} \right)^2.
\end{aligned}
\tag{2.53}
$$

In order to obtain Equation (2.52), we note that the first term in Equation (2.46) is zero for $\gamma = \delta = \tau$, given Equation (2.50), while the second term produces non-zero contributions only for $\alpha \neq \tau$. The third term is zero while the last term requires $\alpha = \epsilon \neq \tau$ for non-zero contributions. To obtain Equation (2.53), the first term requires $\alpha = \tau$, the second term is zero, the third term requires $\alpha = \tau \neq \epsilon$, and the last term requires $\alpha = x$ and $\epsilon = \tau$ or $\alpha = \tau$ and $\epsilon = x$.

The Ricci scalar is

$$R = -\frac{1}{c^2}R_{\tau\tau} + \frac{3}{a^2}R_{xx} = \frac{6}{c^2}\left(\frac{1}{a}\frac{d^2 a}{d\tau^2} + \left(\frac{1}{a}\frac{da}{d\tau} \right)^2 \right), \tag{2.54}$$

in agreement with Equation (2.13) for $N = 1$ and $k = 0$.

2.2.2 Higher-curvature Actions

If we drop the condition that the action depend on the metric through derivatives of order at most two, it is no longer uniquely determined up to the choice of two constants. For instance, instead of R, any function $f(R)$ may be used as a coordinate-independent quantity, subject only to the condition that $\lim_{R\to 0}(f(R)/R) = 1$ in order to have the correct low-curvature behavior well-tested in general relativity. (Even this condition can be weakened because all tests of general relativity so far have been performed in a small range of curvature values that does *not* include $R = 0$. As an example, contributions proportional to $1/R$ are sometimes used in discussion of late-time cosmological effects such as dark energy.) In this way, one obtains a new class of *f(R)-gravity theories* (Buchdahl 1970) with action

$$S[g_{\alpha\beta}, \phi] = \frac{c^4}{16\pi G} \int dt d^3x \sqrt{-\det g} \, f(R) + S_{\text{matter}}[g_{\alpha\beta}, \phi]. \tag{2.55}$$

(The cosmological constant is then simply a constant contribution to $f(R)$.)

If the gravitational action is not required to be linear in curvature, curvature invariants other than functions of the Ricci scalar are possible. As quadratic curvature invariants, for instance, we may consider

$$R_1 = \sum_{\alpha=0}^{3} \sum_{\beta=0}^{3} R_{\alpha\beta} R^{\alpha\beta} = \sum_{\alpha=0}^{3} \sum_{\beta=0}^{3} \sum_{\gamma=0}^{3} \sum_{\delta=0}^{3} R_{\alpha\beta} R_{\gamma\delta} g^{\alpha\gamma} g^{\beta\delta} \tag{2.56}$$

and

$$R_2 = \sum_{\alpha=0}^{3} \sum_{\beta=0}^{3} \sum_{\gamma=0}^{3} \sum_{\delta=0}^{3} R_{\alpha\beta\gamma\delta} R^{\alpha\beta\gamma\delta}$$

$$= \sum_{\alpha_1=0}^{3} \sum_{\alpha_2=0}^{3} \sum_{\beta_1=0}^{3} \sum_{\beta_2=0}^{3} \sum_{\gamma_1=0}^{3} \sum_{\gamma_2=0}^{3} \sum_{\delta_1=0}^{3} \sum_{\delta_2=0}^{3} R_{\alpha\beta_1\gamma_1}{}^{\delta_1} R_{\alpha_2\beta_2\gamma_2}{}^{\delta_2} g^{\alpha_1\alpha_2} g^{\beta_1\beta_2} g^{\gamma_1\gamma_2} g_{\delta_1\delta_2} \tag{2.57}$$

in addition to R^2.

It turns out that

$$\mathcal{G} = R^2 - 4R_1 + R_2, \tag{2.58}$$

called the *Gauss–Bonnet term*, can be added to the classical R without changing the field equations because $\sqrt{-\det g} \, \mathcal{G}$ is a spacetime divergence, such that its integral in an action is reduced to a boundary term which does not affect local field equations. Therefore, only two quadratic curvature invariants are independent, usually taken as R^2 and R_1. (A non-trivial contribution can result from $\mathcal{G}$ if it is multiplied in the action with a scalar field ϕ, adding also the standard scalar action with a kinetic term of first-order in derivatives. Such a theory is called *dynamical Gauss–Bonnet gravity*; see Equation (2.133) below.)

In a curvature expansion, the leading corrections to general relativity are obtained from quadratic curvature invariants in a higher-curvature action

$$S[g_{\alpha\beta}, \phi] = \frac{c^4}{16\pi G} \int dt d^3x \sqrt{-\det g} \, (R - 2\Lambda + AR^2 + BR_1)$$

$$+ S_{\text{matter}}[g_{\alpha\beta}, \phi]. \tag{2.59}$$

The two parameters A and B play the role of new coupling constants in addition to Newton's constant G. Because curvature has units of $1/[L]^2$, the coefficients must have units of $[L]^2$. Given the known constants of nature, a unique length parameter can be defined, the *Planck length*

$$\ell_P = \sqrt{\frac{G\hbar}{c^3}}. \tag{2.60}$$

It requires Planck's constant $\hbar$ which appears only in quantum theories. Therefore, higher-curvature terms in a gravity action are usually considered quantum corrections to general relativity. They are indeed implied by effective theories of perturbative quantum gravity (Donoghue 1994; Burgess 2004).

Because the Planck length evaluates to the tiny $\ell_P = 1.6 \times 10^{-35}$ m, huge curvature is required for significant higher-curvature corrections from quantum effects. Such values may have been realized in the very early universe. If higher-curvature terms are proposed as classical modifications of general relativity, their coefficients are much less constrained and could therefore be larger. However, it is then more difficult to find a convincing explanation of the new length scales implied by the coefficients A and B.

Evaluated for isotropic models, both R^2 and R_1 contain a term quadratic in $\ddot{a}$. For instance, adding a multiple of R^2 to the classical action would contribute a term $\ddot{a}^2/a^2$ which, unlike $\ddot{a}/a$, cannot be reduced to first-order form by integrating by parts. The standard form of Euler–Lagrange equations, (2.2), should then be generalized to the higher-order version

$$\frac{d^2}{dt^2}\frac{\partial L}{\partial \ddot{a}} - \frac{d}{dt}\frac{\partial L}{\partial \dot{a}} + \frac{\partial L}{\partial a} = 0. \tag{2.61}$$

(Equations of even higher order follow the same pattern, with alternating signs.) For $f(R)$ models of isotropic cosmology, for instance, the gravitational action contributes a term

$$6\frac{d^2}{d\tau^2}\left(a^2\frac{df}{dR}\right) - 12\frac{d}{d\tau}\left(a\dot{a}\frac{df}{dR}\right) + 3c^2a^2f(R) - 6\frac{df}{dR}(a\ddot{a} + 2\dot{a}^2) \tag{2.62}$$

to the Friedmann equation (using $N = 1$ and therefore $t = \tau$ for simplicity, as well as $k = 0$). Acting with τ-derivatives on df/dR implies higher than second-order derivatives as long as df/dR depends on R, that is as long as $f(R)$ is not a linear function. The first term in Equation (2.62) is then a time derivative of a of up to fourth order.

Higher-order equations of motion usually modify the coefficients seen in second-order equations and can therefore suggest new physical effects. However, for a physical interpretation it is important to notice that they also imply new degrees of freedom: For second-order equations, all configuration variables and their first time derivatives (velocities) can be chosen freely at an initial time; they represent the degrees of freedom. For higher-order equations, we also have to know at least second-order derivatives (accelerations) at the initial time if we want to produce a unique solution. These are new degrees of freedom which do not appear in lower-order theories.

The differential operator in Equation (2.62) is of fourth order for any non-linear $f(R)$. It therefore requires four initial values, the scale factor and its first three derivatives at an initial time. The same number of initial values would be required if there were two degrees of freedom with standard second-order equations of motion. For inhomogeneous geometries, the counting implies four initial values per spatial point for a fourth-order equation for the metric, or two local degrees of freedom given by two fields with standard second-order field equations.

2.2.3 The Ostrogradsky Problem

General covariance implies that modified action principles for the spacetime metric (without additional fields) can be formulated only by using non-linear functions of curvature invariants. Since curvature depends on up to second-order derivatives of the metric, such action principles lead to higher-derivative equations. In addition to requiring more initial values than second-order equations, higher-derivative theories are often unstable.

For a general demonstration of how instability is implied by higher derivatives, we use a Lagrangian $L(q, \dot{q}, \ddot{q})$ which depends on up to second-order derivatives. Generically, $\ddot{q}$ cannot be removed by integrating by parts. As a formal condition of this implication, we impose that the highest derivative order be non-linear, $\partial^2 L/\partial \ddot{q}^2 \neq 0$. If this condition is violated, as in Equation (2.13), the second-derivative order is spurious because it can be removed through integration by parts.

In the higher-derivative case, the variable q and all its derivatives in the action, except for the highest-derivative one, are considered independent configuration variables because they can be chosen freely as initial values. For instance, $\dot{q}$ is no longer proportional to the momentum of q, but q as well as $\dot{q}$ have their own momenta. We refer to the configuration variables as

$$q_1 = q \quad \text{and} \quad q_2 = \dot{q} \tag{2.63}$$

in order to emphasize this independent nature.

With this definition, we can already see that higher-derivative terms generically imply an unbounded Hamiltonian linear in the momentum p_1 of q_1. Unlike in standard second-order theories, p_1 cannot be proportional to $\dot{q}_1 = \dot{q}$ because it must be independent of the second configuration variable, q_2, while Equation (2.63) requires us to identify $\dot{q}$ with q_2. We will derive the specific expression for p_1 in the next paragraphs, but irrespective of its precise form, Hamilton's equation for q_1 imply that

$$\frac{dq_1}{dt} = \frac{\partial H}{\partial p_1} \tag{2.64}$$

because this equation characterizes p_1 as the momentum of q_1, given a Hamiltonian $H(q_1, p_1, q_2, p_2)$. Since we have defined $q_2 = \dot{q} = \dot{q}_1$, we obtain the partial differential equation $\partial H / \partial p_1 = q_2$, which has the general solution $H = p_1 q_2 + h(q_1, q_2, p_2)$ with an arbitrary function h independent of p_1.

Therefore, the dependence of H on p_1 is linear. Such a Hamiltonian does not determine a stable system because solutions exist for any negative value of the energy, $E = H$. If the system is interacting with other degrees of freedom, even just weekly, it can continually transmit energy to them without reaching a stable ground state.

In more detail, momenta of the configuration variables are derived from partial derivatives of the Lagrangian by suitable time derivatives of q. The specific combinations of partial derivatives are dictated by the higher-order version of Euler–Lagrange equations, such as Equation (2.61), to which the resulting canonical equations must be equivalent. For second-order Lagrangians, the two momenta of q_1 and q_2, respectively, are given by

$$p_1 = \frac{\partial L}{\partial \dot{q}} - \frac{d}{dt} \frac{\partial L}{\partial \ddot{q}} \tag{2.65}$$

$$p_2 = \frac{\partial L}{\partial \ddot{q}}, \tag{2.66}$$

as we will show now by requiring equivalence between the Lagrangian and Hamiltonian pictures.

Since we assumed

$$\frac{\partial^2 L}{\partial \ddot{q}^2} \neq 0 \tag{2.67}$$

to avoid higher-order terms being spurious, it is possible to invert the expression (2.66) for $\ddot{q}$, at least locally. Together with $q = q_1$ and $\dot{q} = q_2$, we can therefore express all variables on which $L(q, \dot{q}, \ddot{q})$ depends in terms of the canonical variables q_1, q_2, p_1 and p_2. Also the Hamiltonian

$$H(q_1, p_1, q_2, p_2) = p_1 \dot{q}_1(q_2) + p_2 \dot{q}_2(q_1, q_2, p_2) - L(q(q_1), \dot{q}(q_2), \ddot{q}(q_1, q_2, p_2)), \tag{2.68}$$

obtained from L by a two-variable Legendre transformation, can be written as such a function. Moreover, the dependence on the various canonical variables is restricted by their definitions. For instance, $\dot{q}_1$ depends only on q_2 because we defined $q_2 = \dot{q} = \dot{q}_1$.

Hamilton's equation for the momentum p_1,

$$\dot{p}_1 = -\frac{\partial H}{\partial q_1}, \tag{2.69}$$

is then equivalent to the required higher-order Euler–Lagrange equation: We have the correct analog of Equation (2.61),

$$0 = \dot{p}_1 + \frac{\partial H}{\partial q_1} = \frac{d}{dt}\frac{\partial L}{\partial \dot{q}} - \frac{d^2}{dt^2}\frac{\partial L}{\partial \ddot{q}} - \frac{\partial L}{\partial q_1}, \qquad (2.70)$$

provided

$$\frac{\partial H}{\partial q_1} = -\frac{\partial L}{\partial q_1}, \qquad (2.71)$$

which indeed follows from the two-variable Legendre transformation (2.68). (Taking a partial derivative of Equation (2.68) by q_1, the two terms that result from the q_1-dependence of $p_2 \dot{q}_2$ and $-L$, respectively, cancel out thanks to Equation (2.76) and $\dot{q}_2 = \ddot{q}$.) Hamilton's equation for p_2,

$$0 = \dot{p}_2 + \frac{\partial H}{\partial q_2} = \frac{d}{dt}\frac{\partial L}{\partial \ddot{q}} + p_1 - \frac{\partial L}{\partial \dot{q}}, \qquad (2.72)$$

is identically satisfied using Equation (2.65), as well as

$$\frac{\partial H}{\partial q_2} = p_1 - \frac{\partial L}{\partial \dot{q}} \qquad (2.73)$$

which is obtained from Equation (2.68) using $\dot{q}_1 = \dot{q} = q_2$ and the same cancellation as described for $\dot{p}_1$.

As indicated in Equation (2.68), $\dot{q}_2 = \ddot{q}$ depends only on q_1, q_2 and p_2 but not on p_1 because it is obtained by inverting Equation (2.76). This inversion can depend only on p_2 on the left-hand side of Equation (2.76) and all variables in L other than the one we are inverting it for, $\ddot{q}$. These other variables are $q = q_1$ and $\dot{q}_2$, without any independent contributions from p_1. The first momentum, p_1, therefore appears only in the first contribution to Equation (2.68), given by $p_1\dot{q}_1(q_2) = p_1 q_2$. Since this term is linear in p_1, the Hamiltonian is unbounded from below, implying instability (Woodard 2007).

2.2.4 R^2 Gravity

Generic higher-derivative theories are unstable, and therefore generic higher-curvature actions of gravity are unstable. However, several possibilities exist to evade this conclusion. We have already seen that degenerate higher-derivative theories need not be unstable because our argument for a Hamiltonian linear in one of the momenta then breaks down. An example for such a theory is general relativity, in which the Ricci scalar (2.13) contains a second-order time derivative of a, but only in linear form. Here, second-order derivatives in the action can be removed through integration by parts.

A second loophole refers to the specific nature of gravity, which is unstable even in its classical form due to gravitational collapse or, in reverse, long-time expansion of the universe. Indeed, the standard gravitational Hamiltonians we have already encountered, such as Equation (1.18) for isotropic models or Equation (1.47) for homogeneous models, are unbounded from below. Unlike Equation (2.68), they are not linear in momenta, but instead have one quadratic momentum term with negative coefficient. The Hamiltonian is unbounded from below also in this case.

For gravity, it is therefore admissible to have one unstable direction in phase space, in which the energy can be lowered indefinitely by increasing a suitable momentum. If the unstable direction implied by higher-derivative terms happens to agree with this unstable direction required for gravity, the model is acceptable.

Canonical Structure
As an example, we consider an isotropic model in which the Ricci scalar R in the Lagrangian is replaced by a quadratic expression, $R + AR^2$, with some constant A with units of $1/[L]^2$. For simplicity, we use spatially flat models, $k = 0$, and choose proper time, such that $N = 1$. The Ricci scalar (2.13) then simplifies to $R = 6c^{-2}(\ddot{a}/a + \dot{a}^2/a^2)$, and our Lagrangian is given by

$$
\begin{aligned}
L &= \frac{V_0 c^4}{16\pi G} a^3 (R + AR^2) + L_{\text{matter}} \\
&= \frac{3V_0}{8\pi G}(6A a \ddot{a}^2 + (c^2 a^2 + 12A\dot{a}^2)\ddot{a} + (c^2 + 6A\dot{a}^2/a^2)a\dot{a}^2) + L_{\text{matter}}.
\end{aligned}
\tag{2.74}
$$

As in our general discussion (2.63), we choose as configuration variables $q_1 = a$ and $q_2 = \dot{a}$. Assuming that L_{matter} depends on a but not on its time derivatives, the momenta can then be derived as

$$
p_1 = \frac{\partial L}{\partial \dot{a}} - \frac{d}{dt}\frac{\partial L}{\partial \ddot{a}} = -\frac{9V_0 A}{2\pi G}\left(a\dddot{a} + \dot{a}\ddot{a} - 2\frac{\dot{a}^3}{a}\right)
\tag{2.75}
$$

and

$$
p_2 = \frac{\partial L}{\partial \ddot{a}} = \frac{3V_0}{8\pi G}(c^2 a^2 + 12A(a\ddot{a} + \dot{a}^2)).
\tag{2.76}
$$

The last equation implies that

$$
\ddot{a} = \dot{q}_2 = \frac{2\pi G}{9V_0 A}\frac{p_2}{q_1} - \frac{c^2}{12A}q_1 - \frac{q_2^2}{q_1}
\tag{2.77}
$$

which can be used to express the Lagrangian and its Legendre transform in terms of canonical variables instead of time derivatives of a:

$$
\begin{aligned}
H(q_1, q_2, p_1, p_2) &= p_1 q_2 + \frac{\pi G}{9V_0 A}\frac{p_2^2}{q_1} - \frac{c^2}{12A}q_1 p_2 \\
&\quad - \frac{q_2^2 p_2}{q_1} + \frac{V_0 c^4}{64\pi G}\frac{q_1^3}{A} + H_{\text{matter}}.
\end{aligned}
\tag{2.78}
$$

The Hamiltonian (2.78) has one unstable direction, given by the term linear in p_1. This direction replaces the unstable direction in the standard isotropic Hamiltonian (1.18) because the dependence on p_2 is now quadratic with *positive* coefficient, and therefore stable. As a model of a gravitational system, the Lagrangian (2.74) is acceptable in spite of its non-degenerate dependence on $\ddot{a}$.

Hamilton's equations generated by Equation (2.78) are

$$\frac{\mathrm{d}q_1}{\mathrm{d}\tau} = q_2 \tag{2.79}$$

$$\frac{\mathrm{d}q_2}{\mathrm{d}\tau} = \frac{2\pi G}{9V_0 A}\frac{p_2}{q_1} - \frac{c^2}{12A}q_1 - \frac{q_2^2}{q_1} \tag{2.80}$$

$$\frac{\mathrm{d}p_1}{\mathrm{d}\tau} = \frac{\pi G}{9V_0 A}\frac{p_2^2}{q_1^2} + \frac{c^2}{12A}p_2 - \frac{q_2^2 p_2}{q_1^2} - \frac{3V_0 c^4}{64\pi G}\frac{q_1^2}{A} - \frac{\partial H_{\text{matter}}}{\partial a} \tag{2.81}$$

$$\frac{\mathrm{d}p_2}{\mathrm{d}\tau} = -p_1 + 2\frac{q_2 p_2}{q_1} \tag{2.82}$$

where

$$\frac{1}{V_0}\frac{\partial H_{\text{matter}}}{\partial a} = 3\rho a^2 + \frac{\partial \rho}{\partial a}a^3 = -3Pa^2 \tag{2.83}$$

with the matter pressure P, using the continuity Equation (1.10).

Equation (2.80) implies that

$$p_2 = \frac{3V_0 c^2}{8\pi G}q_1^2 + \frac{9V_0 A}{2\pi G}\left(q_1\dot{q}_2 + q_2^2\right) \tag{2.84}$$

which is the same as Equation (2.76). Inserting this relationship in Equation (2.82), we can compute

$$p_1 = 2\frac{q_2}{q_1}p_2 - \dot{p}_2 = -\frac{9V_0 A}{2\pi G}\left(q_1\ddot{q}_2 + q_2\dot{q}_2 - 2\frac{q_2^3}{q_1}\right), \tag{2.85}$$

again the same as Equation (2.75). The final equation, (2.81), then takes the form

$$-\frac{9V_0 A}{2\pi G}\frac{\mathrm{d}}{\mathrm{d}\tau}\left(q_1\ddot{q}_2 + q_2\dot{q}_2 - 2\frac{q_2^3}{q_1}\right) = \frac{3V_0 c^2}{8\pi G}(2q_1\dot{q}_2 + q_2^2)$$
$$+ \frac{9V_0 A}{4\pi G}\left(\dot{q}_2^2 - \frac{q_2^4}{q_1^2}\right) + 3V_0 q_1^2 P \tag{2.86}$$

or

$$-2\frac{\dot{q}_2}{q_1} - \frac{\dot{q}_2^2}{q_1^2} = \frac{8\pi G}{c^2}P + \frac{6A}{c^2}\left(2\frac{\ddot{q}_2}{q_1} + 4\frac{\dot{q}_2\ddot{q}_2}{q_1^2} + 3\frac{\dot{q}_2^2}{q_1^2} - 12\frac{\dot{q}_2^2\ddot{q}_2}{q_1^3} + 3\frac{\dot{q}_2^4}{q_1^4}\right). \quad (2.87)$$

For $A = 0$ we obtain the correct Raychaudhuri Equation (1.12), while higher-derivative terms may be viewed as effective pressure terms. In addition, the constraint H = 0 with p_1 and p_2 inserted implies the Friedmann equation

$$\frac{\dot{q}_2^2}{q_1^2} = \frac{8\pi G}{3c^2}\rho - \frac{6A}{c^2}\left(2\frac{\dot{q}_2\ddot{q}_2}{q_1^2} - \frac{\dot{q}_2^2}{q_1^2} + 2\frac{\dot{q}_2^2\ddot{q}_2}{q_1^3} - 3\frac{\dot{q}_2^4}{q_1^4}\right) \quad (2.88)$$

with higher-order corrections.

It is straightforward to confirm that the higher-derivative corrections in both Equations (2.87) and (2.88) vanish for exponential inflation, $a(\tau) \propto \exp(H\tau)$ with $H = \dot{a}/a = q_2/q_1$ constant. Therefore, if the matter contribution is a cosmological constant or dark energy, such that $P = -\rho$, the classical solution can be extended to non-zero A. More importantly, the A-term in Equation (2.87) can, under certain conditions, be viewed as an effective negative pressure, driving inflation even if the matter contribution has zero or positive pressure.

Conformal Transformation
It is easier to see the possibility of inflationary solutions by reformulating the action (Whitt 1984), rather than using equations of motion. Such a reformulation may be viewed as the Lagrangian version of the introduction of auxiliary degrees of freedom as in Equation (2.63). A higher-derivative action which requires multiple derivatives as initial values of the original field can then be mapped to a second-order action in which an enlarged number of fields and their first-order derivatives are prescribed as initial values.

Starting from

$$S[g_{\alpha\beta}] = \frac{c^4}{16\pi G}\int dt d^3x\sqrt{-\det g}\,(R + AR^2), \quad (2.89)$$

we first introduce a non-dynamical auxiliary field, ψ, by writing the action as

$$S[g_{\alpha\beta}, \psi] = \frac{c^4}{16\pi G}\int dt d^3x\sqrt{-\det g}\,(R(1 + 2A\psi) - A\psi^2). \quad (2.90)$$

The auxiliary field ψ is non-dynamical because its time derivative does not appear in the action. Its Euler–Lagrange equation generated by variation with respect to ψ,

$$0 = \frac{\partial L}{\partial\psi} = \frac{c^4 A}{8\pi G}\sqrt{-\det g}\,(R - \psi), \quad (2.91)$$

can be solved algebraically by $\psi = R$. Inserting this solution in Equation (2.90), we obtain Equation (2.89). The new action therefore implies equations of motion equivalent to the original one.

In a second step, a conformal transformation, introducing

$$\tilde{g}_{\alpha\beta} = \frac{1}{1 + 2A\psi}g_{\alpha\beta} \qquad (2.92)$$

as the new metric, eliminates the factor $1 + 2A\psi$ of R in Equation (2.89). In general, a conformal transformation $\tilde{g}_{\alpha\beta} = \Omega^2 g_{\alpha\beta}$ with conformal factor Ω transforms the Ricci scalar R of $g_{\alpha\beta}$ to the new Ricci scalar (Wald 1984)

$$\tilde{R} = \frac{R}{\Omega^2} - \frac{6}{\Omega^2}\sum_{\alpha=0}^{3}\sum_{\beta=0}^{3} g^{\alpha\beta}\left(\nabla_{\alpha}\nabla_{\beta}\log\Omega + \frac{\partial\log\Omega}{\partial x^{\alpha}}\frac{\partial\log\Omega}{\partial x^{\beta}}\right) \qquad (2.93)$$

of $\tilde{g}_{\alpha\beta}$, or

$$R = \Omega^2\tilde{R} + 6\sum_{\alpha=0}^{3}\sum_{\beta=0}^{3}\tilde{g}^{\alpha\beta}\left(\Omega\tilde{\nabla}_{\alpha}\tilde{\nabla}_{\beta}\Omega - 2\frac{\partial\Omega}{\partial x^{\alpha}}\frac{\partial\Omega}{\partial x^{\beta}}\right) \qquad (2.94)$$

where $\tilde{\nabla}$ denotes the covariant derivative determined by the new metric, $\tilde{g}_{\alpha\beta}$.

Together with $\sqrt{-\det g} = \Omega^{-4}\sqrt{-\det\tilde{g}}$, the term $\sqrt{-\det g}\,R(1 + 2A\psi)$ in Equation (2.90) is therefore transformed to

$$\begin{aligned}
\sqrt{-\det g}\,R(1 + 2A\psi) &= \sqrt{-\det\tilde{g}}\,\Omega^{-2}\tilde{R}(1 + 2A\psi) \\
&\quad + 6\sqrt{-\det\tilde{g}}\,(1 + 2A\psi)\sum_{\alpha=0}^{3}\sum_{\beta=0}^{3}\tilde{g}^{\alpha\beta}\Big(\Omega^{-3}\tilde{\nabla}_{\alpha}\tilde{\nabla}_{\beta}\Omega \\
&\quad - 2\Omega^{-4}\frac{\partial\Omega}{\partial x^{\alpha}}\frac{\partial\Omega}{\partial x^{\beta}}\Big).
\end{aligned} \qquad (2.95)$$

The first term equals the standard gravitational Lagrangian provided we choose the conformal factor such that

$$1 + 2A\psi = \Omega^2. \qquad (2.96)$$

The second term, inserted in Equation (2.90), provides a scalar-field action

$$\begin{aligned}
&\frac{3c^2}{8\pi G}\int dt d^3x\sqrt{-\det\tilde{g}}\sum_{\alpha=0}^{3}\sum_{\beta=0}^{3}\tilde{g}^{\alpha\beta}\left(\Omega^{-1}\tilde{\nabla}_{\alpha}\tilde{\nabla}_{\beta}\Omega - 2\Omega^{-2}\frac{\partial\Omega}{\partial x^{\alpha}}\frac{\partial\Omega}{\partial x^{\beta}}\right) \\
&= -\frac{9c^2}{8\pi G}\int dt d^3x\sqrt{-\det\tilde{g}}\,\Omega^{-2}\sum_{\alpha=0}^{3}\sum_{\beta=0}^{3}\tilde{g}^{\alpha\beta}\frac{\partial\Omega}{\partial x^{\alpha}}\frac{\partial\Omega}{\partial x^{\beta}} \\
&= -\frac{9c^2}{8\pi G}A^2\int dt d^3x\frac{\sqrt{-\det\tilde{g}}}{(1 + 2A\psi)^2}\sum_{\alpha=0}^{3}\sum_{\beta=0}^{3}\tilde{g}^{\alpha\beta}\frac{\partial\psi}{\partial x^{\alpha}}\frac{\partial\psi}{\partial x^{\beta}}
\end{aligned} \qquad (2.97)$$

after integration by parts and using Equation (2.96) (as well as Equation (2.35)).

We can transform the derivatives in the last line to standard form

$$S_{\text{kin}}[\phi] = -\frac{1}{2} \int dt d^3x \sqrt{-\det \tilde{g}} \sum_{\alpha=0}^{3} \sum_{\beta=0}^{3} \tilde{g}^{\alpha\beta} \frac{\partial \phi}{\partial x^\alpha} \frac{\partial \phi}{\partial x^\beta} \tag{2.98}$$

by introducing a new scalar field, ϕ, related to ψ by

$$\frac{3cA}{2\sqrt{2\pi G}} \frac{d\psi}{1 + 2A\psi} = d\phi. \tag{2.99}$$

This differential equation is solved by

$$1 + 2A\psi = \exp\left(\frac{2\sqrt{2\pi G}\,\phi}{3cA}\right) \tag{2.100}$$

and thus

$$\psi = \frac{1 - \exp(2\sqrt{2\pi G}\,\phi/(3cA))}{2A}. \tag{2.101}$$

Finally, the quadratic term in Equation (2.90) depending on ψ is transformed to

$$\frac{c^4 A}{16\pi G} \sqrt{-\det g}\, \psi^2 = \frac{c^4 A}{16\pi G} \sqrt{-\det \tilde{g}} \frac{\psi^2}{(1 + 2A\psi)^2}$$
$$= \frac{c^2}{64\pi G A}\left(1 - \exp(-2\sqrt{2\pi G}\,\phi/(3cA))\right)^2. \tag{2.102}$$

Starobinsky Model
Interpreting Equation (2.102) as a potential of the new field ϕ, the quadratic gravity action (2.89) is equivalent to the Einstein–Hilbert action coupled to a standard scalar field with potential

$$W(\phi) = \frac{c^2}{64\pi G A}\left(1 - \exp(-2\sqrt{2\pi G}\,\phi/(3cA))\right)^2. \tag{2.103}$$

The potential and kinetic energies of the scalar field are bounded from below, provided $A > 0$, demonstrating again that there are no instabilities implied by the higher-derivative contributions of AR^2. Higher-derivative terms in this case imply only a single additional degree of freedom, described by the scalar field ϕ.

As shown in Figure 2.2, the potential (2.103) has a long flat range for large ϕ, which can be used to generate a slow-roll regime in which the kinetic energy is much smaller than the potential. The scalar is then very close to a matter source with $P \approx -\rho$, leading to inflationary solutions for the scale factor. This type of inflation, as well as the quadratic action (2.89) it is derived from, are usually referred to as the *Starobinsky model* because inflationary solutions in a quadratic gravity model were first discussed in Starobinsky (1980). However, these solutions were not those

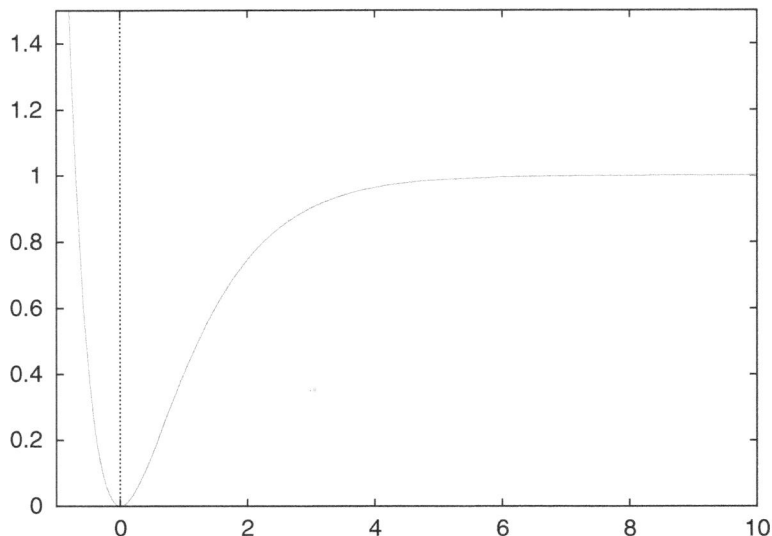

Figure 2.2. The equivalent scalar potential (2.103) of the Starobinsky model.

derived from a potential (2.103). They were rather derived from additional terms of the form Dq_2^4 that could appear in a more general version of Equation (2.87), where D is an independent coefficient in addition to A. Such a term implies that exponential solutions $a(\tau) \propto \exp(H\tau)$ with constant H exist even if there is no matter energy and pressure. These were the original inflationary solutions discussed by Starobinsky.

The D-term does not come from a higher-curvature contribution to the gravitational part, but was instead derived as a possible quantum term in the stress–energy tensor of matter (Davies et al. 1977) that does not have an analog as a higher-curvature term in the action. (The second independent quadratic curvature invariant, R_1 in Equation (2.56), would imply correction terms that are not independent of the A-terms derived from R^2-corrections in isotropic cosmological models.) Moreover, it can appear only in conformally flat spacetimes. The latter include Friedmann–Robertson–Walker models but not all their perturbations. The condition is therefore rather restrictive. Moreover, the constant Hubble parameter H of exponential solutions is proportional to D^{-1}, such that the solutions are not perturbative in D. They are admissible solutions if the dynamics is precisely given by the A and D-corrections, while all higher-curvature terms beyond second order are zero. If some higher-curvature terms are non-zero, even if they are tiny they may significantly contribute to solutions that are not perturbative in the expansion parameters.

Not only the original solutions in Starobinsky's model, but also the refined version based on the potential (2.103) suffer from the drawback of being unreliable in a perturbative context. The potential (2.103) and its exponent are proportional to A^{-1}. It is therefore not guaranteed that higher-curvature terms beyond second order will not change the potential in significant ways (Simon 1990).

Nevertheless, the potential usually referred to in models of Starobinsky inflation seems to be in good agreement with observations of the cosmic microwave background (Ade & Planck Collaboration 2014). Evaluated for Equation (2.103), the number of e-folds (1.110) is given by

$$N_* \approx \frac{4\pi G}{\kappa^2 c^4}(e^{\kappa\varphi_*} - e^{\kappa\varphi_{\text{fin}}}) \approx \frac{4\pi G}{\kappa^2 c^4}e^{\kappa\varphi_*} \tag{2.104}$$

where

$$\kappa = \frac{2\sqrt{2\pi G}}{3cA}. \tag{2.105}$$

The slow-roll parameter (1.99) is then

$$\epsilon \approx \frac{4\pi G}{c^4 \kappa^2 N_*^2} \tag{2.106}$$

which, compared with Equation (1.113) for power-law potentials has an additional small factor of $1/N_*$. Therefore, the tensor-to-scalar ratio (1.115) is small for Starobinsky inflation, and unlike power-law inflation it remains consistent with observations.

A similar result is, of course, obtained from the intermediate action (2.90) used in the derivation of the Starobinsky potential. Instead of higher-curvature terms, one then has a so-called scalar-tensor theory in which the scalar ψ or, in a more common interpretation in this context, $\varphi = \sqrt{\psi}$ is *non-minimally coupled* to gravity through a term $\psi R = \varphi^2 R$ in the action. As an inflation model, this setting is called *Higgs inflation* because the scalar field φ could, under certain conditions, be identified with the observed Higgs particle provided the latter has a non-minimal coupling (Bezrukov & Shaposhnikov 2008; Steinwachs 2020). One should then also add a kinetic term for φ to Equation (2.90), which would only change some numerical coefficients in the potential that results after the conformal transformation (2.92).

2.2.5 Scalar-tensor Theories

As in the equivalent scalar-field theory in R^2-gravity, for any Lagrangian of the form $f(R)$ with some function f it is possible to construct an explicit model with second-order field equations, introducing auxiliary fields that capture the additional degrees of freedom of a higher-derivative theory. For a review with phenomenological applications, see Sotiriou & Faraoni (2010).

f(R)-gravity and Brans–Dicke Models
To this end, we introduce an auxiliary field ψ and rewrite the action as

$$S[g_{\alpha\beta}, \psi] = \frac{c^4}{16\pi G} \int \mathrm{d}t\mathrm{d}^3x\sqrt{-\det g}\left(f(\psi) + \frac{\mathrm{d}f}{\mathrm{d}\psi}(R - \psi)\right). \tag{2.107}$$

(For $f(\psi) = \psi + A\psi^2$, this action is the same as Equation (2.90).) Variation with respect to ψ, which does not appear with time or space derivatives, yields the algebraic Euler–Lagrange equation $(R - \psi)d^2f/d\psi^2 = 0$, which implies $\psi = R$ as long as $f(R)$ is non-linear. The solution $\psi = R$ of an algebraic equation can be reinserted into the action (2.107) without changing its extrema, upon which we gain back the original $f(R)$-theory.

Next, we redefine the auxiliary field by mapping ψ to ϕ such that

$$\phi = \frac{\mathrm{d}f}{\mathrm{d}\psi}. \qquad (2.108)$$

This field is indeed variable as long as $f(R)$ is non-constant. (Because $f(R)$ is linear for general relativity, there is no extra degree of freedom in this case.) The action (2.107) then takes the form

$$S[g_{\alpha\beta}, \phi] = \frac{c^4}{16\pi G} \int \mathrm{d}t\mathrm{d}^3x\sqrt{-\det g}\,(\phi R - V(\phi)) \qquad (2.109)$$

where a new potential has been defined by

$$V(\phi) = \phi\psi(\phi) - f(\psi(\phi)), \qquad (2.110)$$

defining the function $\psi(\phi)$ as obtained by inverting Equation (2.108). This inversion can locally be performed as long as $\mathrm{d}\phi/\mathrm{d}\psi = \mathrm{d}^2f/\mathrm{d}\psi^2 \neq 0$, again as long as f is not linear.

The new action is linear in R, just like the action of general relativity, but the coefficient is not constant and depends on an independent degree of freedom, ϕ. The form (2.109) is a special case of *Brans–Dicke theories* (Brans & Dicke 1961), defined in general by

$$S[g_{\alpha\beta}, \phi] = \frac{c^4}{16\pi G} \int \mathrm{d}t\mathrm{d}^3x\sqrt{-\det g}\left(\phi R - \frac{\omega}{\phi}\sum_{\alpha=0}^{3}\sum_{\beta=0}^{3}g^{\alpha\beta}\frac{\partial\phi}{\partial x^\alpha}\frac{\partial\phi}{\partial x^\beta} - V(\phi)\right) \quad (2.111)$$

with a real parameter ω. The set of $f(R)$-gravity theories is therefore identical with Brans–Dicke theories such that $\omega = 0$, given a suitable potential $V(\phi)$ related to $f(R)$ by Equation (2.110). (If the factor ω/ϕ in Equation (2.111) is replaced by $1/2$, the action is referred to as *dilaton gravity*. The *dilaton potential* $V(\phi)$ is the same used in Section 1.5 in order to generalize spherically symmetric gravity.)

Any Brans–Dicke theory can be mapped by a conformal transformation to general relativity coupled to a scalar field. We redefine the metric used in Equation (2.111) by introducing the conformally related

$$\tilde{g}_{\alpha\beta} = \phi g_{\alpha\beta}. \qquad (2.112)$$

Equation (2.93) implies that this new metric has the Ricci scalar

$$\tilde{R} = \frac{R}{\phi} - \frac{3}{\phi} \sum_{\alpha=0}^{3} \sum_{\beta=0}^{3} g^{\alpha\beta}\left(\nabla_\alpha \nabla_\beta \log\phi + \frac{1}{2}\frac{\partial \log\phi}{\partial x^\alpha}\frac{\partial \log\phi}{\partial x^\beta}\right). \qquad (2.113)$$

Therefore,

$$\sqrt{-\det\tilde{g}}\,\tilde{R} = \phi^2\sqrt{-\det g}\,\tilde{R}$$
$$= \sqrt{-\det g}\left(\phi R - 3\phi \sum_{\alpha=0}^{3} \sum_{\beta=0}^{3} g^{\alpha\beta}\left(\nabla_\alpha \nabla_\beta \log\phi + \frac{1}{2}\frac{\partial \log\phi}{\partial x^\alpha}\frac{\partial \log\phi}{\partial x^\beta}\right)\right). \qquad (2.114)$$

Inserting this equation in an action, we can integrate by parts in the middle term and obtain

$$\int \mathrm{d}t\mathrm{d}^3x\sqrt{-\det\tilde{g}}\,\tilde{R} = \int \mathrm{d}t\mathrm{d}^3x\sqrt{-\det g}\left(\phi R + \frac{3}{2\phi} \sum_{\alpha=0}^{3} \sum_{\beta=0}^{3} g^{\alpha\beta}\frac{\partial \phi}{\partial x^\alpha}\frac{\partial \phi}{\partial x^\beta}\right). \qquad (2.115)$$

The last term depends on the metric only through $\sqrt{-\det g}\,g^{\alpha\beta} = \phi^{-1}\sqrt{-\det\tilde{g}}\,\tilde{g}^{\alpha\beta}$. A suitable contribution to it can therefore be moved to the left, such that in (2.111) we have

$$\frac{c^2}{16\pi G} \int \mathrm{d}t\mathrm{d}^3x\sqrt{-\det g}\left(\phi R - \frac{\omega}{\phi} \sum_{\alpha=0}^{3} \sum_{\beta=0}^{3} g^{\alpha\beta}\frac{\partial \phi}{\partial x^\alpha}\frac{\partial \phi}{\partial x^\beta}\right)$$
$$= \frac{c^2}{16\pi G} \int \mathrm{d}t\mathrm{d}^3x\sqrt{-\det\tilde{g}}\left(\tilde{R} - \frac{\omega + 3/2}{\phi^2} \sum_{\alpha=0}^{3} \sum_{\beta=0}^{3} \tilde{g}^{\alpha\beta}\frac{\partial \phi}{\partial x^\alpha}\frac{\partial \phi}{\partial x^\beta}\right). \qquad (2.116)$$

As long as $\omega > -3/2$, the scalar can be redefined to

$$\tilde{\phi} = \frac{c}{4}\sqrt{\frac{2\omega + 3}{\pi G}} \log\phi \qquad (2.117)$$

to obtain general relativity coupled to a scalar field with standard kinetic term. In particular, $f(R)$-gravity, which is equivalent to a Brans–Dicke theory with $\omega = 0$, can always be written as standard gravity coupled to a scalar field with a potential determined by Equations (2.110) and (2.117).

Palatini $f(R)$-gravity

It is possible to avoid having an extra degree of freedom in $f(R)$-gravity theories if one views R as a function of two independent objects, the metric $g_{\alpha\beta}$ and the symbol $\Gamma^\gamma_{\alpha\beta}$ in Equation (2.46) without assuming the relationship (2.26). One then obtains two sets of Euler–Lagrange equations, one for each index-object. In general relativity (that is, $f(R) = R$), the Euler–Lagrange equations obtained by varying $\Gamma^\gamma_{\alpha\beta}$ are linear in $\Gamma^\gamma_{\alpha\beta}$ and first-order derivatives of the metric. They can be solved algebraically, with the result that $\Gamma^\gamma_{\alpha\beta}$ must be the Christoffel symbol. Inserting this algebraic solution in the original action then implies the usual action, in which R is understood as a function of the metric alone, assuming the usual form of the Christoffel symbol. Therefore, no new theory is obtained in this case, but rather a different formulation, named after Palatini (Palatini 1919; Tsamparlis 1978).

If R is replaced by $f(R)$ in the action, $f(R)$-gravity and Palatini $f(R)$-gravity are no longer equivalent (Buchdahl 1970). In particular, R is now of first order in derivatives because, with unspecified $\Gamma^\gamma_{\alpha\beta}$, first-order partial derivatives of $\Gamma^\gamma_{\alpha\beta}$ count as first-order derivatives of a field independent of the metric, rather than as second-order derivatives of $g_{\alpha\beta}$. Using a non-linear $f(R)$ then means that we just add higher powers of first-order derivatives to the action, but no higher-order derivatives. The standard Euler–Lagrange equations should be used, and they remain of second order. In Palatini $f(R)$-theories one can therefore study modifications of general relativity without having to deal with additional degrees of freedom.

Like $f(R)$-gravity, Palatini $f(R)$-gravity is related to Brans–Dicke theories, but with a non-zero ω (Olmo 2005). We can derive the correct value by following the procedure of introducing auxiliary fields, first writing the action as

$$S[g_{\alpha\beta}, \Gamma^\gamma_{\alpha\beta}, \psi] = \frac{c^2}{16\pi G} \int \mathrm{d}^3 x \sqrt{-\det g} \left(f(\psi) + \frac{\mathrm{d}f}{\mathrm{d}\psi}(R_\Gamma - \psi) \right) \qquad (2.118)$$

where the subscript on R_Γ indicates that R belongs to the Ricci tensor (2.46) for a generic $\Gamma^\gamma_{\alpha\beta}$ independent, at this point, of the metric $g_{\alpha\beta}$ that appears in $\sqrt{-\det g}$. Thus, R_Γ depends on $g_{\alpha\beta}$ only through the inverse metric $g^{\alpha\beta}$ used to compute the trace of the Γ-Ricci tensor, (2.47).

Before we proceed, in order to see how the action may be related to Brans–Dicke gravity we have to understand the relationship of R_Γ with the metric after we solve equations of motion implied by Equation (2.118). Using the same arguments as indicated for Palatini $f(R)$-gravity before we introduced the auxiliary field, varying Equation (2.118) with respect to $\Gamma^\gamma_{\alpha\beta}$ implies an algebraic equation for $\Gamma^\gamma_{\alpha\beta}$ in which the metric-dependent coefficient $g^{\alpha\beta}\sqrt{-\det g}$ in the action now has an additional factor of $\mathrm{d}f/\mathrm{d}\psi$. Since the conformal transformation $\tilde{g}_{\alpha\beta} = (\mathrm{d}f/\mathrm{d}\psi)g_{\alpha\beta}$ implements just the same factor, we see that Euler–Lagrange equations of (2.118) imply that $\Gamma^\gamma_{\alpha\beta}$ used in R_Γ is the Christoffel symbol of the metric $\tilde{g}_{\alpha\beta}$. However, R_Γ is *not* equal to the Ricci scalar $\tilde{R}$ of $\tilde{g}_{\alpha\beta}$ because R_Γ in Equation (2.118) is understood as being obtained from the Ricci tensor using the inverse metric $g^{\alpha\beta}$ rather than $\tilde{g}^{\alpha\beta}$: We have $\tilde{R} = \sum_{\alpha=0}^3 \sum_{\beta=0}^3 \tilde{g}^{\alpha\beta} \tilde{R}_{\alpha\beta}$ but

$$R_\Gamma = \sum_{\alpha=0}^3 \sum_{\beta=0}^3 g^{\alpha\beta} \tilde{R}_{\alpha\beta} = \frac{\mathrm{d}f}{\mathrm{d}\psi} \tilde{R}. \qquad (2.119)$$

As before, there is the relationship (2.93) between R and $\tilde{R}$, which implies a relationship between R_Γ and R. First, we write

$$\frac{\mathrm{d}f}{\mathrm{d}\psi}\tilde{R} = R - 3\sum_{\alpha=0}^{3}\sum_{\beta=0}^{3}g^{\alpha\beta}\left(\nabla_\alpha\nabla_\beta\log\frac{\mathrm{d}f}{\mathrm{d}\psi} + \frac{1}{2}\frac{\partial\log(\mathrm{d}f/\mathrm{d}\psi)}{\partial x^\alpha}\frac{\partial\log(\mathrm{d}f/\mathrm{d}\psi)}{\partial x^\beta}\right)$$

$$= R - \frac{3}{\mathrm{d}f/\mathrm{d}\psi}\sum_{\alpha=0}^{3}\sum_{\beta=0}^{3}g^{\alpha\beta}\nabla_\alpha\nabla_\beta\frac{\mathrm{d}f}{\mathrm{d}\psi} \tag{2.120}$$

$$+ \frac{3}{2(\mathrm{d}f/\mathrm{d}\psi)^2}\sum_{\alpha=0}^{3}\sum_{\beta=0}^{3}g^{\alpha\beta}\frac{\partial(\mathrm{d}f/\mathrm{d}\psi)}{\partial x^\alpha}\frac{\partial(\mathrm{d}f/\mathrm{d}\psi)}{\partial x^\beta}$$

using the product rule in the middle term. For the action (2.118), we need $(\mathrm{d}f/\mathrm{d}\psi)R_\Gamma = (\mathrm{d}f/\mathrm{d}\psi)^2\tilde{R}$, given by

$$\frac{\mathrm{d}f}{\mathrm{d}\psi}R_\Gamma = \frac{\mathrm{d}f}{\mathrm{d}\psi}R - 3\sum_{\alpha=0}^{3}\sum_{\beta=0}^{3}g^{\alpha\beta}\nabla_\alpha\nabla_\beta\frac{\mathrm{d}f}{\mathrm{d}\psi}$$

$$+ \frac{3}{2\mathrm{d}f/\mathrm{d}\psi}\sum_{\alpha=0}^{3}\sum_{\beta=0}^{3}g^{\alpha\beta}\frac{\partial(\mathrm{d}f/\mathrm{d}\psi)}{\partial x^\alpha}\frac{\partial(\mathrm{d}f/\mathrm{d}\psi)}{\partial x^\beta}. \tag{2.121}$$

The term in the middle is now a total derivative, which vanishes when integrated over after multiplication with $\sqrt{-\det g}$ in Equation (2.118) (or produces a boundary term that does not affect local field equations). Identifying $\mathrm{d}f/\mathrm{d}\psi$ with a new auxiliary field ϕ, the action therefore takes the form

$$S[g_{\alpha,\beta}, \phi] = \frac{c^2}{16\pi G}\int \mathrm{d}^3x\left(\phi R + \frac{3}{2\phi}\sum_{\alpha=0}^{3}\sum_{\beta=0}^{3}g^{\alpha\beta}\frac{\partial\phi}{\partial x^\alpha}\frac{\partial\phi}{\partial x^\beta} - V(\phi)\right) \tag{2.122}$$

with a potential $V(\phi)$ defined as in Equation (2.110). This action corresponds to a Brans–Dicke theory with $\omega = -3/2$.

A second conformal transformation $\tilde{g}_{\alpha\beta} = \phi g_{\alpha\beta}$ maps this theory to standard general relativity. We can simply use the result (2.116) with $\omega = -3/2$. For this value, the kinetic term of the scalar drops out, and only the potential remains. As a consequence, the scalar does not represent a propagating degree of freedom but instead is directly related to the metric through an algebraic Euler–Lagrange equation. This result is consistent with the fact that $f(R)$-gravity in a Palatini formulation is of first order and does not imply dynamical degrees of freedom other than the metric.

Horndeski Theories
The most general scalar-tensor theory with second-order equations of motion, as determined by Horndeski (1974) and reviewed by Kobayashi (2019), has a complicated action of the form

$$S[g_{\alpha\beta}, \phi] = \frac{c^2}{16\pi G} \int dt d^3 x \sqrt{-\det g} \left(H_1(\phi, X) + H_2(\phi, X) \sum_{\alpha=0}^{3} \sum_{\beta=0}^{3} g^{\alpha\beta} \nabla_\alpha \nabla_\beta \phi \right.$$

$$+ H_3(\phi, X) R + \frac{\partial H_3}{\partial X} \sum_{\alpha=0}^{3} \sum_{\beta=0}^{3} \sum_{\gamma=0}^{3} \sum_{\delta=0}^{3} (g^{\alpha\beta} g^{\gamma\delta} - g^{\alpha\gamma} g^{\beta\delta})(\nabla_\alpha \nabla_\beta \phi)(\nabla_\gamma \nabla_\delta \phi)$$

$$+ H_4(\phi, X) \sum_{\alpha=0}^{3} \sum_{\beta=0}^{3} \sum_{\gamma=0}^{3} \sum_{\delta=0}^{3} g^{\alpha\gamma} g^{\beta\delta} \left(R_{\alpha\beta} - \frac{1}{2} R g_{\alpha\beta} \right) \nabla_\gamma \nabla_\delta \phi \qquad (2.123)$$

$$- \frac{1}{6} \frac{\partial H_4}{\partial X} \sum_{\alpha=0}^{3} \sum_{\beta=0}^{3} \sum_{\gamma=0}^{3} \sum_{\delta=0}^{3} \sum_{\epsilon=0}^{3} \sum_{\eta=0}^{3} (g^{\alpha\beta} g^{\gamma\epsilon} g^{\delta\eta} + 2 g^{\alpha\gamma} g^{\beta\delta} g^{\epsilon\eta} - 3 g^{\alpha\gamma} g^{\beta\epsilon} g^{\delta\eta})$$

$$\left. \times (\nabla_\alpha \nabla_\beta \phi)(\nabla_\gamma \nabla_\epsilon \phi)(\nabla_\delta \nabla_\eta \phi) \right)$$

where

$$X = -\frac{1}{2} \sum_{\alpha=0}^{3} \sum_{\beta=0}^{3} g^{\alpha\beta} \frac{\partial \phi}{\partial x^\alpha} \frac{\partial \phi}{\partial x^\beta} \qquad (2.124)$$

is the standard kinetic term of a scalar field. The four free functions of ϕ and X, H_1, H_2, H_3 and H_4, parameterize all possible scalar-tensor theories with second-order equations of motion.

The action (2.123) depends on up to second-order derivatives of the scalar field and, through the Ricci tensor, of the metric. Variation by ϕ or $g_{\alpha\beta}$ and subsequent integrations by parts therefore could imply higher than second-order derivatives of ϕ in Euler–Lagrange equations. However, the coefficient functions in the action are arranged such that all higher than second-order derivatives cancel out in equations of motion.

This property is relatively easy to see for the coefficient function H_2, which implies two terms in Euler–Lagrange equations that can lead to third-order derivatives. After integrating by parts, we have

$$-2 \sum_{\alpha=0}^{3} \sum_{\beta=0}^{3} \sum_{\gamma=0}^{3} \sum_{\delta=0}^{3} g^{\alpha\beta} \nabla_\beta \left(\frac{\partial H_2}{\partial X} (\nabla_\alpha \phi) g^{\gamma\delta} \nabla_\gamma \nabla_\delta \phi \right) + \sum_{\gamma=0}^{3} \sum_{\delta=0}^{3} \nabla_\gamma \nabla_\delta \left(g^{\gamma\delta} H_2(\phi, X) \right). \quad (2.125)$$

The first term is obtained by varying ϕ in the X-dependence of H_2, while the second term is obtained by varying ϕ in the second-order derivative of the contribution $\sum_{\alpha=0}^{3} \sum_{\beta=0}^{3} g^{\alpha\beta} \nabla_\alpha \nabla_\beta \phi$. (To derive these terms from variations, functional derivatives are useful which are explained and applied in more detail in Section 2.3.4.) Any other ϕ-dependence in the H_2-term leads to low-order derivatives. By taking the derivatives in Equation (2.125), using the product rule and also applying Equation (2.35) and renaming suitable indices, one can see that all terms of precisely third order in Equation (2.125) cancel out.

As we noticed in our general discussion of higher-order Lagrangians in Section 2.2.3, Ostrogradsky instabilities can be avoided even in higher-order equations provided the non-degeneracy condition (2.67) is violated. For multiple degrees of freedom in a Lagrangian L, this equation takes on a matrix form, stating that the matrix with components

$$\frac{\partial^2 L}{\partial \psi_i \partial \psi_j} \tag{2.126}$$

is invertible in a non-degenerate theory, where ψ_i represents all components of the fields, here $g_{\alpha\beta}$ and ϕ. If the matrix is not invertible, it may be possible to evade Ostrogradsky instabilities. This outcome has been confirmed in Gleyzes et al. (2015), provided the higher-order contribution in the action is a combination of terms of the form

$$bH_3(\phi, X) \sum_{\alpha=0}^{3} \sum_{\beta=0}^{3} \sum_{\gamma=0}^{3} \sum_{\delta=0}^{3} \sum_{\alpha'=0}^{3} \sum_{\beta'=0}^{3} \sum_{\gamma'=0}^{3} \sum_{\delta'=0}^{3} g_{\delta\delta'} \epsilon^{\alpha\beta\gamma\delta} \epsilon^{\alpha'\beta'\gamma'\delta'}$$
$$(\nabla_\alpha \phi)(\nabla_{\alpha'} \phi)(\nabla_\beta \nabla_{\beta'} \phi)(\nabla_\gamma \nabla_{\gamma'} \phi) \tag{2.127}$$

with a free function $bH_3(\phi, X)$, and

$$bH_4(\phi, X) \sum_{\alpha=0}^{3} \sum_{\beta=0}^{3} \sum_{\gamma=0}^{3} \sum_{\delta=0}^{3} \sum_{\alpha'=0}^{3} \sum_{\beta'=0}^{3} \sum_{\gamma'=0}^{3} \sum_{\delta'=0}^{3} \epsilon^{\alpha\beta\gamma\delta} \epsilon^{\alpha'\beta'\gamma'\delta'}$$
$$(\nabla_\alpha \phi)(\nabla_{\alpha'} \phi)(\nabla_\beta \nabla_{\beta'} \phi)(\nabla_\gamma \nabla_{\gamma'} \phi)(\nabla_\delta \nabla_{\delta'} \phi) \tag{2.128}$$

with a free function $bH_4(\phi, X)$. The contributions in Equation (2.123) plus one or both of the terms (2.127) and (2.128) then define stable scalar-tensor theories beyond Horndeski. Such theories have been explored in more detail in Langlois & Noui (2016a, 2016b), Motohashi et al. (2016) Langlois (2017), where they have been named degenerate higher-order scalar-tensor (DHOST) theories.

Special Cases. As simple examples, we have already seen Brans–Dicke theories (2.111), which are Horndeski theories with a function H_3 that depends only on ϕ, while H_2 and H_4 are zero. An action of a similar form but with X-dependent H_3 is given by

$$S_X[g_{\alpha\beta}, \phi] = \frac{c^2}{16\pi G} \int \mathrm{d}t \mathrm{d}^3 x \sqrt{-\det g} \left(XR + \frac{3}{2X} \sum_{\alpha=0}^{3} \sum_{\beta=0}^{3} g^{\alpha\beta} \frac{\partial X}{\partial x^\alpha} \frac{\partial X}{\partial x^\beta} \right). \tag{2.129}$$

A conformal transformation to a new metric

$$\tilde{g}_{\alpha\beta} = X g_{\alpha\beta}, \tag{2.130}$$

using Equation (2.116), maps this action to

$$S[\tilde{g}_{\alpha\beta}] = \frac{c^2}{16\pi G} \int \mathrm{d}t \mathrm{d}^3 x \sqrt{-\det \tilde{g}} \, \tilde{R} \Big|_{C=0} \tag{2.131}$$

which, at first sight, looks like standard general relativity. However, as indicated in Equation (2.131) by the subscript $C = 0$, the conformal transformation (2.130) together with the definition (2.124) of X implies that $\tilde{g}_{\alpha\beta}$ satisfies the constraint

$$C = 1 + \frac{1}{2} \sum_{\alpha=0}^{3} \sum_{\beta=0}^{3} \tilde{g}^{\alpha\beta} \frac{\partial\phi}{\partial x^\alpha} \frac{\partial\phi}{\partial x^\beta} = 1 + \frac{1}{2X} \sum_{\alpha=0}^{3} \sum_{\beta=0}^{3} g^{\alpha\beta} \frac{\partial\phi}{\partial x^\alpha} \frac{\partial\phi}{\partial x^\beta} = 0. \qquad (2.132)$$

Therefore, $\tilde{g}_{\alpha\beta}$ in Equation (2.131) cannot be varied freely, which implies that the resulting theory, called *mimetic gravity* (Chamseddine & Mukhanov 2013; Sebastiani et al. 2017), has additional solutions compared with general relativity: A constrained variation leads to a smaller number of Euler–Lagrange equations to be satisfied by solutions of the theory. Generalized versions of mimetic gravity, in which additional terms depending on ϕ and X are added to the original action, are related to DHOST theories (Langlois et al. 2019).

Another example is *Einstein-dilaton–Gauss–Bonnet gravity*, which introduces a dynamical scalar field by coupling it to the Gauss–Bonnet term $\mathcal{G}$, defined in Equation (2.58):

$$S[g_{\alpha\beta}, \phi] = \frac{c^2}{16\pi G} \int \mathrm{d}t\mathrm{d}^3x \sqrt{-\det g} \left(R - \frac{1}{2} \sum_{\alpha=0}^{2} \sum_{\beta=0}^{3} g^{\alpha\beta} \frac{\partial\phi}{\partial x^\alpha} \frac{\partial\phi}{\partial x^\beta} + f(\phi)\mathcal{G} \right) \qquad (2.133)$$

with some function $f(\phi)$. Suitable integrations by parts can be used to write the $\mathcal{G}$-term, which is quadratic in curvature components, in Horndeski form with functions (Kobayashi et al. 2011)

$$H_1(\phi, X) = 8X^2 \frac{\mathrm{d}^4 f}{\mathrm{d}\phi^4} (3 - \log|X|) \qquad (2.134)$$

$$H_2(\phi, X) = -4X \frac{\mathrm{d}^3 f}{\mathrm{d}\phi^3} (7 - 3\log|X|) \qquad (2.135)$$

$$H_3(\phi, X) = 4X \frac{\mathrm{d}^2 f}{\mathrm{d}\phi^2} (2 - \log|X|) \qquad (2.136)$$

$$H_4(\phi, X) = -4 \frac{\mathrm{d}f}{\mathrm{d}\phi} \log|X|. \qquad (2.137)$$

Non-hyperbolicity. While Horndeski and DHOST theories evade the Ostrogradsky instability, their equations of motion in many cases (but not always) fail to be sufficiently hyperbolic (Papallo & Reall 2017; Papallo 2017; Ripley & Pretorius 2019a, 2019b, 2020; Kovacs & Reall 2020). Therefore, they do not always imply well-posed initial-value problems that could be used for deterministic evolution. In some regions of spacetime, the scalar field may even be subject to an elliptic rather than hyperbolic partial differential equation. While such an equation is well-posed with boundary conditions surrounding a four-dimensional region, its solutions would not be stable with initial conditions. For instance, the wave equation,

$$\beta \frac{1}{c^2} \frac{\partial^2 \phi}{\partial t^2} + \frac{\partial^2 \phi}{\partial x^2} = 0 \tag{2.138}$$

in two dimensions has stable (that is, bounded) plane-wave solutions $\phi(t, x) = A \exp(i(x - ct)) + B \exp(-i(x + ct))$ for $\beta = -1$, in which case (2.138) is hyperbolic. It has unbounded solutions $\phi(t, x) = A \exp(x - ct) + B \exp(-(x + ct))$ for $\beta = 1$, in which case (2.138) is elliptic.

Another example in which cosmological perturbation equations can become unstable is an extended version of mimetic gravity, in which a term $f(\phi)$ is added to the action (2.131), subject to the usual constraint (2.132). As derived in Haro et al. (2019), scalar perturbations of the metric on a spatially flat isotropic background, combined with the scalar field ϕ as an analog of the Bardeen potential (1.73), obey a second-order partial differential equation with highest-order term

$$-\frac{1}{c^2} \frac{\partial^2 \Phi}{\partial t^2} + \frac{\beta(\phi)}{a^2} \left(\frac{\partial^2 \Phi}{\partial x^2} + \frac{\partial^2 \Phi}{\partial y^2} + \frac{\partial^2 \Phi}{\partial z^2} \right) \tag{2.139}$$

with

$$\frac{1}{\beta(\phi)} = 1 - \frac{3}{2} \frac{d^2 f}{d\phi^2}. \tag{2.140}$$

Depending on the function f, $\beta(\phi)$ may change sign in certain ranges of ϕ, in which case the scalar mode is subject to an elliptic equation in some regions of spacetime, and to a hyperbolic equation in others. The partial differential equation is then of mixed type and requires a combination of initial and boundary values (Tricomi 1968). Tensor modes in this theory, by contrast, always obey a hyperbolic partial differential equation

$$-\frac{1}{c^2} \frac{\partial^2 h}{\partial t^2} - 3H \frac{\partial h}{\partial t} + \frac{1}{a^2} \left(\frac{\partial^2 \Phi}{\partial x^2} + \frac{\partial^2 \Phi}{\partial y^2} + \frac{\partial^2 \Phi}{\partial z^2} \right) = 0. \tag{2.141}$$

Gravity therefore remains stable in such theories, but scalar fields may become destabilized.

Observational Constraints. In general, Horndeski or DHOST theories may even lead to modifications of the wave Equation (2.141) for tensor modes, which, unlike Equation (2.141), change the speed of gravitational waves (De Felice & Tsujikawa 2012; Bellini & Sawicki 2014; Gleyzes et al. 2015; Langlois et al. 2017). This possibility has become a pressing issue since the direct detection of gravitational waves from a merger of neutron stars, which put stringent constraints on the speed of gravitational waves relative to the speed of light (Abbott et al. 2017a, 2017b). In particular, the speed c_{GW} of gravitational waves relative to the speed of light, c, must obey the inequalities

$$-3 \times 10^{-15} \leqslant \frac{c_{GW}}{c} - 1 \leqslant 7 \times 10^{-16}. \tag{2.142}$$

While this constraint on its own can be met by several scalar-tensor theories, it is often in conflict with theories that have been proposed to explain dark energy through the extra scalar field of a scalar-tensor theory (Ezquiaga & Zumalacárregui 2017; Creminelli & Vernizzi 2017; Sakstein & Jain 2017; Baker et al. 2017; de Rham & Melville 2018).

2.2.6 Hořava–Lifshitz and Einstein Aether Theories

It is possible to evade the Ostrogradsky instability while allowing higher-order derivatives if the latter are only with respect to spatial coordinates. This possibility, however, comes at the price of breaking general covariance either explicitly or dynamically, through the introduction of a fixed vector field in the action.

Hořava–Lifshitz gravity (Hořava 2009) can be defined in different versions, which lead to an action of the form

$$S[g_{ij}] = \frac{c^2}{16\pi G} \int dt d^3 x \sqrt{-\det g} \left(f_1 \left(\partial g_{ij}/\partial t \right) + f_2 \left({}^{(3)}R_{ijk}{}^l \right) \right). \tag{2.143}$$

While the function f_2 can include spatial higher derivatives through the spatial Riemann tensor ${}^{(3)}R_{ijk}{}^l$, computed as in Equation (2.45) but only for the spatial part g_{ij} of the metric, the function f_1 may add higher powers of the time derivative of g_{ij} but not higher-order time derivatives. Therefore, there are no new degrees of freedom and no Ostrogradsky instability. Higher-order spatial curvature terms can nevertheless imply interesting cosmological consequences, such as dark energy (Calcagni 2009; Kiritsis & Kofinas 2009).

The general form is not compatible with covariance, but it is possible to implement a subset of coordinate transformations. In most cases, time-dependent transformations of the spatial coordinates are allowed, which are of the form $x_t^{i'}(x^i)$, while time can be reparameterized, that is, mapped to $t'(t)$, but not in a position-dependent way. In addition, most theories of this form obey a scaling symmetry with equations invariant under $x^i = \lambda x^i$ while $t' = \lambda^3 t$ for any constant $\lambda \neq 0$. The exponent of three in the time scaling again illustrates the different treatment of time and space in this non-covariant theory.

Horařava–Lifshitz theories are closely related (Blas et al. 2011) to Einstein ether theories (Jacobson & Mattingly 2001) with an action of the form

$$S[g_{\alpha\beta}, v^\gamma] = \frac{c^2}{16\pi G} \int dt d^3 x \sqrt{-\det g}$$
$$\times \left(R + \sum_{\alpha=0}^{3} \sum_{\beta=0}^{3} \left(A \sum_{\gamma=0}^{3} \sum_{\delta=0}^{3} g_{\gamma\delta} v^\alpha v^\beta (\nabla_\alpha v^\gamma)(\nabla_\beta v^\delta) \right. \right.$$
$$\left. \left. + B (\nabla_\alpha v^\beta)(\nabla_\beta v^\alpha) + C (\nabla_\alpha v^\alpha)(\nabla_\beta v^\beta) \right) \right) \tag{2.144}$$

with constants A, B and C. As shown in Gumrukcuoglu et al. (2018), these theories have modified speeds of gravitational waves (at least if $B \neq 0$) and are therefore subject to observational constraints.

2.3 Canonical Formulation

Any action principle can be transformed to a canonical formulation by first computing momenta (derivatives of the Lagrangian by the velocities) and then applying a Legendre transformation of the Lagrangian to the Hamiltonian, schematically $H(q, p) = \dot{q}(p)p - L(q, \dot{q}(p))$. In the second step, we have to eliminate the velocities in favor of momenta, which requires an inversion of the relationships found in the first step. However, invertibility is not always guaranteed, in particular when the action principle enjoys gauge symmetries. In general relativity, these symmetries are given by general covariance.

Transforming the action of general relativity to a canonical formulation is therefore quite involved at a technical level. Fortunately, it is possible to construct the canonical formulation of general relativity from scratch, without using an action principle. In this construction, the canonical formulation of general relativity, or any generally covariant theory of spacetime, appears as a generalization of the better known spacetime diagrams of special relativity to non-linear transformations.

2.3.1 Spacetime in Special Relativity

A common set of symmetry transformations of Euclidean space is given by rotations around a given axis, by a fixed angle φ. If we describe space by Cartesian coordinates, the axes of a coordinate system are transformed by a rotation. Figure 2.3 shows a two-dimensional illustration.

One of the main insights into the structure of spacetime and its differences from Euclidean geometry comes from an analysis of the corresponding set of transformations. Mathematically, these transformations are given by the equations of Lorentz boosts, in which the rotation angle φ is replaced by the relative velocity v of a new observer moving with respect to one represented by the original coordinate system,

$$t' = \frac{t - vx/c^2}{\sqrt{1 - v^2/c^2}} \tag{2.145}$$

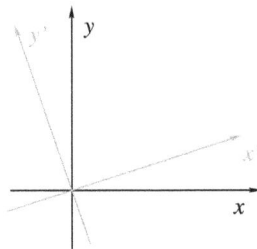

Figure 2.3. A rotation of a two-dimensional Cartesian coordinate system.

$$x' = \frac{x - vt}{\sqrt{1 - v^2/c^2}} \tag{2.146}$$

for a boost in the x-direction, keeping $y' = y$ and $z' = z$ unchanged. More precisely, the rotation angle in the transformations $x' = \cos(\varphi)x + \sin(\varphi)y$ and $y' = -\sin(\varphi)x + \cos(\varphi)y$ is replaced by the rapidity, or hyperbolic angle, ρ defined by $\tanh \rho = v/c$. With this parameter, Equations (2.145) and (2.146) take the form

$$ct' = \cosh(\rho)ct - \sinh(\rho)x \quad \text{and} \quad x' = -\sinh(\rho)ct + \cosh(\rho)x. \tag{2.147}$$

The transformation equations are such that the speed of light, c, takes the same value for all observers. That is, the transformations leave the wave equation

$$\left(-\frac{1}{c^2}\frac{\partial^2}{\partial t^2} + \frac{\partial^2}{\partial x^2} + \frac{\partial^2}{\partial y^2} + \frac{\partial^2}{\partial z^2} \right) f(t, x, y, z) = 0 \tag{2.148}$$

invariant. Geometrically, the Minkowski line element

$$ds^2 = -c^2 dt^2 + dx^2 + dy^2 + dz^2 \tag{2.149}$$

is preserved by Lorentz boosts.

The constant speed of light also determines how the axes of a Cartesian coordinate system in spacetime should be transformed if a new observer is moving at relative speed v with respect to the original one. As shown in Figure 2.4, the new time axis must move to the right (into the first quadrant): The new x'-axis, at an angle with the x-axis that is related to the speed v, determines all events in spacetime at which $t' = 0$. Two light rays emitted at $t' = 0$ but on opposite sides of the center (that is, with $x_1'(t' = 0) = -x_2'(t' = 0)$) and aimed toward the center, should then meet at the center, or at $x' = 0$. The set of all points with $x' = 0$ is just the definition of the t'-axis. Since an observer moving to the right moves toward the light source on the right, the right light ray has to travel a shorter distance to reach the observer than the left light ray. Therefore, the t'-axis is moved into the first quadrant, in contrast to the result of a Euclidean rotation.

The main difference between space and spacetime, or in terms of geometrical properties between Euclidean and Lorentzian signature, is therefore how one

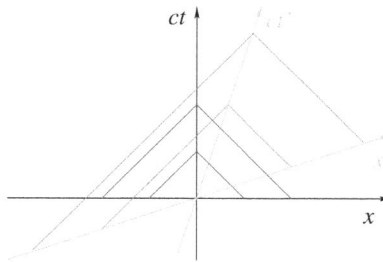

Figure 2.4. A boost of a two-dimensional Cartesian coordinate system in spacetime. The preservation of the speed of light by a boost implies that both axes are moved into the same quadrant. Blue lines show light rays in the original system (ct, x), while red lines show light rays in the new system, (ct', x').

determines the *normal* direction to hypersurfaces of constant time. The normal direction is defined as the direction in which one must move away from the hypersurface such that all position coordinates in a Cartesian system, in which the line element is diagonal and has constant coefficients, are held fixed. (The scalar product of the normal vector and a vector tangential to the hypersurface is then zero, which would be a coordinate-independent definition.) In Euclidean space, the normal directions are standard orthogonal directions. In Minkowski spacetime, the normal directions are orthogonal in a modified, non-Euclidean geometry, as given by the Minkowski line element. However, they do not look orthogonal in an illustration drawn on a sheet of paper (or a computer screen), which always retains its Euclidean geometry even if we try to visualize the Minkowski geometry of spacetime. For this reason, the normal direction in Figure 2.4, or the angle between the x'-axis and the t'-axis, does not look orthogonal.

Normal directions are the main characteristic of the structure of space or spacetime, as well as their differences. It is therefore sufficient to analyze the geometry by drawing a selection of spatial hypersurfaces and their normal directions. For this purpose, we do not need the full Cartesian system with its axes and marks. Focusing on the relevant structure in this way will allow us to generalize results from the linear setting of special relativity to non-linear transformations in general relativity.

2.3.2 Symmetries

In an abstract way, we are able to probe spacetime just as we probe matter: By pushing and deforming the object to be analyzed and seeing how it reacts, we can determine its structural properties. In spacetime, this process is described by hypersurface deformations. Performing some of them in a row in different orderings and checking for differences in the final outcomes, we can find their algebraic properties. These properties encode how an underlying symmetry group is realized, corresponding to general covariance. Making use of spatial hypersurfaces implies that we are working in a canonical formulation, in which field equations are not specified for spacetime tensors through their partial derivatives, but rather for the spatial geometry at any given time and its rate of change determined by an independent spatial field given by the momentum.

Special relativity makes use of linear deformations (2.145) and (2.146). These transformations, by virtue of being linear, correspond to a transformations of entire axes in a Cartesian diagram as shown in Figure 2.4. By considering their implications for subsets of spacetime defined by $ct' = $ const, they show how spatial hypersurfaces take on different forms after a transformation.

Infinitesimally, a given spatial hypersurface in one frame is deformed by a transformation to a new frame. Figure 2.5 illustrates the implications for deformations of spatial hypersurfaces, expressed through a normal deformation of the original hypersurface, defined by constant ct in the original frame. By this normal deformation, each point $\vec{x}$ is moved a distance $N(\vec{x})$ in the normal direction $\vec{n}$, where $N(\vec{x})$ is a linear function

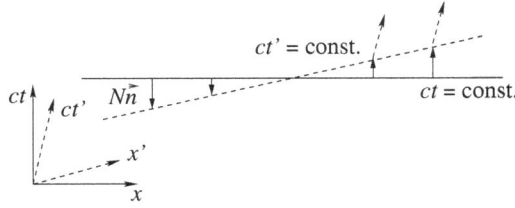

Figure 2.5. Linear deformation of a spatial hypersurface. The initial surface defined by ct = const is deformed by moving each point $\vec{x}$ in spacetime a distance $N(\vec{x})$ in the normal direction, given by a vector $\vec{n}$. The deformation is linear provided the function $N(\vec{x})$ is linear. Normal directions are determined according to the directions of axes in spacetime diagrams.

$$N(\vec{x}) = c\Delta t + \frac{\vec{v}}{c} \cdot \vec{x} \qquad (2.150)$$

with Δt a time displacement and $\vec{v}$ the boost velocity. This linear deformation preserves the flat nature of spatial hypersurfaces, drawn as straight lines in a two-dimensional illustration.

In four dimensions, there are four parameters, given by the time displacement Δt and the three components of the velocity vector $\vec{v}$. We will denote any transformation in the normal direction by H[N], the letter H standing for "Hamiltonian" because in a canonical formulation the Hamiltonian generates transformations in the time direction. In addition, we are able to perform linear transformations of space, within a ct = const hypersurface. Taking into account the Euclidean geometry of space, these transformations are determined by spatial vectors

$$\vec{w}(\vec{x}) = \Delta\vec{x} + \mathbf{R}\vec{x} \qquad (2.151)$$

at each point $\vec{x}$, with a spatial translation $\Delta\vec{x}$ and a rotation matrix $\mathbf{R}$. Any spatial deformation will be denoted as D[$\vec{w}$], the letter D standing for "diffeomorphism."

The basic algebraic properties of transformations are determined by their commutators, for which we pick two transformations, T_1 and T_2, and compute the difference of the outcomes from combinations in two different orderings, $T_1 T_2$ and $T_2 T_1$. Using our diagrams derived for single transformations, the combinations are illustrated in Figure 2.6. In this example, the first transformation, $T_1 = $ H[N_1] is a normal deformation by $N_1(x) = vx/c$ (a Lorentz boost), and the second transformation, $T_2 = $ H[N_2], is a normal deformation by $N_2(x) = c\Delta t - vx/c$ (the reverse Lorentz boost after waiting some time Δt). These two transformations do not commute with each other because we end up at different points on the final hypersurface. Their commutator,

$$[\mathrm{H}[N_1], \mathrm{H}[N_2]] = \mathrm{D}[w] \qquad (2.152)$$

with the spatial displacement $w(x) = \Delta x = v\Delta t$, is determined by hyperbolic geometry of the triangle indicated in the diagram.

As a first application of these methods to understanding the structure of space-time, we repeat the commutator construction in Euclidean space, as shown in

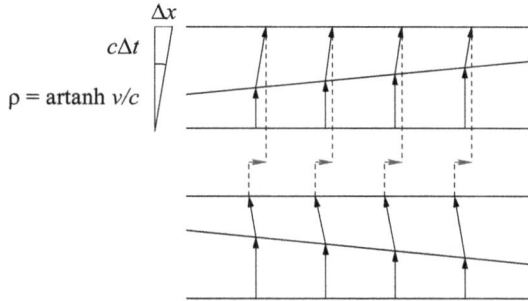

Figure 2.6. The graphical commutator of a time translation and a boost. The sharp hyperbolic angle ρ at the bottom of the triangle is such that $\tanh \rho = v/c$. Therefore, the spatial displacement, given by the top side of the triangle (blue side, equal to length of blue arrows), equals $\Delta x = c\Delta t \tanh \rho = v\Delta t$.

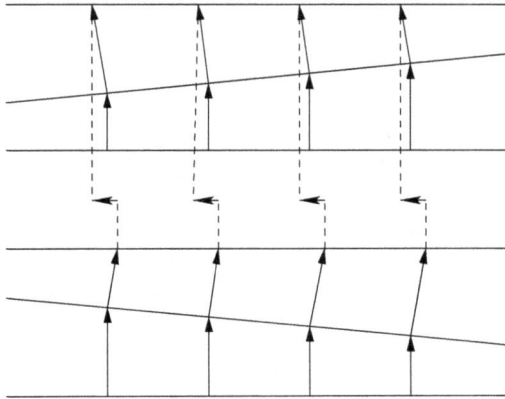

Figure 2.7. The graphical commutator of a spatial translation and a rotation in Euclidean space. The direction of the final spatial displacement is reversed compared with Figure 2.6.

Figure 2.7. Since there is no time in Euclidean space, we have to relabel some of our parameters in the deformation functions N_1 and N_2: We use $N_1(x) = \tan(\varphi)x$ with an angle φ, and $N_2(x) = \Delta y - \tan(\varphi)x$ with the same angle and a spatial displacement, Δy. In this case, geometry determines almost the same commutator,

$$[H[N_1], H[N_2]] = -D[w], \tag{2.153}$$

but with the opposite sign. The final spatial displacement is now given by $w(x) = -\tan(\varphi)\Delta y$, which is the same as a rotation by an angle φ. The geometrical property called *signature*, that is whether we have space or spacetime, is therefore encoded in the sign of commutators of hypersurface deformations.

2.3.3 Spacetime in General Relativity

The transition to general relativity is now straightforward, at least in terms of diagrams: In general relativity we replace the special, linear Lorentz transformations

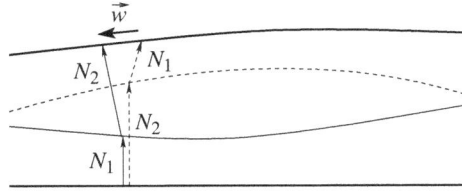

Figure 2.8. Non-linear hypersurface deformations, with non-linear lapse functions N_1 and N_2. They imply a commutator (2.154) with final spatial displacement (2.155).

of special relativity with arbitrary, non-linear coordinate changes. Cartesian coordinate axes are no longer meaningful in this context because they would not remain straight, but we can easily replace linear hypersurfaces and their linear deformations with non-linear versions. Their commutators can be shown as well, as in Figure 2.8. Two arbitrary normal deformations then have the commutator

$$[\mathrm{H}[N_1], \mathrm{H}[N_2]] = \mathrm{D}[\vec{w}(\vec{x})] \tag{2.154}$$

where the vector field $\vec{w}(\vec{x})$ has components

$$w^i = \sum_{j=1}^{3} q^{ij}\left(N_1\frac{\partial N_2}{\partial x^j} - N_2\frac{\partial N_1}{\partial x^j}\right) \tag{2.155}$$

and q^{ij} is the inverse of the spatial metric on the initial hypersurface. The previous relation (2.152) is a special case of this general relationship, using the linear functions N_1 and N_2 specified for boosts in Equation (2.150). (In four-dimensional Euclidean space, w^i would be replaced by $-w^i$.)

The commutator of two spatial deformations depends only on the geometry of space and is given by

$$[\mathrm{D}[\vec{v}], \mathrm{D}[\vec{w}]] = \mathrm{D}[\vec{u}] \tag{2.156}$$

where $\vec{u}$ has components

$$u^i = \sum_{j=1}^{3}\left(v^j\frac{\partial w^i}{\partial x^j} - w^j\frac{\partial v^i}{\partial x^j}\right). \tag{2.157}$$

Moreover, a normal and a spatial deformation have the commutator

$$[\mathrm{H}[N], \mathrm{D}[\vec{v}]] = \mathrm{H}[M] \tag{2.158}$$

with

$$M = -\sum_{j=1}^{3} v^j\frac{\partial N}{\partial x^j}. \tag{2.159}$$

The hypersurface-deformation bracket (2.154) was first derived (in a different way) by Dirac (1958b), and usefully rewritten in Arnowitt et al. (1962), introducing normal deformations and the displacement function N (called the *lapse function*). More recently (Blohmann et al. 2013), the bracket has been found to define a Lie algebroid, a generalization of a Lie algebra (Pradines 1967).

The brackets determine the possible equations of motion compatible with them, provided one specifies the evolving field (the metric) and the order of derivatives that can appear (two in the classical case). Only two free parameters then remain for consistent classical equations (Hojman et al. 1976), corresponding to Newton's constant and the cosmological constant. This result is therefore equivalent to the discussion of higher-curvature actions, leading to Equation (2.48). We will first derive the bracket (2.154) with (2.155) from spacetime geometry, and then demonstrate the uniqueness of equations of motion for spherically symmetric spacetimes and for cosmological perturbations around a spatially flat isotropic background, also deriving the required Hamiltonians of these models.

Derivation of the Commutator of Normal Hypersurface Deformations
We consider two transversal deformations with lapse functions N_1 and N_2, performed one after the other in the two different orderings. We call the initial hypersurface in Figure 2.8 S_{in}, and the two intermediate ones S_1, obtained by deforming S_{in} with N_1 in the normal direction, and S_2, obtained by deforming S_{in} with N_2 in the normal direction. The intermediate hypersurfaces imply two final hypersurfaces, $S_{\text{fin}}^{(1)}$ by deforming S_1 with N_2 in the new normal direction of S_1, and $S_{\text{fin}}^{(2)}$ by deforming S_2 with N_1 in the new normal direction of S_2.

By comparing the two final hypersurfaces we can derive the commutator of normal deformations. The metric tensor, which should appear in the final result (2.155), enters the calculation because it determines the normals of S_{in}, S_1 and S_2. Based on Equation (2.155), we expect that only the spatial metric is relevant, and most of the spacetime components will indeed cancel out. However, the result depends on the signature of spacetime, given by the sign of the determinant of the spacetime metric. We will refer to this signature as $\sigma = \pm 1$, where $\sigma = -1$ indicates Lorentzian signature.

Our calculation will be performed in a fixed coordinate system, which we may adjust to the initial slice, such that S_{in} is given by constant time, $t = t_{\text{in}}$, in this system. The hypersurface is then simply characterized by a mapping $x^\alpha(y^i) = (t_{\text{in}}, y^1, y^2, y^3)$ from surface coordinates y^i to spacetime coordinates x^α, such that $x^i = y^i$ and $x^0 = t_{\text{in}}$. For a general embedding $x^\alpha(y^i)$, the unit normal to the hypersurface is determined by

$$n^\alpha = \frac{\sigma}{\|\cdot\|} \sum_{\alpha'=0}^{3} \sum_{\beta=0}^{3} \sum_{\gamma=0}^{3} \sum_{\delta=0}^{3} g^{\alpha\alpha'} \epsilon_{\alpha'\beta\gamma\delta} \frac{\partial x^\beta}{\partial y^1} \frac{\partial x^\gamma}{\partial y^2} \frac{\partial x^\delta}{\partial y^3} \tag{2.160}$$

where normalization is imposed by dividing by the norm $\|\cdot\|$ of the multiple sum. The tensor $\epsilon_{\alpha'\beta\gamma\delta}$ is, by definition, completely antisymmetric in all four indices, with $\epsilon_{0123} = 1$. The factor of σ (just defined as the spacetime signature) ensures that the normal vector points to the future in Lorentzian spacetime. For an embedding with constant t_{in} and $x^i = y^i$, the general expression simplifies to $n_{\text{in}}^\alpha = \sigma g^{\alpha 0}/\sqrt{|g^{00}|}$.

Moving points a distance $N_1(y^i)$ and $N_2(y^i)$, respectively, along the normal, we obtain the intermediate hypersurfaces,

$$S_1: \; x_1^\alpha(y^i) = x^\alpha(y^i) + N_1(y^i)n_{\text{in}}^\alpha = x^\alpha(y^i) + \sigma N_1(y^i)\frac{g^{\alpha 0}(y^i)}{\sqrt{|g^{00}|}} \tag{2.161}$$

and

$$S_2: \; x_2^\alpha(y^i) = x^\alpha(y^i) + N_2(y^i)n_{\text{in}}^\alpha = x^\alpha(y^i) + \sigma N_2(y^i)\frac{g^{\alpha 0}(y^i)}{\sqrt{|g^{00}|}}. \tag{2.162}$$

The general Equation (2.160) applied to these expressions yields the two new normal directions. Expanded to first order in the lapse functions (suitable for infinitesimal deformations that appear in a commutator), they are

$$
\begin{aligned}
n_1^\alpha &= \sigma\frac{g^{\alpha 0}}{\sqrt{|g^{00}|}} + \sum_{j=1}^{3}\left(-\sigma g^{\alpha j} + \frac{g^{\alpha 0}g^{j0}}{|g^{00}|}\right)\frac{\partial N_1}{\partial y^j} + N_1 X + O\left(N_1^2\right) \\
&= \sigma\frac{g^{\alpha 0}}{\sqrt{|g^{00}|}} - \sigma\sum_{i=1}^{3} q^{\alpha i}\frac{\partial N_1}{\partial y^i} + N_1 X + O\left(N_1^2\right)
\end{aligned}
\tag{2.163}
$$

and

$$
\begin{aligned}
n_2^\alpha &= \sigma\frac{g^{\alpha 0}}{\sqrt{|g^{00}|}} + \sum_{j=1}^{3}\left(-\sigma g^{\alpha j} + \frac{g^{\alpha 0}g^{j0}}{|g^{00}|}\right)\frac{\partial N_2}{\partial y^j} + N_2 X + O\left(N_2^2\right) \\
&= \sigma\frac{g^{\alpha 0}}{\sqrt{|g^{00}|}} - \sigma\sum_{i=1}^{3} q^{\alpha i}\frac{\partial N_2}{\partial y^i} + N_2 X + O\left(N_2^2\right).
\end{aligned}
\tag{2.164}
$$

The coefficient X denotes a combination of metric components and their derivatives. We do not need its precise form because it depends only on N_1 and N_2 but not on their derivatives, and will therefore cancel out in the final commutator. In the second step of each equation, we have introduced the inverse spatial metric $q^{\alpha\beta} = g^{\alpha\beta} - \sigma n_{\text{in}}^\alpha n_{\text{in}}^\beta$ on the initial slice. (Even though it is defined here as a tensor with spacetime indices α and β, its independent contributions result only from spatial components because its normal projections vanish:

$$\sum_{\alpha=0}^{3} n_{\text{in}}^\alpha q_{\alpha\beta} = n_{\text{in}\beta}\left(1 - \sigma\sum_{\alpha=0}^{3} n_{\text{in}}^\alpha n_{\text{in}\alpha}\right) = 0$$

because $\sigma\sum_{\alpha=0}^{3} n_{\text{in}}^\alpha n_{\text{in}\alpha} = \sigma^2 = 1$.)

The final hypersurfaces obtained with the two orderings of lapse functions are

$$
\begin{aligned}
S_{\text{fin}}^{(1)}: x_{12}^{\alpha}(y^i) &= x^{\alpha}(y^i) + N_1(y^i)n_{\text{in}}^{\alpha} + N_2(y^i)n_1^{\alpha} \\
&= x^{\alpha}(y^i) + \sigma N_1(y^i)\frac{g^{\alpha 0}}{\sqrt{|g^{00}|}} \\
&\quad + \sigma N_2(y^i)\left(\frac{g^{\alpha 0}}{\sqrt{|g^{00}|}} - \sum_{j=1}^{3} q^{\alpha j}\frac{\partial N_1}{\partial y^j}\right) \\
&\quad + N_1 N_2 X + O(N_1^2)
\end{aligned}
\tag{2.165}
$$

and

$$
\begin{aligned}
S_{\text{fin}}^{(2)}: x_{21}^{\alpha}(y^i) &= x^{\alpha}(y^i) + N_2(y^i)n_{\text{in}}^{\alpha} + N_1(y^i)n_2^{\alpha} \\
&= x^{\alpha}(y^i) + \sigma N_2(y^i)\frac{g^{\alpha 0}}{\sqrt{|g^{00}|}} \\
&\quad + \sigma N_1(y^i)\left(\frac{g^{\alpha 0}}{\sqrt{|g^{00}|}} - \sum_{j=1}^{3} q^{\alpha j}\frac{\partial N_2}{\partial y^j}\right) \\
&\quad + N_2 N_1 X + O(N_2^2).
\end{aligned}
\tag{2.166}
$$

Subtracting the final positions from each other, we have

$$
\delta S^{\alpha}(y^i) = x_{21}^{\alpha}(y^i) - x_{12}^{\alpha}(y^i) = -\sigma \sum_{i=1}^{3} q^{\alpha i}\left(N_1\frac{\partial N_2}{\partial y^i} - N_2\frac{\partial N_1}{\partial y^i}\right).
\tag{2.167}
$$

This expression depends only on the spatial metric and is therefore a spatial deformation along the final hypersurface: $\delta S^0(y^i) = 0$ because $q^{0i} = 0$. It is equivalent to Equation (2.154) with Equation (2.155) and $\sigma = -1$. All information about the spacetime metric that is not already determined by the spatial metric is contained in the signature σ.

2.3.4 Spherical Symmetry

In order to demonstrate the close relationship between spacetime symmetries encoded in Equation (2.154) and generally covariant field equations, we consider spherically symmetric general relativity. The methods applied in Hojman et al. (1976) to the full theory allow us to derive the Hamiltonian of these systems which we already used in Section 1.5.

Any such spacetime has a line element of the form

$$
\begin{aligned}
\mathrm{d}s^2 = {}&- N(t, x)^2 c^2 \mathrm{d}t^2 + L(t, x)^2(\mathrm{d}x + M(t, x)c\mathrm{d}t)^2 \\
&+ S(t, x)^2(\mathrm{d}\vartheta^2 + \sin^2\vartheta \mathrm{d}\varphi^2).
\end{aligned}
\tag{2.168}
$$

As we have seen in cosmological models, the lapse function N, and similarly the function M, does not have a momentum, while L and S have momenta p_L and p_S. The momenta represent time derivatives of the metric components, while their spatial derivatives $L' = \partial L/\partial x$, $S' = \partial S/\partial x$ and higher orders appear directly in the Hamiltonian.

As it turns out, it is convenient to express these spatial derivatives through the combinations

$$\Gamma = \frac{S'}{L}, \quad \Gamma' = \frac{S''}{L} - \frac{S'L'}{L^2} \tag{2.169}$$

which have the derivatives

$$\frac{\delta\Gamma(y)}{\delta S(x)} = \frac{1}{L(y)}\frac{\partial\delta(x, y)}{\partial y}, \quad \frac{\delta\Gamma(y)}{\delta L(x)} = -\frac{S'(y)}{L(y)^2}\delta(x, y) \tag{2.170}$$

$$\frac{\delta\Gamma'(y)}{\delta S(x)} = \frac{1}{L(y)}\frac{\partial^2\delta(x, y)}{\partial y^2} - \frac{L'(y)}{L(y)^2}\frac{\partial\delta(x, y)}{\partial y} \tag{2.171}$$

$$\frac{\delta\Gamma'(y)}{\delta L(x)} = -\frac{S''(y)}{L(y)^2}\delta(x, y) - \frac{S'(y)}{L(y)^2}\frac{\partial\delta(x, y)}{\partial y} + 2\frac{S'(y)L'(y)}{L(y)^3}\delta(x, y). \tag{2.172}$$

(The combination Γ, defined in the first equation of (2.169), is related to a certain component of the Christoffel symbol in spherically symmetric spacetimes.)

In the last three equations, the symbol δ denotes a *functional derivative*, which is a derivative by the value of a field, such as $p_S(x)$, taken at a fixed point x. Since there is a continuum of infinitely many independent x, the functional derivative is a derivative in an infinite-dimensional space, but in practice it obeys the usual rules of a partial derivative. One should only keep in mind that the functional derivative by, say, $p_S(x)$ is completely independent of a usual partial derivative by the coordinate x: We can independently change the position x and the value the function p_S assigns to x. Therefore, $\delta/\delta p_S(x)$ commutes with $\partial/\partial x$, as well as with an integration over x.

Moreover, since field values at different points can be changed independently, we have $\delta p_S(y)/\delta p_S(x) = 0$ if $x \neq y$. If the case of $x = y$ is included, the basic functional derivative is

$$\frac{\delta p_S(y)}{\delta p_S(x)} = \delta(x - y) \tag{2.173}$$

because the delta distribution replaces the unit matrix in a function space:

$$\int_{-\infty}^{\infty} f(x)\delta(x - y)\mathrm{d}x = f(y). \tag{2.174}$$

Similarly,

$$\frac{\delta S(y)}{\delta S(x)} = \delta(x - y),$$ (2.175)

while

$$\frac{\delta p_S(y)}{\delta S(x)} = 0 = \frac{\delta S(y)}{\delta p_S(x)}$$ (2.176)

because two different functions can be changed independently at any point.

Spatial Deformations
The usefulness of the ratio Γ can be seen from transformation properties of the components L and S. If we infinitesimally transform the only spatial coordinate in which we have inhomogeneity, $\tilde{x}(x) = x - \epsilon$, the component S is simply changed by a Taylor expansion:

$$\tilde{S}(\tilde{x}) = S(x(\tilde{x})) = S(\tilde{x}+\epsilon) = S(\tilde{x}) + \epsilon S'(\tilde{x}) + O(\epsilon^2).$$ (2.177)

Therefore,

$$\delta_\epsilon S = \tilde{S} - S = \epsilon S',$$ (2.178)

which characterizes the transformation of a scalar function.

The component L, however, transforms differently because ds^2 must be invariant, while both L and dx transform in $L^2 dx^2$. The invariance condition of the line element implies that L must transform such that $L dx = \tilde{L} d\tilde{x}$, or

$$\tilde{L}(\tilde{x}) = \frac{dx}{d\tilde{x}} L(x(\tilde{x})).$$ (2.179)

Any function L with such a transformation behavior is called a scalar with *density weight*, in this case with weight plus one according to the power by which the factor of $dx/d\tilde{x}$ appears. (The function L transforms like a one-dimensional version of the integration measure, where the determinant shown in Equation (1.22) is replaced by a single factor.) While S is a scalar without density weight, its spatial derivative, S', transforms with density weight plus one owing to the chain rule. Therefore, the ratio S'/L, used in the definition of Γ, has zero density weight and is therefore a simple scalar function.

For our infinitesimal transformation, we have

$$\tilde{L}(\tilde{x}) = \frac{1}{1 - \epsilon'} L(\tilde{x}+\epsilon) = L(\tilde{x}) + \epsilon' L(\tilde{x}) + \epsilon L'(\tilde{x}) + O(\epsilon^2)$$
$$= L(\tilde{x}) + (\epsilon L)'(\tilde{x}) + O(\epsilon^2)$$ (2.180)

or

$$\delta_\epsilon L = (\epsilon L)'.$$ (2.181)

Together with Equation (2.178), these equations determine how the two dynamical components of the metric with momenta p_S and p_L, respectively, transform with respect to an infinitesimal change of the spatial coordinate x in which we have inhomogeneity. An infinitesimal change of a spatial coordinate is nothing but a spatial deformation, described by our diffeomorphism generator D[w]. If we can find a phase-space expression for D[w]—that is, an expression for this generator as a function not only of the displacement w but also of the phase-space degrees of freedom S, L, and their momenta p_S and p_L—the transformations (2.178) and (2.181) should equal a D-version of Hamiltons equations:

$$\delta_\epsilon S(x) = \frac{\delta D[\epsilon]}{\delta p_S(x)}, \quad \delta_\epsilon L(x) = \frac{\delta D[\epsilon]}{\delta p_L(x)}. \tag{2.182}$$

The diffeomorphism generator D[w] then generates spatial transformations just as the Hamiltonian H of a dynamical system generates time translations.

Combining Equation (2.182) with Equations (2.178) and (2.181), we obtain the first-order functional differential equations

$$\frac{\delta D[\epsilon]}{\delta p_S(x)} = \epsilon S' \quad \text{and} \quad \frac{\delta D[\epsilon]}{\delta p_L(x)} = (\epsilon L)' \tag{2.183}$$

which must be true for any ϵ. Since the right-hand sides do not depend on p_S or p_L, the solution D[ϵ] is linear in the momenta or their spatial derivatives. The coefficients in

$$D[\epsilon] = \int_{-\infty}^{\infty} \epsilon(y)\big(p_S S' - Lp_L'\big)dy \tag{2.184}$$

then follow uniquely from the functional differential equations, and we discard integration constants. Indeed, the correct expression for $\delta_\epsilon S(x)$ follows immediately, while

$$\delta_\epsilon L(x) = -\int_{-\infty}^{\infty} \epsilon(y)L(y)\frac{\partial \delta(x-y)}{\partial y}dy = \frac{d(\epsilon L)}{dx} \tag{2.185}$$

is obtained after integrating by parts. The different appearance of S and L in Equation (2.184) reflects their density weights.

In what follows we will work directly with the integrand in Equation (2.184), given by

$$D(x) = p_S(x)S(x)' - L(x)p_L(x)'. \tag{2.186}$$

Moreover, as in classical mechanics, it is convenient to express various versions of Hamilton's equations or their generalizations, such as Equation (2.182), in terms of Poisson brackets. The *Poisson bracket* of two functions on a functional phase space, $f(S, L, p_S, p_L)$ and $g(S, L, p_S, p_L)$, is defined as

$$\{f, g\} = \int_{-\infty}^{\infty} \left(\frac{\delta f}{\delta S(x)} \frac{\delta g}{\delta p_S(x)} - \frac{\delta f}{\delta p_S(x)} \frac{\delta g}{\delta S(x)} \right.$$

$$\left. + \frac{\delta f}{\delta L(x)} \frac{\delta g}{\delta p_L(x)} - \frac{\delta f}{\delta p_L(x)} \frac{\delta g}{\delta L(x)} \right) \mathrm{d}x. \qquad (2.187)$$

If f and g are defined through integrations, as in $D[\epsilon]$, the two delta functions in each product in Equation (2.187), which arise from applying Equation (2.173) or (2.175) as well as the corresponding L-versions, are integrated three times, such that the Poisson bracket $\{D[\epsilon_1], D[\epsilon_2]\}$ is also an integrated expression. The Poisson bracket of two local functions, which are not integrated but rather depend on x through the fields, is a combination of the delta function and its partial derivatives by position coordinates.

Two Normal Deformations
The Poisson bracket is a classical or phase-space analog of the commutator. The commutator relationship (2.154) with (2.155) should therefore take the form

$$\{H[N_1], H[N_2]\} = D[L^{-2}(N_1 N_2' - N_1' N_2)] \qquad (2.188)$$

where the inverse spatial metric of spherically symmetric spacetimes is taken only in the inhomogeneous x-direction, $q^{xx} = 1/L^2$. The generator of normal deformations of spatial hypersurfaces, $H[N] = \int_{-\infty}^{\infty} N(x)H(x)\mathrm{d}x$, can also be written with a local integrand, $H(x)$, which depends on the position x through the fields S, L and their momenta. For the local integrands, the commutator (2.154) looks like

$$\{H(x), H(y)\} = \int \mathrm{d}z \left(\frac{\delta H(x)}{\delta S(z)} \frac{\delta H(y)}{\delta p_S(z)} - \frac{\delta H(y)}{\delta p_S(z)} \frac{\delta H(x)}{\delta S(z)} \right.$$

$$\left. + \frac{\delta H(x)}{\delta L(z)} \frac{\delta H(y)}{\delta p_L(z)} - \frac{\delta H(y)}{\delta p_L(z)} \frac{\delta H(x)}{\delta L(z)} \right) \qquad (2.189)$$

$$= -\frac{D(x)}{L(x)^2} \frac{\partial \delta(x - y)}{\partial y} + \frac{D(y)}{L(y)^2} \frac{\partial \delta(x - y)}{\partial x}.$$

The first line simply inserts the definition of the Poisson bracket, while the last line results in Equation (2.188) upon integration with multipliers N_1 and N_2:

$$\int_{-\infty}^{\infty} \int_{-\infty}^{\infty} N_1(x)N_2(y)\left(-\frac{D(x)}{L(x)^2} \frac{\partial \delta(x - y)}{\partial y} + \frac{D(y)}{L(y)^2} \frac{\partial \delta(x - y)}{\partial x} \right) \mathrm{d}x\mathrm{d}y$$

$$= \int_{-\infty}^{\infty} \int_{-\infty}^{\infty} \left(\frac{D(x)}{L(x)^2} N_1(x) \frac{\mathrm{d}N_2(y)}{\mathrm{d}y} - \frac{D(y)}{L(y)^2} \frac{\mathrm{d}N_1(x)}{\mathrm{d}x} N_2(y) \right) \delta(x - y)\mathrm{d}x\mathrm{d}y$$

$$= \int_{-\infty}^{\infty} \frac{D}{L^2}\left(N_1 \frac{\mathrm{d}N_2}{\mathrm{d}x} - \frac{\mathrm{d}N_1}{\mathrm{d}x} N_2 \right)\mathrm{d}x = D[L^{-2}(N_1 N_2' - N_1' N_2)].$$

Our aim is to construct classical spherically symmetric theories compatible with the symmetry of hypersurface deformations in spacetime. Since we already know the phase-space expression of D, it remains to construct a suitable H. (Our derivation here follows Bojowald & Paily 2012.) At this time, we are interested in classical theories, and therefore assume that the phase-space expression of H(x) depends quadratically on the momenta p_L and p_S, and contains spatial derivatives (up to second order) only of L and S in the "potential"-like terms independent of momenta. Any such expression can be expanded as

$$H(x) = H_{00}(x) + H_{11}(x)p_S(x)p_L(x) + H_{20}(x)p_S(x)^2 + H_{02}(x)p_L(x)^2 \qquad (2.190)$$

with coefficients H_{IJ} that depend on x through the metric components L and S and their spatial derivatives. Time-reversal symmetry excludes the presence of terms linear in the momenta. Moreover, only H_{00} is allowed to depend on spatial derivatives of L and S, or on Γ and Γ', because the remaining terms implicitly contain two derivatives (by time) through the momenta.

We therefore have linear derivatives

$$\frac{\delta H(x)}{\delta p_S(z)} = \big(A_1(x)p_S(x) + B_1(x)p_L(x)\big)\delta(x - z) \qquad (2.191)$$

$$\frac{\delta H(x)}{\delta p_L(z)} = \big(A_2(x)p_S(x) + B_2(x)p_L(x)\big)\delta(x - z) \qquad (2.192)$$

where $A_1(x) = 2H_{20}(x)$, $A_2(x) = B_1(x) = H_{11}(x)$ and $B_2(x) = 2H_{02}(x)$. Therefore,

$$\begin{aligned}
\{H(x), H(y)\} &= \frac{\delta H(y)}{\delta S(x)}\big(A_1(x)p_S(x) + B_1(x)p_L(x)\big) \\
&\quad + \frac{\delta H(y)}{\delta L(x)}\big(A_2(x)p_S(x) + B_2(x)p_L(x)\big) - (x \leftrightarrow y) \qquad (2.193) \\
&= \frac{p_S(x)S'(x) - L(x)p_L{}'(x)}{L(x)^2}\frac{\partial\delta(x - y)}{\partial y} - (x \leftrightarrow y),
\end{aligned}$$

indicating antisymmetrization in x and y by "$-(x \leftrightarrow y)$".

This equation can be evaluated by comparing coefficients of p_L and p_S. For $p_L = 0$ and $p_S = 0$, the equation is automatically satisfied. The left-hand side contains terms of up to third order in p_L and p_S, while the right-hand side is linear. The quadratic and cubic terms do not receive contributions from H_{00}, which is the only coefficient of H that may depend on spatial derivatives. Therefore, all quadratic and cubic terms are multiplied by $\delta(x - y)$ without spatial derivatives: they vanish identically upon antisymmetrization. It remains to evaluate the linear terms on both sides of the equation.

We extract first-order coefficients of p_S by taking a derivative $\delta/\delta p_S(z)$ followed by setting $p_S = 0 = p_L$:

$$\left(\frac{\delta H_{00}(y)}{\delta S(x)} A_1(x) + \frac{\delta H_{00}(y)}{\delta L(x)} A_2(x) \right) \delta(x - z) - (x \leftrightarrow y)$$

$$= \frac{S'(x)}{L(x)^2} \frac{\partial \delta(x - y)}{\partial y} \delta(x - z) - (x \leftrightarrow y). \tag{2.194}$$

Similarly, we extract first-order coefficients of p_L by taking a derivative $\delta/\delta p_L(z)$ followed by setting $p_S = 0 = p_L$:

$$\left(\frac{\delta H_{00}(y)}{\delta S(x)} B_1(x) + \frac{\delta H_{00}(y)}{\delta L(x)} B_2(x) \right) \delta(x - z) - (x \leftrightarrow y)$$

$$= -\frac{1}{L(x)} \frac{\partial \delta(x - y)}{\partial y} \frac{\partial \delta(x - z)}{\partial x} - (x \leftrightarrow y). \tag{2.195}$$

The coefficient H_{00} can depend directly on S, or indirectly through spatial derivatives of S which we have parameterized in terms of Γ and Γ'. By the chain rule, therefore,

$$\frac{\delta H_{00}(y)}{\delta S(x)} = \frac{\partial H_{00}(y)}{\partial S(y)} \frac{\delta S(y)}{\delta S(x)} + \frac{\partial H_{00}(y)}{\partial \Gamma(y)} \frac{\delta \Gamma(y)}{\delta S(x)} + \frac{\partial H_{00}(y)}{\partial \Gamma'(y)} \frac{\delta \Gamma'(y)}{\delta S(x)} \tag{2.196}$$

and a similar relation for $\delta H_{00}(y)/\delta L(x)$.

In order to proceed, we eliminate delta functions by multiplying our relationship with arbitrary *test functions* $a(x)$, $b(y)$, and $c(z)$, and then integrate over x, y, and z. The z-integration is straightforward in Equation (2.194) because there are no derivatives of delta functions by z. We obtain

$$\int_{-\infty}^{\infty} \int_{-\infty}^{\infty} dx\, dy\, a(x) b(y) c(x) \left(\frac{\partial H_{00}(y)}{\partial S(y)} \delta(x - y) \right.$$

$$+ \frac{\partial H_{00}(y)}{\partial \Gamma(y)} \frac{1}{L(y)} \frac{\partial \delta(x - y)}{\partial y}$$

$$+ \frac{\partial H_{00}(y)}{\partial \Gamma'(y)} \left(\frac{1}{L(y)} \frac{\partial^2 \delta(x - y)}{\partial y^2} - \frac{L'(y)}{L(y)^2} \frac{\partial \delta(x - y)}{\partial y} \right) \bigg) A_1(x)$$

$$+ \left(\frac{\partial H_{00}(y)}{\partial L(y)} \delta(x - y) - \frac{\partial H_{00}(y)}{\partial \Gamma(y)} \frac{S'(y)}{L(y)^2} \delta(x - y) \right. \tag{2.197}$$

$$- \frac{\partial H_{00}(y)}{\partial \Gamma'(y)} \left(\frac{S''(y)}{L(y)^2} \delta(x - y) \right.$$

$$\left. + \frac{S'(y)}{L(y)^2} \frac{\partial \delta(x - y)}{\partial y} - 2 \frac{S'(y)L'(y)}{L(y)^3} \delta(x - y) \right) \bigg) A_2(x) - (x \leftrightarrow y) \bigg)$$

$$= \int_{-\infty}^{\infty} \int_{-\infty}^{\infty} dx\, dy\, a(x) b(y) c(x) \left(\frac{S'(x)}{L(x)^2} \frac{\partial \delta(x - y)}{\partial y} - (x \leftrightarrow y) \right).$$

Any term that contains a delta function $\delta(x - y)$ without derivatives cancels out with the corresponding term upon antisymmetrization. Only five terms with derivatives of delta functions are then left in the equation. Their evaluations require integrations by parts. Also here, depending on the term, there may be cancellations thanks to antisymmetrization. In particular, any terms in which we take a derivative of a coefficient not equal to one of the test functions a, b, or c cancels out with such a corresponding term. If we integrate by parts over y, therefore, we need to take into account only terms in which integration by parts produces at least one derivative of $b(y)$. We have to be careful only when integrating by parts in the one term that contains a second derivative, $\partial^2\delta(x - y)/\partial y^2$, which we have to integrate by parts twice. After the first integration by parts, we produce a coefficient b', which after the second integration by parts leads to a non-zero contribution with b'' as well as a non-zero contribution from $2b'(L^{-1}\partial H_{00}/\partial \Gamma')'$, taking a derivative of the whole coefficient by y: With generic coefficients, we have

$$\int_{-\infty}^{\infty} dy Q(y) R(y) \frac{\partial^2 \delta(x - y)}{\partial y^2} = -\int_{-\infty}^{\infty} dy (QR)' \frac{\partial \delta(x - y)}{\partial y}$$

$$= \int_{-\infty}^{\infty} dy (Q''R + 2Q'R' + QR'')\delta(x - y). \tag{2.198}$$

Since the cross-term appears twice, a factor of 2 must be inserted in one of the b'-terms

In this way, we arrive at

$$\int_{-\infty}^{\infty} dx \left(ab'cA_1 \left(-\frac{1}{L} \frac{\partial H_{00}}{\partial \Gamma} + 2\left(\frac{1}{L} \frac{\partial H_{00}}{\Gamma'} \right)' + \frac{L'}{L^2} \frac{\partial H_{00}}{\partial \Gamma'} \right) \right.$$

$$\left. + ab''cA_1 \frac{1}{L} \frac{\partial H_{00}}{\Gamma'} + ab'cA_2 \frac{S'}{L^2} \frac{\partial H_{00}}{\partial \Gamma'} + ab'c \frac{S'}{L^2} \right) - (a \leftrightarrow b) = 0 \tag{2.199}$$

where we brought the right-hand side of Equation (2.197) to the left. There are now two kinds of coefficients, those of $c(ab'' - a''b)$ and those of $c(ab' - a'b)$, respectively, which must vanish independently of each other:

$$\frac{A_1}{L} \frac{\partial H_{00}}{\partial \Gamma'} = 0 \tag{2.200}$$

from $c(ab'' - a''b)$ and

$$A_1 \left(\frac{1}{L} \frac{\partial H_{00}}{\partial \Gamma} - \frac{L'}{L^2} \frac{\partial H_{00}}{\partial \Gamma'} - 2\left(\frac{1}{L} \frac{\partial H_{00}}{\partial \Gamma'} \right)' \right) - A_2 \frac{S'}{L^2} \frac{\partial H_{00}}{\partial \Gamma'} - \frac{S'}{L^2} = 0 \tag{2.201}$$

from $c(ab' - a'b)$.

Similarly, Equation (2.195) first implies

$$\int_{-\infty}^{\infty} \int_{-\infty}^{\infty} dx dy a(x) b(y) c(x) \left(\left(\frac{\partial H_{00}(y)}{\partial S(y)} \delta(x-y) + \frac{\partial H_{00}(y)}{\partial \Gamma(y)} \frac{1}{L(y)} \frac{\partial \delta(x-y)}{\partial y} \right. \right.$$

$$+ \frac{\partial H_{00}(y)}{\partial \Gamma'(y)} \left(\frac{1}{L(y)} \frac{\partial^2 \delta(x-y)}{\partial y^2} - \frac{L'(y)}{L(y)^2} \frac{\partial \delta(x-y)}{\partial y} \right) \right) B_1(x)$$

$$+ \left(\frac{\partial H_{00}(y)}{\partial L(y)} \delta(x-y) - \frac{\partial H_{00}(y)}{\partial \Gamma(y)} \frac{S'(y)}{L(y)^2} \delta(x-y) - \frac{\partial H_{00}(y)}{\partial \Gamma'(y)} \left(\frac{S''(y)}{L(y)^2} \delta(x-y) \right. \right. \quad (2.202)$$

$$\left. \left. + \frac{S'(y)}{L(y)^2} \frac{\partial \delta(x-y)}{\partial y} - 2 \frac{S'(y) L'(y)}{L(y)^3} \delta(x-y) \right) \right) B_2(x) - (x \leftrightarrow y) \right)$$

$$= -\int_{-\infty}^{\infty} \int_{-\infty}^{\infty} dx dy \left(a(x) b(y) c'(x) \frac{1}{L(x)} \frac{\partial \delta(x-y)}{\partial y} - (x \leftrightarrow y) \right)$$

where we have integrated by parts with respect to z in the last term before integrating over z, using the identity

$$\frac{\partial \delta(x-z)}{\partial x} = -\frac{\partial \delta(x-z)}{\partial z}. \quad (2.203)$$

The y-integration then produces

$$\int_{-\infty}^{\infty} dx \left[ab' c B_1 \left(-\frac{1}{L} \frac{\partial H_{00}}{\partial \Gamma} + \frac{L'}{L^2} \frac{\partial H_{00}}{\partial \Gamma'} + 2 \left(\frac{1}{L} \frac{\partial H_{00}}{\partial \Gamma'} \right)' \right) + ab' c B_2 \frac{S'}{L^2} \frac{\partial H_{00}}{\partial \Gamma'} \right.$$

$$\left. + ab'' c B_1 \frac{1}{L} \frac{\partial H_{00}}{\partial \Gamma'} + ab' c \left(\frac{1}{L} \right)' + ab'' c \frac{1}{L} \right) - (a \leftrightarrow b) = 0. \quad (2.204)$$

The last two terms result after integrating by parts one last time to remove a derivative from c.

Again collecting the coefficients of $c(ab'' - a''b)$ and $c(ab' - a'b)$, respectively, we obtain

$$\frac{B_1}{L} \frac{\partial H_{00}}{\partial \Gamma'} + \frac{1}{L} = 0 \quad (2.205)$$

and

$$B_1 \left(-\frac{1}{L} \frac{\partial H_{00}}{\partial \Gamma} + \frac{L'}{L^2} \frac{\partial H_{00}}{\partial \Gamma'} + 2 \left(\frac{1}{L} \frac{\partial H_{00}}{\partial \Gamma'} \right)' \right) + B_2 \frac{S'}{L^2} \frac{\partial H_{00}}{\partial \Gamma'} + \left(\frac{1}{L} \right)' = 0. \quad (2.206)$$

It is convenient to rewrite the last equation as

$$B_1\left(-\frac{1}{L}\frac{\partial H_{00}}{\partial \Gamma} + \left(\frac{1}{L}\right)'\left(\frac{\partial H_{00}}{\partial \Gamma'} + \frac{1}{B_1}\right) + \frac{2}{L}\left(\frac{\partial H_{00}}{\partial \Gamma'}\right)'\right) + B_2\frac{S'}{L^2}\frac{\partial H_{00}}{\partial \Gamma'} = 0. \quad (2.207)$$

As a result of Equation (2.205), $\delta H_{00}/\delta \Gamma'$ cannot be zero. Therefore, Equation (2.200) implies $A_1 = 0$, and from Equations (2.201) and (2.205) we obtain

$$A_2 = -\left(\frac{\partial H_{00}}{\partial \Gamma'}\right)^{-1} = B_1, \quad (2.208)$$

such that Equation (2.207) takes the form

$$\frac{B_1}{L}\frac{\partial H_{00}}{\partial \Gamma} + \frac{2}{L}B_1\left(\frac{1}{B_1}\right)' + \frac{B_2}{B_1}\frac{S'}{L^2} = 0. \quad (2.209)$$

We can write the last two equations as partial differential equations for the dependence of H_{00} on derivatives of S and L, parameterized through $\Gamma = S'/L$ and Γ':

$$\frac{\partial H_{00}}{\partial \Gamma} = 2\frac{B_1'}{B_1^2} - \frac{B_2}{B_1^2}\frac{S'}{L} = \left(2\frac{L}{B_1^2}\frac{dB_1}{dS} - \frac{B_2}{B_1^2}\right)\Gamma \quad (2.210)$$

$$\frac{\partial H_{00}}{\partial \Gamma'} = -\frac{1}{B_1}. \quad (2.211)$$

Since L' canceled out in Equation (2.207) upon using Equation (2.205), the last two equations, (2.210) and (2.211), are decoupled and can easily be integrated, with the result

$$H_{00}(S, L, \Gamma, \Gamma') = \left(\frac{L}{B_1^2}\frac{dB_1}{dS} - \frac{1}{2}\frac{B_2}{B_1^2}\right)\Gamma^2 - \frac{1}{B_1}\Gamma' + H_0(S, L). \quad (2.212)$$

There are now three free functions of S and L, $B_1 = H_{11}$, $B_2 = 2H_{02}$ and H_0, which appear in the full Hamiltonian

$$H = H_{00} + H_{11}p_S p_L + H_{02}p_L^2 \quad (2.213)$$

$$= B_1 p_S p_L + \frac{1}{2}B_2 p_L^2 + \left(\frac{L}{B_1^2}\frac{dB_1}{dS} - \frac{1}{2}\frac{B_2}{B_1^2}\right)\Gamma^2 - \frac{1}{B_1}\Gamma' + H_0 \quad (2.214)$$

using $H_{20} = \frac{1}{2}A_1 = 0$.

Normal and Spatial Deformations
In addition to Equation (2.154), the bracket

$$\{H[N], D[M]\} = -H[MN'] \quad (2.215)$$

or

$$\{H(x), D(y)\} = H(x)\frac{\partial \delta(x - y)}{\partial x} + \frac{\partial H}{\partial x}\delta(x - y) \tag{2.216}$$

must also be satisfied, which ensures that H transforms properly by spatial deformations; see Equation (2.158).

We can check this equation independently for each of the terms that sum up to form the Hamiltonian constraint, in particular for the possible "kinetic" contributions $L^{q_1}S^{q_2}p_S p_L$ and $L^{p_1}S^{p_2}p_L^2$ with arbitrary exponents q_1, q_2, p_1 and p_2. For the first term of this generic form, we obtain

$$\begin{aligned}
\{L(x)^{q_1}S(x)^{q_2}p_S(x)p_L(x), D(y)\} = {} & q_2 L(x)^{q_1}S(x)^{q_2-1}S'(y)p_S(x)p_L(x)\delta(x - y) \\
& - L(x)^{q_1}S(x)^{q_2}p_L(x)p_S(y)\frac{\partial \delta(x - y)}{\partial y} \\
& - q_1 L(x)^{q_1-1}S(x)^{q_2}L(y)p_S(x)p_L(x)\frac{\partial \delta(x - y)}{\partial y} \\
& + L(x)^{q_1}S(x)^{q_2}p_S(x)p_L(y)'\delta(x - y).
\end{aligned} \tag{2.217}$$

We may now use the identity

$$\begin{aligned}
A(x)B(y)\frac{\partial \delta(x - y)}{\partial y} & = A(x)\frac{\partial}{\partial y}(B(y)\delta(y - x)) - A(x)B'(y)\delta(x - y) \\
& = A(x)\frac{\partial}{\partial y}(B(x)\delta(y - x)) - A(x)B'(x)\delta(x - y) \tag{2.218} \\
& = A(x)B(x)\frac{\partial \delta(x - y)}{\partial y} - A(x)B'(x)\delta(x - y)
\end{aligned}$$

and rewrite the terms containing a derivative of the delta function in such a way that their coefficients depend only on x:

$$\begin{aligned}
\{L(x)^{q_1}S(x)^{q_2}p_S(x)p_L(x), D(y)\} = {} & \big(q_2 L(x)^{q_1}S(x)^{q_2-1}S'(x)p_S(x)p_L(x) \\
& + L(x)^{q_1}S(x)^{q_2}p_L(x)p_S(x)' \\
& + q_1 L(x)^{q_1-1}S(x)^{q_2}L(x)'p_S(x)p_L(x) \\
& + L(x)^{q_1}S(x)^{q_2}p_S(x)p_L(x)'\big)\delta(x - y) \tag{2.219} \\
& - \big(L(x)^{q_1}S(x)^{q_2}p_L(x)p_S(x) \\
& + q_1 L(x)^{q_1}S(x)^{q_2}p_S(x)p_L(x)\big)\frac{\partial \delta(x - y)}{\partial y}
\end{aligned}$$

$$= (L(x)^{q_1} S(x)^{q_2} p_S(x) p_L(x))' \delta(x-y)$$

$$- L(x)^{q_1} S(x)^{q_2} p_S(x) p_L(x) \frac{\partial \delta(x-y)}{\partial y} \tag{2.220}$$

$$- q_1 L(x)^{q_1} S(x)^{q_2} p_S(x) p_L(x) \frac{\partial \delta(x-y)}{\partial y}.$$

This bracket is of the required form provided $q_1 = 0$, while q_2 is unrestricted. Similarly, $p_1 = 1$ while p_2 is unrestricted in any expression of the form $L^{p_1} S^{p_2} p_L^2$. More generally, any term must be linear in L and p_S, while the dependence on S and p_L is unrestricted. A shortcut through this argument uses the density weight of the basic fields, which is one for L and p_S and zero for S and p_L. The density weights add up in a product, and each term in $H(x)$ must have density weight one for it to have a well-defined and coordinate-independent spatial integral.

Applied to the current form of the Hamiltonian, we can write the free functions as

$$B_1 = b_1(S), \quad B_2 = L b_2(S), \quad H_0 = \frac{1}{4} L S V(S), \tag{2.221}$$

introducing $V(S)$ in this specific form so as to match with conventions about the dilaton potential. Therefore,

$$H = b_1(S) p_S p_L + \frac{1}{2} b_2(S) L p_L^2$$

$$+ L \left(\frac{1}{b_1(S)^2} \frac{db_1}{dS} - \frac{1}{2} \frac{b_2(S)}{b_1(S)^2} \right) \Gamma^2 - \frac{1}{b_1(S)} \Gamma' + \frac{1}{4} L S V(S). \tag{2.222}$$

In the Γ-dependent terms, we can combine

$$-\frac{1}{b_1^2} \frac{db_1}{dS} L \Gamma^2 + \frac{\Gamma'}{b_1} = -\frac{1}{b_1^2} b_1' \frac{S'}{L} + \frac{1}{b_1} \left(\frac{S'}{L} \right)' = \left(\frac{S'}{b_1 L} \right)'. \tag{2.223}$$

If $f(S)$ is such that $b_1(S) = (df/dS)^{-1}$, this expression equals $(f(S)/L)' = \Gamma_f'$, where $\Gamma_f = f(S)'/L$ is defined just like Γ but after transforming S to $f(S)$. The same transformation implies $\Gamma^2/b_1^2 = \Gamma_f^2$. It is also part of a canonical transformation, mapping (S, p_S) to $(\tilde{S}, \tilde{p}_S)$ with $\tilde{S} = f(S)$ and $\tilde{p}_S = b_1(S) p_S$ and preserving the Poisson brackets of basic fields.

After this canonical transformation, the Hamiltonian is simplified to

$$\tilde{H} = \tilde{p}_S p_L + \frac{1}{2} b_2(S) L p_L^2 - \frac{1}{2} L b_2(S) \Gamma_f^2 - \Gamma_f' + \frac{1}{4} L S V(S). \tag{2.224}$$

The free function b_1 can therefore be eliminated by a canonical transformation, or mapped to any other non-zero function. Moreover, b_2 can be absorbed in the lapse function since only $H[N] = \int N(x) H(x) dx$ with arbitrary N has meaning as a spacetime generator. As a result, only the dilaton potential $V(S)$ can be chosen freely.

For $b_1(S) = -1/S$ and $b_2(S) = 1/S^2$, we have

$$\mathrm{H} = -\frac{p_L p_S}{S} + \frac{Lp_L^2}{2S^2} + \frac{1}{2}\frac{(S')^2}{L} + \frac{SS''}{L} - \frac{SS'L'}{L^2} + \frac{1}{4}LSV(S) \qquad (2.225)$$

which, when set equal to zero, is identical with the constraint (1.141) used in Section 1.5. The spatial generator D, when set equal to zero, is identical with the constraint (1.142), while the equations of motion (1.143) and (1.144) as well as the momentum relationships (1.140) are identical with Hamilton's equations for p_S, p_L, L, and S generated by the combined $\mathrm{H}[N] + \mathrm{D}[M]$.

It is therefore possible to derive the dynamics of two-dimensional gravity models from the required behavior of hypersurface deformations, imposing the canonical version of general covariance. Only the dilaton potential remains free in classical-type theories with second-order field equations. In this sense, the dilaton potential in two-dimensional models generalizes the cosmological constant by replacing a free constant in the Einstein–Hilbert action for gravity in four-dimensional spacetimes with a free function for gravity in two-dimensional spacetimes.

2.3.5 Perturbative Cosmological Inhomogeneity

Like the equations of spherically symmetric gravity, the classical dynamics of perturbative inhomogeneity in cosmology is also determined by general covariance. We will assume an isotropic background spacetime with vanishing spatial curvature, described by a time-dependent scale factor a subject to the Friedmann equation with some matter source. In our derivation of generally covariant Hamiltonians, it is convenient to use the variable $q = a^2$ to describe the background, which is directly related to a spatial metric component.

Canonical Fields
Transforming the canonical variables V and p_V found in Chapter 1, we see that, up to a constant factor, q has the momentum $p = -da/d\tau$: These two variables have the Poisson bracket

$$\{q, p\} = \frac{8\pi G}{3c^2 V_0}. \qquad (2.226)$$

As in Chapter 1, V_0 is the averaging volume. The background Hamiltonian

$$\bar{\mathrm{H}} = -\frac{3c^2 V_0}{8\pi G}\sqrt{q}\, p^2 \qquad (2.227)$$

determines the correct Friedmann equation when set equal to zero after adding a matter Hamiltonian. Although q and p are not strictly canonical because their Poisson bracket does not equal one, the bracket is constant and therefore leads to the usual form of Hamilton's equations just with a constant rescaling. We will use q and p because these variables imply more convenient coefficients in the following expressions.

In order to introduce perturbative inhomogeneity, we consider the background variables to be spatial averages of fields q_{ij} and p^{kl}, respectively. The fields q_{ij} should just be the components of the spatial metric, while the precise relationship between momentum fields p^{kl} and time derivatives of the metric is left open at this point as it will be determined by equations of motion generated by the Hamiltonian still to be derived. For now, we will only require that the momentum fields have the right structure for the formulation of a canonical theory: We have written them with superscript indices to make sure that no extra metric factor is required in a term

$$\int \mathrm{d}^3 x \sum_{i=1}^{3} \sum_{j=1}^{3} \dot{q}_{ij} p^{ij} \tag{2.228}$$

used in a Legendre transformation. Moreover, for the integration to be valid, p^{ij} should have density weight one. No factor of $\sqrt{\det q}$ is therefore necessary in Equation (2.228) because the density weight of p^{ij} makes sure that this field alone transforms like $\sqrt{\det q}\, \tilde{p}^{ij}$ where $\tilde{p}^{ij}$ has the same index structure as p^{ij} but no density weight.

In Equation (2.228) and in what follows, spatial integrations with unspecified limits will be over the averaging region of coordinate size $V_0 = \int \mathrm{d}^3 x$. Perturbative inhomogeneity is defined in the same averaging region, but we will keep up to second-order perturbations in our Hamiltonian, extending the averaged homogeneous dynamics.

We then formulate the Poisson bracket

$$\{q_{ij}(x), p^{kl}(y)\} = \frac{4\pi G}{c^2}\left(\delta_i^{\,k}\delta_j^{\,l} + \delta_j^{\,k}\delta_i^{\,l}\right)\delta(x - y). \tag{2.229}$$

The right-hand side is determined by analogy with Equation (2.226), adjusting the numerical factor by symmetrizing on the right because $q_{ij} = q_{ji}$ is always symmetric, as well as $p^{kl} = p^{lk}$ in order to ensure the same number of degrees of freedom in q_{ij} and p^{kl}. The missing factor of 1/3 compared with Equation (2.226) is explained by an implied average in Equation (2.226) over independent spatial directions: If we use isotropic and spatially constant $q_{ij} = q\delta_{ij}$ and $p^{kl} = p\delta^{kl}$, the left-hand side of Equation (2.229) is averaged to

$$\sum_{i=1}^{3} \sum_{j=1}^{3} \sum_{k=1}^{3} \sum_{l=1}^{3} \delta^{ij}\delta_{kl} \int \mathrm{d}^3 x \int \mathrm{d}^3 y \{q\delta_{ij}, p\delta^{kl}\} = 9 V_0^2 \{q, p\} \tag{2.230}$$

while the right-hand side reads

$$\frac{4\pi G}{c^2} \sum_{i=1}^{3} \sum_{j=1}^{3} \sum_{k=1}^{3} \sum_{l=1}^{3} \delta^{ij}\delta_{kl}\left(\delta_i^{\,k}\delta_j^{\,l} + \delta_j^{\,k}\delta_i^{\,l}\right) \int \mathrm{d}^3 x \int \mathrm{d}^3 y\, \delta(x - y) = \frac{24\pi G}{c^2} V_0. \tag{2.231}$$

These two equations are equivalent provided Equation (2.226) holds.

As used in this calculation, the fields q_{ij} and p^{kl} contain background contributions and not just inhomogeneity: They do not average to zero. We define pure inhomogeneity fields, δq_{ij} and δp^{kl}, by splitting the fields into background and perturbations,

$$q_{ij} = q\delta_{ij} + \delta q_{ij}, \quad p^{kl} = p\delta^{kl} + \delta p^{kl} \tag{2.232}$$

such that

$$\int d^3x\,\delta q_{ij} = 0 \quad \text{and} \quad \int dy\,\delta p^{kl} = 0. \tag{2.233}$$

These equations ensure that spatial averages of q_{ij} and p^{kl}, respectively, indeed produce q and p:

$$\frac{1}{V_0}\int d^3x\,q_{ij} = q\delta_{ij}, \quad \frac{1}{V_0}\int d^3y\,p^{kl} = p\delta^{kl}. \tag{2.234}$$

The pure perturbations then have the Poisson bracket

$$\{\delta q_{ij}(x),\, \delta p^{kl}(y)\} = \frac{4\pi G}{c^2}\left(\delta_i{}^k\delta_j{}^l + \delta_j{}^k\delta_i{}^l\right)\left(\delta(x-y) - \frac{1}{V_0}\right). \tag{2.235}$$

Subtracting $1/V_0$ on the right implements the averaging condition (2.233), but it will not be relevant for the derivations that follow.

Spatial Deformations

Next, we will determine the form of the generator $\mathrm{D}[\vec{\xi}]$ of spatial deformations. It should be a phase-space function depending on q, p, δq_{ij} and δp^{kl}, such that the Poisson brackets $\{\delta q_{ij}, \mathrm{D}[\vec{\xi}]\}$ and $\{\delta p^{kl}, \mathrm{D}[\vec{\xi}]\}$ are identical with metric and momentum perturbations implied by adding the components ξ^i to the original spatial coordinates x^i. We can easily derive the required transformation of δq_{ij} by inserting the coordinate change in the spatial line element $ds^2 = \sum_{i=1}^{3}\sum_{j=1}^{3}q_{ij}dx^idx^j$ which, by definition, is invariant under a change of coordinates.

For the leading transformation of δq_{ij}, it is sufficient to assume that $q_{ij} = q\delta_{ij}$ before the coordinate change. Thus,

$$ds^2 = \sum_{i=1}^{3}\sum_{j=1}^{3} q\delta_{ij}d(x^i + \xi^i)d(x^j + \xi^j)$$

$$= \sum_{i=1}^{3}\sum_{j=1}^{3} q\left(\delta_{ij} + \sum_{k=1}^{3}\delta_{kj}\frac{\partial\xi^k}{\partial x^i} + \sum_{k=1}^{3}\delta_{ik}\frac{\partial\xi^k}{\partial x^j}\right)dx^idx^j \tag{2.236}$$

$$= \sum_{i=1}^{3}\sum_{j=1}^{3}\left(q\delta_{ij} + \Delta_\xi\delta q_{ij}\right)dx^idx^j$$

where we Taylor expand ξ^i and ignore terms quadratic in the small change ξ^i. Comparing the last two lines, we read off the change

$$\Delta_\xi \delta q_{ij} = q \sum_{k=1}^{3} \left(\delta_{kj} \frac{\partial \xi^k}{\partial x^i} + \delta_{ik} \frac{\partial \xi^k}{\partial x^j} \right) \tag{2.237}$$

implied by a small spatial deformation. (This change of δq_{ij} is called the *Lie derivative* of δq_{ij}.) The same expression can be computed from the Poisson bracket

$$\Delta_\xi \delta q_{ij} = \{\delta q_{ij}, D[\vec{\xi}]\} \tag{2.238}$$

provided $D[\vec{\xi}]$ contains a term

$$D_1[\vec{\xi}] = -\frac{c^2}{4\pi G} \int d^3x q \sum_{i=1}^{3}\sum_{j=1}^{3}\sum_{k=1}^{3} \xi^i \delta_{ij} \frac{\partial \delta p^{jk}}{\partial x^k}. \tag{2.239}$$

The remaining terms of $D[\vec{\xi}]$ should be proportional to first-order derivatives of δq_{ij}, and generate a small change of δp^{kl}. Since we have not fully determined the meaning of p^{kl} yet, we cannot directly infer how it changes under a coordinate transformation. However, from the canonical structure we deduce that the change of δp^{kl} has to fulfill two conditions: The components δp^{kl} must be "dual" to those of δq_{ij}, in the sense that the combination (2.228) used in a Legendre transformation must be invariant. Applying the product rule to $\delta \dot{q}_{ij} \delta p^{ij}$, therefore, for each term in $\Delta_\xi \delta q_{ij}$ there must be a corresponding term in $\Delta_\xi \delta p^{ij}$ with the opposite sign. Second, δp^{ij} has density weight one: It implicitly provides a factor of $\sqrt{\det q}$ required for an integration in the Legendre transformation.

The square root of the determinant of the metric changes by

$$\Delta_\xi \sqrt{\det q} = \frac{1}{2\sqrt{\det q}} \Delta_\xi \det q$$

$$= \frac{1}{12q^{3/2}} \Delta_\xi \left(\sum_{i=1}^{3}\sum_{j=1}^{3}\sum_{k=1}^{3}\sum_{m=1}^{3}\sum_{n=1}^{3}\sum_{o=1}^{3} \epsilon^{ijk} \epsilon^{mno} q_{im} q_{jn} q_{ko} \right) \tag{2.240}$$

$$= \frac{1}{2}\sqrt{q} \sum_{i=1}^{3}\sum_{j=1}^{3} \delta^{ij} \Delta_\xi \delta q_{ij},$$

where we used the completely antisymmetric tensor ϵ^{ijk} in order to express the determinant of q_{ij} in terms of its components. Using Equation (2.237),

$$\frac{\Delta_\xi \sqrt{\det q}}{q^{3/2}} = \sum_{i=1}^{3} \frac{\partial \xi^i}{\partial x^i}. \tag{2.241}$$

Combining the two conditions required for a coordinate transformation of δp^{ij}, we need

$$\Delta_\xi \delta p^{ij} = -p \sum_{k=1}^{3} \left(\delta_{kj} \frac{\partial \xi^k}{\partial x^i} + \delta_{ik} \frac{\partial \xi^k}{\partial x^j} \right) + p \sum_{i=1}^{3} \frac{\partial \xi^i}{\partial x^i}. \tag{2.242}$$

The first term is determined by the condition that δp^{ij} is dual to δq_{ij} in terms of its index structure and should therefore transform with a term opposite in sign to Equation (2.237). The second term implements the density weight of δp^{ij}, adding p times Equation (2.241) to the transformation. The same terms are generated by $\mathrm{D}[\vec{\xi}]$ provided we extend Equation (2.239) to

$$\mathrm{D}^{(1)}[\vec{\xi}] = -\frac{c^2}{4\pi G} \int \mathrm{d}^3x \sum_{i=1}^{3} \sum_{j=1}^{3} \xi^i \left(\delta_{ij} q \sum_{k=1}^{3} \frac{\partial \delta p^{jk}}{\partial x^k} + p \frac{\partial \delta q_{ij}}{\partial x^j} - \frac{1}{2} p \frac{\partial \delta q_{ij}}{\partial x^i} \right). \tag{2.243}$$

(The full expression $\mathrm{D}[\xi]$ also contains terms quadratic in the perturbations, but they will not be required for our derivation. We will need only the first-order generator $\mathrm{D}^{(1)}[\xi]$ given in Equation (2.243).)

Generator of Normal Deformations

Normal deformations are sensitive to the dynamics of spacetime. Nevertheless, in a generally covariant theory of the metric, their form is completely determined by hypersurface-deformation commutators if we are looking for classical dynamics governed by second-order field equations and assume the background Hamiltonian given in Chapter 1. The normal generator $\mathrm{H}[N]$ can then depend only on spatial derivatives of δq_{ij} of up to second order, together with terms quadratic in p and δp_{ij} because classical momenta contain a first-order time derivative.

First Order. Given the averaging conditions (2.233), any first-order perturbation of the Hamiltonian vanishes. However, in addition to δq_{ij} and δp^{ij}, we have the lapse function N in the generator $\mathrm{H}[N]$, which we should also expand for a perturbative analysis: $N = \bar{N} + \delta N$. We can then obtain a second-order contribution to $\mathrm{H}[N]$ which is of first order in δq_{ij} and δp^{ij} and does not average to zero thanks to a factor of δN from the lapse function. The most general such contribution to $\mathrm{H}[N]$ allowed in a classical metric theory with quadratic momentum dependence and up to second-order derivatives is given by

$$\mathrm{H}^{(1)}[\delta N] = \frac{c^2}{8\pi G} \int \mathrm{d}^3y \, \delta N(y) \sum_{i=1}^{3} \sum_{j=1}^{3} \left(A_1 p \sqrt{q} \, \delta_{ij} \delta p^{ij} + B_1 \frac{p^2}{\sqrt{q}} \delta^{ij} \delta q_{ij} \right.$$
$$\left. + C_1 \frac{1}{\sqrt{q}} \frac{\partial^2 \delta q_{ij}}{\partial y^i \partial y^j} + D_1 \frac{1}{\sqrt{q}} \frac{\partial^2 \delta q_{ii}}{\partial y^j \partial y^j} \right). \tag{2.244}$$

The coefficients A_1, B_1, C_1, and D_1 are constants and cannot depend on q or p for the following reasons: The p-dependence has explicitly been spelled out, using the condition that all momentum terms must appear raised to even powers in a theory that is time-reversal symmetric, and can be at most quadratic in a standard theory of classical type. The q-dependence, as given in the four terms, is determined by the behavior with respect to isotropic rescalings of the coordinates, $y^i \rightarrow \lambda y^i$, which according to the invariance of the spatial line element must transform $q \to \lambda^{-2}q$ and $\delta q_{ij} \to \lambda^{-2}\delta q_{ij}$. The same rescaling must transform $p \to \lambda^{-1}p$ and $\delta p^{ij} \to \lambda^{-1}\delta p^{ij}$ to ensure invariant Poisson brackets, or invariant terms (2.228) for perturbations and $V_0\dot{q}p$ for the background in a Legendre transformation.

All but one of the four coefficients can be uniquely determined by imposing two conditions: (1) The contribution $H^{(1)}[\delta N]$ should apply in particular to isotropic and homogeneous perturbations $\delta q_{ij} = \delta_{ij}\delta q$, $\delta p^{kl} = \delta^{kl}\delta p$ where δq, δp as well as δN are spatially constant. The terms containing C_1 and D_1 then do not contribute, while the terms containing A_1 and B_1 must agree with a first-order expansion of the background generator $(\bar{N} + \delta N)\bar{H}(q + \delta q, p + \delta p)$ using Equation (2.227). A quick calculation implies that

$$A_1 = -2 \quad \text{and} \quad B_1 = -\frac{1}{2}. \tag{2.245}$$

(2) Spatial covariance determines the relationship between C_1 and D_1. The metric terms in $H^{(1)}[\delta N]$ have a Poisson bracket with $D^{(1)}[\vec{\xi}]$ given by

$$\left\{ \int dy \delta N(y) \sum_{i=1}^{3} \sum_{j=1}^{3} \left(C_1 \frac{\partial^2 \delta q_{ij}}{\partial y^i \partial y^j} + D_1 \frac{\partial^2 \delta q_{ii}}{\partial y^j \partial y^j} \right), D^{(1)}[\vec{\xi}] \right\}$$

$$= -2q \int d^3x \sum_{i=1}^{3} \sum_{j=1}^{3} \xi^j (C_1 + D_1) \frac{\partial^3 \delta N}{(\partial x^i)^2 \partial x^j} \tag{2.246}$$

after integrations by parts. Since this result is not of the form seen in the right-hand side of Equation (2.158), which requires an integration of a phase-space function times $\sum_{j=1}^{3} \xi^j \partial \delta N / \partial x^j$, it must vanish, such that $D_1 = -C_1$. With this condition, the derivative terms in Equation (2.244) can easily be recognized as being proportional to the first-order perturbation of

$$\sqrt{\det q} \sum_{i=1}^{3} \sum_{j=1}^{3} q^{ij} R_{ij} \tag{2.247}$$

where R_{ij} is the Ricci tensor (2.46) evaluated for the spatial metric q_{ij}. This expression is indeed invariant under spatial coordinate changes as we saw in Section 2.2.1, but the canonical derivation does not require knowledge of curvature invariants.

The remaining coefficient, C_1, cannot be independently determined because one parameter can always be changed by applying a canonical transformation to q_{ij} and p^{ij} with a constant factor, together with a suitable rescaling of N to keep $\bar{H}$ unchanged. We therefore fix C_1 by appealing to the usual convention of general relativity, according to which the spatial Ricci scalar R contributes a term $-c^2(16\pi G)^{-1}\sqrt{\det q}\,R$ to the Hamiltonian constraint, derived from the action (2.48). Thus,

$$C_1 = -\frac{1}{2} \tag{2.248}$$

and therefore

$$D_1 = \frac{1}{2}. \tag{2.249}$$

Using $\bar{H}$ and $H^{(1)}[\delta N]$, the full normal generator up to first order in δq_{ij} and δp^{ij} is given by

$$\delta H[\bar{N} + \delta N] = \bar{H}\bar{N} + H^{(1)}[\delta N]. \tag{2.250}$$

The Poisson bracket $\{\bar{H}\bar{N}, H^{(1)}[\delta N]\}$ therefore provides a contribution to the first-order term of the Poisson bracket (2.154). It can easily be evaluated:

$$\{\bar{H}\bar{N}, H^{(1)}[\delta N]\} = \frac{c^2}{8\pi G} \int d^3x \bar{N}\delta N$$
$$\times \sum_{i=1}^{3} \left(\frac{1}{2} A_1 p^2 \delta p^{ii} - 2B_1 \frac{p^3}{q} \delta q_{ii} - \frac{p}{q} \sum_{j=1}^{3} \left(C_1 \frac{\partial^2 \delta q_{ij}}{\partial x^i \partial x^j} + D_1 \frac{\partial^2 \delta q_{ii}}{\partial x^j \partial x^j} \right) \right). \tag{2.251}$$

This result is in general non-zero, and it is *not* of the form in which D or $D^{(1)}$ should appear according to Equation (2.154). In order to produce Equation (2.154), we should add to $\bar{H}\bar{N}$ the correct contribution to $H[N]$ of second order in δq_{ij} and δp^{ij}. Second-order contributions are relevant in this case because the Poisson bracket of a second-order term with a first-order term from $H^{(1)}[\delta N]$ contains first-order terms which are of the same order as Equation (2.251).

Second Order. The general form of a contribution $H^{(2)}[\bar{N}]$ to $H[N]$ of second order in δq_{ij} and δp^{ij} can be deduced up to constant coefficients using the same principles applied to the first-order contribution: time-reversal symmetry and spatial scaling invariance. However, the larger number of indices in combinations of two perturbation fields implies a larger number of possible terms in the second-order generator:

$$H^{(2)}[\bar{N}] = \frac{c^2 \bar{N}}{8\pi G} \int d^3x \sum_{i=1}^{3} \sum_{j=1}^{3} \left(\sqrt{q} \sum_{k=1}^{3} \sum_{l=1}^{3} \delta p^{ij} \delta p^{kl} \left(A_2 \delta_{ik}\delta_{jl} + B_2 \delta_{ij}\delta_{kl} \right) \right.$$
$$+ \frac{p}{\sqrt{q}} \left(C_2 \delta p^{ij} \delta q_{ij} + D_2 \sum_{k=1}^{3} \sum_{l=1}^{3} \delta p^{ij} \delta q_{kl} \delta_{ij} \delta^{kl} \right)$$
$$+ \frac{p^2}{q^{3/2}} \sum_{k=1}^{3} \sum_{l=1}^{3} \delta q_{ij} \delta q_{kl} (E_2 \delta^{ik}\delta^{jl} + F_2 \delta^{ij}\delta^{kl}) \tag{2.252}$$
$$+ \frac{1}{q^{3/2}} \sum_{k=1}^{3} \left(G_2 \frac{\partial \delta q_{ij}}{\partial x^i} \frac{\partial \delta q_{kj}}{\partial x^k} + H_2 \frac{\partial \delta q_{ii}}{\partial x^k} \frac{\partial \delta q_{jj}}{\partial x^k} \right.$$
$$\left. \left. + I_2 \frac{\partial \delta q_{ij}}{\partial x^k} \frac{\partial \delta q_{ij}}{\partial x^k} + J_2 \frac{\partial \delta q_{ij}}{\partial x^i} \frac{\partial \delta q_{kk}}{\partial x^j} \right) \right).$$

There is no need to include second-order derivatives of δq_{ij} because, at the given order, they can always be reduced to terms in the last line by integrating by parts.

A longer calculation with, as usual, several integrations by parts implies the Poisson bracket

$$\{H^{(2)}[\bar{N}], H^{(1)}[\delta N]\} = \frac{c^2 \bar{N}}{8\pi G} \int d^3x \sum_{i=1}^{3} \sum_{j=1}^{3} (p^2 \delta N \delta_{ij} \delta p^{ij} (A_1 C_2$$
$$+ 3A_1 D_2 - 2B_1 A_2 - 6B_1 B_2)$$
$$+ \frac{p^3}{q} \delta N \delta^{ij} \delta q_{ij} (2A_1 E_2 + 6A_1 F_2 - B_1 C_2 - 3B_1 D_2)$$
$$- \frac{p}{q} \frac{\partial \delta N}{\partial x^i} \frac{\partial \delta q_{ij}}{\partial x^j} (2A_1 G_2 + 3A_1 J_2 + C_1 C_2)$$
$$+ \frac{p}{q} \frac{\partial \delta N}{\partial x^j} \frac{\partial \delta q_{ii}}{\partial x^j} (6A_1 H_2 + 2A_1 I_2$$
$$+ A_1 J_2 + C_1 D_2 + D_1 C_2 + 3D_1 D_2)$$
$$+ \frac{\partial \delta N}{\partial x^i} \frac{\partial \delta p^{ij}}{\partial x^j} \cdot 2C_1 A_2$$
$$+ \frac{\partial \delta N}{\partial x^j} \frac{\partial \delta p^{ii}}{\partial x^j} (2C_1 B_2 + 2D_1 A_2 + 6D_1 B_2) \Bigg)$$

$$(2.253)$$

which can indeed help to produce terms of the correct form for Equations (2.154) and (2.155) to hold.

In the perturbative context, we need

$$\{H[\bar{N}], H^{(1)}[\delta N]\} = D[q^{-1} \bar{N} \nabla \delta N] \tag{2.254}$$

where $H[\bar{N}] = \bar{H}\bar{N} + H^{(2)}[\bar{N}]$. Combining the preceding $\{H^{(2)}[\bar{N}], H^{(1)}[\delta N]\}$ with Equation (2.251) and using $D^{(1)}[\vec{\xi}]$ in Equation (2.243), the perturbative commutator (2.254) implies the conditions

$$C_1 A_2 = -1 \tag{2.255}$$

$$2C_1 B_2 + 2D_1 A_2 + 6D_1 B_2 = 0 \tag{2.256}$$

$$A_1 \left(\frac{1}{2} + C_2 \right) + 3A_1 D_2 - 2B_1 A_2 - 6B_1 B_2 = 0 \tag{2.257}$$

$$-2B_1 + 2A_1 E_2 + 6A_1 F_2 - B_1 C_2 - 3B_1 D_2 = 0 \tag{2.258}$$

$$C_1 + 2A_1 G_2 + 3A_1 J_2 + C_1 C_2 = -2 \tag{2.259}$$

$$D_1 + 6A_1 H_2 + 2A_1 I_2 + A_1 J_2 + C_1 D_2 + D_1 C_2 + 3D_1 D_2 = 1. \tag{2.260}$$

Since we already know $B_1 = -1/2 = C_1$ and $D_1 = 1/2$, the first two equations can easily be solved for

$$A_2 = 2 \quad \text{and} \quad B_2 = -1. \tag{2.261}$$

There are four remaining equations for eight undetermined coefficients. We therefore need additional relationships, which we can implement as in the first-order generator. In particular, we should make sure that the second-order expansion of the background generator $(\bar{N} + \delta N)\bar{H}(q + \delta q, p + \delta p)$ agrees with the second-order generator evaluated for isotropic and homogeneous perturbations, $\delta q_{ij} = \delta_{ij}\delta q$ and $\delta p^{ij} = \delta^{ij}\delta p$ with constant δq, δp and δN. Second, we should ensure spatial covariance by computing the Poisson bracket of $H^{(2)}[\bar{N}]$ with $D^{(1)}[\vec{\xi}]$ and making sure that it is consistent with Equation (2.158).

The first condition implies the relationships

$$A_2 + 3B_2 = -1 \tag{2.262}$$

$$C_2 + 3D_2 = -1 \tag{2.263}$$

$$E_2 + 3F_2 = \frac{1}{8}, \tag{2.264}$$

which are, however, not independent of what we already found. The first equation is consistent with Equation (2.261) but does not determine any further coefficients. Moreover, using Equations (2.245), (2.248), and (2.249), Equation (2.257) is equivalent to Equation (2.263) and Equation (2.258) to Equation (2.264). Although the background expansion does not provide independent equations at second order, it shows that the existence of covariant generators is non-trivial because the same equations can be obtained in different ways, implying additional consistency conditions which are fulfilled in our derivation based on the bracket (2.254).

Implementing spatial covariance turns out to be more productive. The required bracket (2.158), evaluated with our parameterized second-order $H^{(2)}$, leads to several terms on the left which are not identically zero but should not appear on the right. One such term contains first-order derivatives of the spatial metric, which can be produced from the Poisson bracket in three different ways: (i) using the background generator $\bar{H}$ and the second or third term in Equation (2.243), (ii) using the terms with coefficients C_2 or D_2 in Equation (2.252) and the second or third term in Equation (2.243), and (iii) using the terms with coefficients E_2 or F_2 in Equation (2.252) and the first term in Equation (2.243). After several integrations by parts, these terms add up to

$$\{\bar{H}\bar{N} + H^{(2)}[\bar{N}], D^{(1)}[\vec{\xi}]\} = -\frac{c^2}{4\pi G}\bar{N}p \int d^3x \sum_{i=1}^{3} \sum_{j=1}^{3} \xi^i \left(\left(2E_2 - C_2 - \frac{1}{2}\right)\frac{\partial \delta q_{ij}}{\partial x^j} \right.$$
$$\left. + \left(2F_2 + \frac{1}{2}D_2 + \frac{1}{2}C_2 + \frac{1}{4}\right)\frac{\partial \delta q_{jj}}{\partial x^i} \right). \tag{2.265}$$

For the correct commutator as required by Equation (2.158), the right-hand side must vanish because $\sum_{j=1}^{3} \xi^j \partial \bar{N} / \partial x^j = 0$. The two resulting conditions,

$$2E_2 - C_2 = \frac{1}{2} \tag{2.266}$$

$$2F_2 + \frac{1}{2}D_2 + \frac{1}{2}C_2 = -\frac{1}{4}, \tag{2.267}$$

together with Equations (2.257) and (2.258) (or, equivalently, Equations (2.263) and (2.264)), imply

$$C_2 = 2, \quad D_2 = -1, \quad E_2 = \frac{5}{4}, \quad F_2 = -\frac{3}{8}. \tag{2.268}$$

The remaining four coefficients—G_2, H_2, I_2, and J_2—are only partially restricted by the two Equations (2.259) and (2.260) not yet evaluated. Since these coefficients determine terms in Equation (2.252) with spatial derivatives of the metric, they should add up to a spatial curvature invariant, just as we saw for the terms with coefficients C_1 and D_1 in $H^{(1)}[\delta N]$. At second order, it is however very tedious to expand $\sqrt{\det q} \sum_{i=1}^{3}\sum_{j=1}^{3} q^{ij} R_{ij}$ with the spatial Ricci tensor R_{ij}. The canonical procedure, computing the Poisson bracket of these terms with $D^{(1)}$ in Equation (2.243), is also tedious, but much faster and less prone to calculational mistakes. Again after several integrations by parts, we obtain

$$\left\{ \int d^3x \sum_{i=1}^{3}\sum_{j=1}^{3}\sum_{k=1}^{3} \left(G_2 \frac{\partial \delta q_{ij}}{\partial x^i}\frac{\partial \delta q_{kj}}{\partial x^k} + H_2 \frac{\partial \delta q_{ii}}{\partial x^k}\frac{\partial \delta q_{jj}}{\partial x^k} + I_2 \frac{\partial \delta q_{ij}}{\partial x^k}\frac{\partial \delta q_{ij}}{\partial x^k} + J_2 \frac{\partial \delta q_{ij}}{\partial x^i}\frac{\partial \delta q_{kk}}{\partial x^j} \right), D^{(1)} \right\}$$

$$= 2q \int d^3x \sum_{i=1}^{2}\sum_{j=1}^{3}\sum_{k=1}^{3} \xi^i \left((G_2 + 2I_2) \frac{\partial^3 \delta q_{ij}}{\partial x^j (\partial x^k)^2} + (G_2 + J_2) \frac{\partial^3 \delta q_{jk}}{\partial x^i \partial x^j \partial x^k} \right. \tag{2.269}$$

$$\left. + (2H_2 + J_2) \frac{\partial^3 \delta q_{kk}}{\partial x^i (\partial x^j)^2} \right).$$

There are three independent terms, which should all vanish according to Equation (2.158). Therefore, only one of the four coefficients, such as G_2, remains independent:

$$H_2 = \frac{1}{2}G_2, \quad I_2 = -\frac{1}{2}G_2, \quad J_2 = -G_2. \tag{2.270}$$

Together with Equation (2.259) or (2.260), we obtain the precise values of all remaining coefficients:

$$G_2 = -\frac{1}{4}, \quad H_2 = -\frac{1}{8}, \quad I_2 = \frac{1}{8}, \quad J_2 = \frac{1}{4}. \tag{2.271}$$

Summary. With only one free choice given by the value of C_1, the background Hamiltonian together with the symmetry conditions of spatial covariance and the correct hypersurface-deformation behavior (2.154) together with Equation (2.158) have completely determined the perturbed Hamiltonian up to second order. The resulting expression is given by

$$H[\bar{N} + \delta N] = \bar{H}\bar{N} + H^{(1)}[\delta N] + H^{(2)}[\bar{N}] \qquad (2.272)$$

with

$$\bar{H} = -\frac{3c^2 V_0}{8\pi G}\sqrt{q}\, p^2 \qquad (2.273)$$

for the background Hamiltonian,

$$H^{(1)}[\delta N] = -\frac{c^2}{16\pi G}\int d^3 y\, \delta N(y) \sum_{i=1}^{3}\sum_{j=1}^{3}\left(4p\sqrt{q}\,\delta_{ij}\delta p^{ij} + \frac{p^2}{\sqrt{q}}\delta^{ij}\delta q_{ij}\right.$$
$$\left. + \frac{1}{\sqrt{q}}\frac{\partial^2 \delta q_{ij}}{\partial y^i \partial y^j} - \frac{1}{\sqrt{q}}\frac{\partial^2 \delta q_{ii}}{\partial y^j \partial y^j}\right) \qquad (2.274)$$

at first order in phase-space variables, and

$$H^{(2)}[\bar{N}] = \frac{c^2 \bar{N}}{8\pi G}\int d^3 x \sum_{i=1}^{3}\sum_{j=1}^{3}\left(\sqrt{q}\sum_{k=1}^{3}\sum_{l=1}^{3}\delta p^{ij}\delta p^{kl}\left(2\delta_{ik}\delta_{jl} - \delta_{ij}\delta_{kl}\right)\right.$$
$$+ \frac{p}{\sqrt{q}}\left(2\delta p^{ij}\delta q_{ij} - \sum_{k=1}^{3}\sum_{l=1}^{3}\delta p^{ij}\delta q_{kl}\delta_{ij}\delta^{kl}\right)$$
$$+ \frac{1}{8}\frac{p^2}{q^{3/2}}\sum_{k=1}^{3}\sum_{l=1}^{3}\delta q_{ij}\delta q_{kl}(10\delta^{ik}\delta^{jl} - 3\delta^{ij}\delta^{kl})$$
$$\left. - \frac{1}{8}\frac{1}{q^{3/2}}\sum_{k=1}^{3}\left(2\frac{\partial \delta q_{ij}}{\partial x^i}\frac{\partial \delta q_{kj}}{\partial x^k} + \frac{\partial \delta q_{ii}}{\partial x^k}\frac{\partial \delta q_{jj}}{\partial x^k} - \frac{\partial \delta q_{ij}}{\partial x^k}\frac{\partial \delta q_{ij}}{\partial x^k} - 2\frac{\partial \delta q_{ij}}{\partial x^i}\frac{\partial \delta q_{kk}}{\partial x^j}\right)\right) \qquad (2.275)$$

at second order.

These expressions can be used in multiple ways. First, they generate Hamilton's equations of motion

$$\frac{d\delta q_{ij}}{dt} = \frac{\delta(H[\bar{N} + \delta N] + D^{(1)}[\vec{\xi}])}{\delta p^{ij}},$$
$$\frac{d\delta p^{ij}}{dt} = -\frac{\delta(H[\bar{N} + \delta N] + D^{(1)}[\vec{\xi}])}{\delta q_{ij}} \qquad (2.276)$$

where t is the time variable determined by a choice of the lapse function $\bar{N} + \delta N$ together with a spatial shift of $\vec{\xi}$ per time. Because $H[\bar{N} + \delta N] + D^{(1)}[\vec{\xi}]$ is quadratic in δp^{ij}, the first equation yields a linear relationship between δp^{ij} and $\mathrm{d}\delta q_{ij}/\mathrm{d}t$, which determines the geometrical meaning of the momentum fields δp^{ij}. The second equation then implies second-order evolution equations for δq_{ij} in time.

For this dynamics, we use hypersurface deformations generated by $\mathrm{H}[\bar{N} + \delta N] + \mathrm{D}^{(1)}[\vec{\xi}]$ as time evolution through spacetime. The same generator can be used to compute implications of a coordinate transformation on the canonical fields. To linear order in perturbations, Bardeen potentials can then be determined as combinations of the fields that are independent under inhomogeneous coordinate changes. Here, the distinction between homogeneous and inhomogeneous coordinate changes, observed in Section 1.4.2, is more clear than in our previous derivation because they correspond to two independent generators, $\mathrm{H}^{(2)}$ in the former case and $\mathrm{H}^{(1)}$ in the latter. The canonical equations of motion for δq_{ij}, as well as the constraint equations which require that $\mathrm{H}^{(1)}[\delta N]$ and $\mathrm{D}^{(1)}[\partial w/\partial x^i]$ (plus suitable matter terms) vanish, then imply the differential equations quoted in Section 1.4.3 for scalar modes.

2.3.6 Modified Canonical Gravity

In the spacetime setting, the form of curvature invariants has led the way to possible modifications of general relativity such as $f(R)$-gravity or different versions of scalar-tensor theories. In canonical gravity, the transformation behavior of spacetime tensors is replaced by transformations obeying the hypersurface-deformation commutators (2.154), (2.158), and (2.156). As we have seen in several detailed examples, these commutators uniquely determine the dynamics of covariant classical theories with second-order field equations, that is, general relativity.

Higher-curvature actions allow modified versions of general relativity by dropping the condition of having second-order field equations. (As shown explicitly in Deruelle et al. 2010, all higher-curvature actions of the standard form obey the classical hypersurface-deformation brackets.) However, the higher-derivative nature often leads to instabilities.

As an alternative, one may work with second-order theories and modify the hypersurface-deformation commutators. Such an approach to modified gravity does not introduce new degrees of freedom unless one explicitly extends the phase space of perturbations δq_{ij} and δp^{ij}. Instead of modifying the dynamics by rearranging degrees of freedom, one rather implements a new spacetime structure by modifying the relationships between generators of hypersurface deformations compared with general relativity, or any other generally covariant theory of classical type. However, a priori the existence of non-trivial but consistent canonical theories for a given form of modified hypersurface-deformation relations is an open question. As we have seen in our canonical derivations of spherically symmetric and perturbative Hamiltonians, the set of all equations implied by hypersurface-deformation relations is over-determined and implies consistency conditions for the available parameters.

A general class of modified brackets is given by the example of $f(\mathrm{H})$ theories (Mukohyama & Noui 2019). Given a generally covariant theory with Hamiltonian H, a consistent $f(\mathrm{H})$-theory is defined by using the Hamiltonian

$\tilde{H} = \sqrt{\det q}\, f(H/\sqrt{\det q})$ instead of H (where H is defined as the integrand in $H[N] = \int d^3x N H$). The square root of $\det q$ with the spatial metric q is used in such a way that the Hamiltonian retains a density weight of one, which implies that the bracket (2.158) remains intact. It follows directly from the chain rule that the Poisson bracket (2.154) is replaced by

$$\{\tilde{H}[N_1], \tilde{H}[N_2]\} = D[(d\tilde{H}/dH)^2 \vec{w}] \qquad (2.277)$$

with the same $\vec{w}$ as in Equation (2.155), where $d\tilde{H}/dH = df(H/\sqrt{\det q})/d(H/\sqrt{\det q})$ does not have a density weight. Given the dependence of $\vec{w}$ on N_1 and N_2, it is possible to transform (2.277) back to the original (2.154) by redefining the lapse functions as $\tilde{N}_1 = (d\tilde{H}/dH)N_1$ and $\tilde{N}_2 = (d\tilde{H}/dH)N_2$, such that

$$\{H[\tilde{N}_1], H[\tilde{N}_2]\} = D[\vec{w}]. \qquad (2.278)$$

However, the bracket $\{H[\tilde{N}], D[\vec{\xi}]\}$ then depends on $d\tilde{H}/dH$ and differs from Equation (2.158), and therefore the theory remains modified even after a redefinition of the lapse function.

Although $f(H)$ theories at first sight look similar to $f(R)$-gravity, they are conceptually quite different because they modify the spacetime structure: While all higher-curvature actions are subject to the same hypersurface-deformation brackets as in general relativity, some of the brackets are modified in $f(H)$ theories. Such theories therefore differ from general relativity not only in their dynamics but also in the structure of spacetime they imply. For instance, the perturbation δq_{ij} that appears in the Hamiltonian of an $f(H)$-theory, inherited from δq_{ij} in the original theory with Hamiltonian H, cannot directly be used as a spatial metric in a line element: Modified hypersurface-deformation relations imply that canonically generated transformations $\Delta \delta q_{ij} = \{\delta q_{ij}, \tilde{H}[N] + D[\vec{\xi}]\}$ are no longer related to coordinate changes such that $\sum_{\alpha=0}^{3}\sum_{\beta=0}^{3} g_{\alpha\beta} dx^\alpha dx^\beta$ with q_{ij} as the spatial part of $g_{\alpha\beta}$ is invariant. Riemannian geometry is then inapplicable in general, and physical equations of motion can only be derived canonically, as indicated at the end of the preceding section.

Another difference between $f(H)$ and $f(R)$-theories is that the function f of $f(H)$ theories with matter must be applied to the full Hamiltonian, obtained by adding the matter Hamiltonian to the gravitational part. If only the gravitational Hamiltonian were used in f, the bracket $\{\tilde{H}[N_1], \tilde{H}[N_2]\}$ would not be a valid generator of hypersurface deformations because the modified right-hand side of Equation (2.277) could not properly add up with an unmodified matter generator to form a full generator. In an $f(H)$-theory, or any other theory with modified hypersurface deformations, it is therefore impossible to modify only gravity and couple it to standard matter.

The bracket (2.277) is a special case of a bracket

$$\{H[N_1], H[N_2]\} = D[\beta \vec{w}] \qquad (2.279)$$

where $\vec{w}$ is again related to N_1 and N_2 as in Equation (2.155), but with a phase-space function β on the right-hand side, depending on the canonical fields q_{ij} or p^{ij} of gravity. Even for linear functions N_1 and N_2, the bracket then differs from the classical possibilities (2.152) and (2.153), indicating a new version of non-classical geometry. Unlike in Equation (2.278), the function β may in general become negative in some regions of phase-space, for instance at large curvature where significant deviations from general relativity could occur. In this case, it is no longer possible to redefine the lapse functions and obtain the form (2.278): Because the lapse functions must be real, a transformation $\tilde{N} = \sqrt{\beta}N$ is not allowed. Geometrically, regions in which β has the opposite sign correspond to signature change from Lorentzian spacetime to four-dimensional Euclidean space: As our derivation of the hypersurface-deformation commutator (2.154) in Section 2.3.3 showed, the classical signature determines the sign of the right-hand side of Equation (2.154). In a modified spacetime theory in which Equation (2.279) is consistently realized, continuous signature change is possible by β moving through zero. Gravitational and scalar-tensor theories of this form have been analyzed in Reyes (2009), Cailleteau et al. (2014), Bojowald et al. (2014), Cuttell & Sakellariadou (2014), Cuttell (2019).

The possibility of continuous signature change is of interest in particular because it shows that canonical formulations allow for modified spacetime structures in a way that cannot be modeled by higher-curvature actions. Such theories therefore provide explicit examples for a possible inequivalence between Lagrangian and Hamiltonian formulations in the spacetime setting, discussed at the beginning of this chapter. Modified canonical gravity indeed extends the possible modifications included by higher-curvature actions. All those latter actions, as shown in Deruelle et al. (2010), have the same classical brackets with $\beta = 1$. If $\beta \neq 1$ in a canonical formulation, additional models of modified gravity can be analyzed, vindicating Dirac's independent conviction (Dirac 1958b) that "It would be permissible to look upon the Hamiltonian form as the fundamental one, and there would then be no fundamental four-dimensional symmetry in the theory."

As a simple model that illustrates how modified generators can lead to new spacetime structures, we take a scalar field $\phi(x)$ in one spatial dimension, with momentum $p(x)$, and define the generators

$$H[N] = \int dx N \left(f(p) - \frac{1}{4}\left(\frac{d\phi}{dx}\right)^2 - \frac{1}{2}\phi\frac{d^2\phi}{dx^2} \right), \quad D[w] = -\int dx w \phi \frac{dp}{dx}. \quad (2.280)$$

The dependence of $H[N]$ on ϕ and its spatial derivatives is modeled on curvature expressions that appear in gravitational theories, depending on up to second-order derivatives of the metric. The numerical coefficients of 1/4 and 1/2 ensure that the bracket is closed, as shown below. The function $f(p)$ is free and can encode modifications. In a classical version of the model, $f(p)$ would be quadratic for the standard form of kinetic contributions to the energy.

The dependence of $D[w]$ on the fields is strictly determined by the spatial coordinate changes that this function should generate, given by

$$\delta_w \phi = \{\phi, \mathrm{D}[w]\} = \frac{\mathrm{d}(w\phi)}{\mathrm{d}x}, \quad \delta_w p = \{p, \mathrm{D}[w]\} = w\frac{\mathrm{d}p}{\mathrm{d}x}. \quad (2.281)$$

As we have already seen in the example of spherical symmetry in Section 2.3.4; the two transformations are slightly different because one of the canonical fields, here ϕ, has a density weight. Since these transformations are linear in canonical fields, the generator must be quadratic and of the form shown above. As in Section 2.3.4, D[w] can be derived from partial differential equations implied by the Poisson brackets in Equation (2.281).

A longer calculation determines the bracket

$$\{\mathrm{H}[N], \mathrm{H}[M]\} = \mathrm{D}\left[\frac{1}{2}\frac{\mathrm{d}^2 f}{\mathrm{d}p^2}\left(N\frac{\mathrm{d}M}{\mathrm{d}x} - \frac{\mathrm{d}N}{\mathrm{d}x}M\right)\right] \quad (2.282)$$

of two normal deformations. In its dependence on N, M and their spatial derivatives, we recognize the general structure of Equation (2.154). However, unless $\frac{1}{2}\mathrm{d}^2 f/\mathrm{d}p^2 = 1$, the spatial displacement differs from the classical case obtained if $f(p) = p^2$. As a non-quadratic example, the function $f(p) = p_0^2 \sin^2(p/p_0)$ with some parameter p_0 implies a factor of $\cos(2p/p_0)$ on the right-hand side of the bracket. This factor is not only non-constant, it can also change sign: We have $\cos(2p/p_0) \approx 1$ for small $p/p_0 \ll 1$, while $\cos(2p/p_0) = -1$ for $p = \frac{1}{2}\pi p_0$. In this simple model of spacetime in a canonical formulation, there is therefore a possibility of signature change.

This model is not complete because the bracket $\{\mathrm{H}[N], \mathrm{D}[w]\}$ is not of the form (2.158) and is not even a consistent modification of this classical bracket: The function determined by this bracket does not vanish when both H[N] and D[w] are set equal to zero for all N and w. One can explain this failure by the density weight of H[N] defined in Equation (2.280), which is not the same for all three terms and therefore does not lead to a consistent spatial transformation of H[N]. Therefore, the model does not provide a consistent set of constraints on the phase-space functions. In Chapter 6 we will see how the model can be extended to a fully consistent modified canonical theory with signature change. This version introduces an additional field, such that the model takes the form of a consistent modification of spherically symmetric canonical gravity, where we have two fields, $L(x)$ and $S(x)$, instead of a single $\phi(x)$. Since one of the spherically symmetric fields, $L(x)$, has a density weight, it can be used to bring the density weight of $(\mathrm{d}S/\mathrm{d}x)^2$ or $\mathrm{d}^2 S/\mathrm{d}x^2$ to the correct value, one, such that the new H[N] has the required Poisson bracket with D[w]. We will derive such a theory in Chapter 6.

References

Abbott, B. P., Abbott, R., Abbott, T. D., et al. 2017a, ApJ, 848, L12

Abbott, B. P., Abbott, R., Abbott, T. D., et al. 2017b, ApJ, 848, L13

Ade, P. A. R., & Planck Collaboration, 2014, A&A, 571, A22

Anderson, J. L., & Bergmann, P. G. 1951, PhRv, 83, 1018

Arnowitt, R., Deser, S., & Misner, C. W. 1962, in Gravitation: An Introduction to Current Research, ed. L. Witten (New York: Wiley) Reprinted in Belinskii, V. A., Khalatnikov, I. M., and Lifschitz, E. M. 1982, Adv. Phys., 13, 639

Arnowitt, R., Deser, S., & Misner, C. W. 2008, GReGr, 40, 1997

Baker, T., Bellini, E., Ferreira, P. G., et al. 2017, PhRvL, 119, 251301

Bellini, E., & Sawicki, I. 2014, JCAP, 07, 050

Bergmann, P. G. 1949, PhRv, 75, 680

Bezrukov, F. L., & Shaposhnikov, M. E. 2008, PhLB, 659, 703

Blas, D., Pujolas, O., & Sibiryakov, S. 2011, JHEP, 04, 018

Blohmann, C., Barbosa Fernandes, M. C., & Weinstein, A. 2013, CConMa, 15, 1250061

Bojowald, M. 2010, Canonical Gravity and Applications: Cosmology, Black Holes, and Quantum Gravity (Cambridge: Cambridge University Press)

Bojowald, M., & Paily, G. M. 2012, PhRvD, 86, 104018

Bojowald, M., Paily, G. M., & Reyes, J. D. 2014, PhRvD, 90, 025025

Brans, C. H., & Dicke, R. H. 1961, PhRv, 124, 925

Buchdahl, H. A. 1970, MNRAS, 150, 1

Burgess, C. P. 2004, LRR, 7, 5 http://www.livingreviews.org/lrr-2004-5

Cailleteau, T., Linsefors, L., & Barrau, A. 2014, CQGra, 31, 125011

Calcagni, G. 2009, JHEP, 09, 112

Chamseddine, A. H., & Mukhanov, V. 2013, JHEP, 11, 135

Creminelli, P., & Vernizzi, F. 2017, PhRvL, 119, 251302

Cuttell, R. 2019, PhD thesis, King's College London

Cuttell, R., & Sakellariadou, M. 2014, PhRvD, 90, 104026

Davies, P. C. W., Fulling, S. A., Christensen, S. M., & Bunch, T. S. 1977, AnPh, 109, 108

De Felice, A., & Tsujikawa, S. 2012, JCAP, 02, 007

de Rham, C., & Melville, S. 2018, PhRvL, 121, 221101

Deruelle, N., Sasaki, M., Sendouda, Y., & Yamauchi, D. 2010, PThPh, 123, 169

Dirac, P. A. M. 1950, Canadian Journal of Mathematics, 2, 129

Dirac, P. A. M. 1958a, RSPSA, 246, 333

Dirac, P. A. M. 1958b, RSPSA, 246, 326

Dirac, P. A. M. 1969, Lectures on Quantum Mechanics (New York: Yeshiva)

Donoghue, J. F. 1994, PhRvD, 50, 3874

Ezquiaga, J. M., & Zumalacárregui, M. 2017, PhRvL, 119, 251304

Gleyzes, J., Langlois, D., Piazza, F., & Vernizzi, F. 2015, PhRvL, 114, 211101

Gleyzes, J., Langlois, D., Piazza, F., & Vernizzi, F. 2015, JCAP, 02, 018

Gumrukcuoglu, A. E., Saravani, M., & Sotiriou, T. P. 2018, PhRvD, 97, 024032

Haro, J., Aresté Saló, L., & Pan, S. 2019, GReGr, 51, 49

Hojman, S. A., Kuchař, K., & Teitelboim, C. 1976, AnPh, 96, 88

Horndeski, G. W. 1974, IJTP, 10, 363

Hořava, P. 2009, PhRvD, 79, 084008

Jacobson, T., & Mattingly, D. 2001, PhRvD, 64, 024028

Kiritsis, E., & Kofinas, G. 2009, NuPhB, 821, 467

Kobayashi, T. 2019, RPPh, 82, 086901

Kobayashi, T., Yamaguchi, M., & Yokoyama, J. 2011, PThPh, 126, 511

Kovacs, A. D., & Reall, H. S. 2020, PhRvD, 101, 124003

Langlois, D. 2017, arXiv:1707.03625

Langlois, D., Mancarella, M., Noui, K., & Vernizzi, F. 2017, JCAP, 05, 033

Langlois, D., Mancarella, M., Noui, K., & Vernizzi, F. 2019, JCAP, 02, 036

Langlois, D., & Noui, K. 2016a, JCAP, 02, 034

Langlois, D., & Noui, K. 2016b, JCAP, 07, 016

Motohashi, H., Noui, K., Suyama, T., Yamaguchi, M., & Langlois, D. 2016, JCAP, 07, 033

Mukohyama, S., & Noui, K. 2019, JCAP, 07, 049

Olmo, G. J. 2005, PhRvL, 95, 261102

Palatini, A. 1919, Rendiconti del Circolo Matematico di Palermo, 43, 203

Papallo, G. 2017, PhRvD, 96, 124036

Papallo, G., & Reall, H. S. 2017, PhRvD, 96, 044019

Pradines, J. 1967, CRAS, 264, A245

Reyes, J. D. 2009, PhD thesis, Pennsylvania State University

Ripley, J. L., & Pretorius, F. 2019a, CQGra, 36, 134001

Ripley, J. L., & Pretorius, F. 2019b, PhRvD, 99, 084014

Ripley, J. L., & Pretorius, F. 2020, CQGra, 37, 155003

Sakstein, J., & Jain, B. 2017, PhRvL, 119, 251303

Sebastiani, L., Vagnozzi, S., & Myrzakulov, R. 2017, AdHEP, 2017, 3156915

Simon, J. Z. 1990, PhRvD, 41, 3720

Sotiriou, T. P., & Faraoni, V. 2010, RvMPh, 82, 451

Starobinsky, A. A. 1980, PhLB, 91, 99

Steinwachs, C. 2020, in 100 Years of Gauge Theory. Past, present and future perspectives, ed. S. De Bianchi, & C. Kiefer (Berlin: Springer)

Tricomi, F. G. 1968, Repertorium der Theorie der Differentialgleichungen (Berlin: Springer)

Tsamparlis, M. 1978, JMaPh, 19, 555

Wald, R. M. 1984, General Relativity (Chicago, IL: The University of Chicago Press)

Whitt, B. 1984, PhLB, 145, 176

Woodard, R. P. 2007, LNP, 720, 403

Chapter 3

Quantum Corrections

Quantum corrections are usually formulated for material objects such as elementary point particles or their compounds. Since this is the only realm in which quantum corrections have been tested by experiments, we will begin by developing those of their features that are relevant for cosmology. Chapter 4 will then show how the formalism can be adapted to the volume or scale factor of the universe. Although the general mathematical formalism of quantum mechanics can still be used in cosmology, several crucial differences in the interpretation of some terms are implied by scaling transformations. They foreshadow the special role played by the condition of general covariance in quantum gravity.

3.1 Algebra

In quantum mechanics, a point particle appears to be extended even if it has no internal structure. Moreover, interference phenomena can be measured for particles, suggesting that they share some properties with the behavior of waves. Because the force acting on a quantum-mechanical object then depends on several positions at once, the theory is non-local. A mathematical consequence of this observation, with important physical implications, is the fact that generic operations that can be applied in the mathematical description of a particle do not have the commutative behavior of classical mechanics: If we consider a point particle as the limiting case of a wave which is non-zero only at a single point, it can be modified only by multiplying the non-zero value with some real or complex number, constituting a commutative operation. For an extended wave, by contrast, we have a variety of mathematical operations, including multiplication with a function as well as differentiation. These two operations do not commute with each other.

3.1.1 Commutator

Multiplication can, for instance, be used to narrow a plane wave $\psi_k(x) = \exp(ikx)$ with some real wave number k down to a smaller range, as in $\psi_{k,\sigma}(x) = \exp(-x^2/4\sigma^2)\psi_k(x)$

doi:10.1088/2514-3433/ab9c98ch3

which implies a finite variance σ of the new wave packet $\psi_{k,\sigma}(x)$. Differentiation can be used to extract the wave number,

$$-i\frac{\mathrm{d}\psi_k(x)}{\mathrm{d}x} = k\psi_k(x). \tag{3.1}$$

Both operations are useful, but they give different results depending on the order in which they are applied. If we start with a generic function $\psi(x)$, we have

$$\frac{\mathrm{d}}{\mathrm{d}x}(x\psi(x)) = \psi(x) + x\psi'(x) \tag{3.2}$$

while

$$x\frac{\mathrm{d}}{\mathrm{d}x}\psi(x) = x\psi'(x). \tag{3.3}$$

The difference, given by the *commutator*

$$\left[\frac{\mathrm{d}}{\mathrm{d}x}, x\right] = \frac{\mathrm{d}}{\mathrm{d}x}x - x\frac{\mathrm{d}}{\mathrm{d}x} = 1 \tag{3.4}$$

(to be applied to a generic function which is no longer spelled out explicitly), is therefore non-zero. Relations such as Equation (3.4) define a non-commutative algebra.

The commutator, in general defined as

$$[\hat{A}, \hat{B}] = \hat{A}\hat{B} - \hat{B}\hat{A} \tag{3.5}$$

for elements $\hat{A}$ and $\hat{B}$ of a non-commutative algebra, enjoys useful mathematical properties that provide it with the important role of a quantum version of the Poisson bracket encountered in Hamilton's equations—or, in a special cases, of a quantum version of partial derivatives. One can directly confirm that the commutator is antisymmetric,

$$[\hat{A}, \hat{B}] = -[\hat{B}, \hat{A}]. \tag{3.6}$$

Therefore, any transformation of the operators that reverses their order in products changes the sign of the commutator. An example of such a transformation is taking the adjoint of a matrix, defined as a combination of transposition with complex conjugation, for which $(\hat{A}\hat{B})^\dagger = \hat{B}^\dagger\hat{A}^\dagger$ and therefore

$$[\hat{A}, \hat{B}]^\dagger = [\hat{B}^\dagger, \hat{A}^\dagger] = -[\hat{A}^\dagger, \hat{B}^\dagger]. \tag{3.7}$$

Such transformations are important as generalizations of the conjugate of complex numbers. In this generalized sense, therefore, the commutator of two self-adjoint operators ($\hat{A}^\dagger = \hat{A}$ and $\hat{B}^\dagger = \hat{B}$) plays the role of an imaginary number:

$$[\hat{A}, \hat{B}]^\dagger = -[\hat{A}, \hat{B}]. \tag{3.8}$$

In addition, the commutator is linear in each argument:

$$[\hat{A}_1 + \hat{A}_2, \hat{B}] = [\hat{A}_1, \hat{B}] + [\hat{A}_2, \hat{B}] \tag{3.9}$$

and

$$[\hat{A}, \hat{B}_1 + \hat{B}_2] = [\hat{A}, \hat{B}_1] + [\hat{A}, \hat{B}_2]. \tag{3.10}$$

The next property of the commutator reveals its close relationship with derivatives. If we consider the product of two operators, $\hat{B}\hat{C}$, its commutator with another operator, $\hat{A}$, obeys a product rule

$$\begin{aligned}
[\hat{A}, \hat{B}\hat{C}] &= \hat{A}\hat{B}\hat{C} - \hat{B}\hat{C}\hat{A} \\
&= \hat{A}\hat{B}\hat{C} - (\hat{B}\hat{A}\hat{C} - \hat{B}\hat{A}\hat{C}) - \hat{B}\hat{C}\hat{A} \\
&= [\hat{A}, \hat{B}]\hat{C} + \hat{B}[\hat{A}, \hat{C}].
\end{aligned} \tag{3.11}$$

The final basic property of the commutator also involves three operators, $\hat{A}$, $\hat{B}$ and $\hat{C}$, but in a double commutator:

$$[\hat{A}, [\hat{B}, \hat{C}]] + [\hat{C}, [\hat{A}, \hat{B}]] + [\hat{B}, [\hat{C}, \hat{A}]] = 0, \tag{3.12}$$

which can be seen by explicitly writing all twelve terms generated by the commutators. Equation (3.12) is called the *Jacobi identity*.

Antisymmetry, linearity, the product rule and the Jacobi identity are fulfilled also by the classical Poisson bracket of two phase-space functions. (In fact, these conditions on a combination of two functions define a Poisson bracket in general mathematical terms.) This relationship helps us to select *canonical operators* in a non-commutative algebra by analogy with canonical variables in classical mechanics. In classical mechanics, canonical variables, such as x and p in one spatial dimension, are such that they have Poisson brackets

$$\{x, H\} = \frac{\partial H}{\partial p} \quad \text{and} \quad \{p, H\} = -\frac{\partial H}{\partial x} \tag{3.13}$$

with any other function $H(x, p)$, usually used as the Hamiltonian of the system. Canonical, or basic, variables directly reveal the relationship between the Poisson bracket and partial derivatives.

3.1.2 Basic Operators

We therefore introduce canonical, or basic, operators $\hat{x}$ and $\hat{p}$ such that they model Hamilton's equations:

$$\frac{[\hat{x}, \hat{H}]}{i\hbar} = \frac{\partial \hat{H}}{\partial p}, \quad \frac{[\hat{p}, \hat{H}]}{i\hbar} = -\frac{\partial \hat{H}}{\partial x} \tag{3.14}$$

for any function $H(x, p)$, which on the right-hand sides of these two equations determines operators through its derivatives. (For instance, if $H = \frac{1}{2}p^2/m$,

$\partial \hat{H}/\partial p = \hat{p}/m$.) On the left-hand sides, the commutators are divided by i because, according to Equation (3.8), they are "generalized imaginary." Moreover, we should divide them by a constant $\hbar$ that provides the correct units of a p-derivative to x, and an x-derivative to p. Therefore, $\hbar$ should have the units of an action (or of the product of x and p). In classical physics, there is no need for such a parameter because Hamilton's equations relate partial derivatives of H to time derivatives of x and p, which already have matching units provided H is an energy. (The product of energy with time has the units of an action, just like the product of position with momentum.) This parameter is therefore a new fundamental constant, *Planck's constant*.

We still have to make sure that a commutator with $\hat{x}$ indeed works like a partial derivative by p, and vice versa. Since Equation (3.14) is required to work for any H, we may make the special choices H = p in the commutator with $\hat{x}$, and H = x in the commutator with $\hat{p}$. Both equations in (3.14) are then fulfilled, provided the basic operators have the commutator

$$[\hat{x}, \hat{p}] = i\hbar. \tag{3.15}$$

(Both equations in Equation (3.14) are reproduced by this choice thanks to the antisymmetry property of the commutator, 3.6.) Using the product rule (3.11), we can determine the Poisson bracket of any two polynomials in $\hat{x}$ and $\hat{p}$ from the basic commutator 3.15. The commutator algebra of such polynomials is therefore determined. Formally, we can extend these commutators to power series, such as Taylor expansions of non-polynomial functions. In order to determine whether there is a specific element in the algebra that equals the power series, we need to study the latter's convergence, which requires a topology on the algebra. Such questions can be analyzed more conveniently if we know a *representation* of the algebra elements as matrices acting on vectors (if finitely many components are sufficient) or as differential operations on functions. The non-commutativity of the matrices or derivatives selected in a representation should then agree with the commutator required for a specific algebra, in particular with Equation (3.15) for basic operators.

3.1.3 Representations

It is useful to know not just the commutator relationships but also specific representations of operators $\hat{x}$ and $\hat{p}$ as mathematical operations such that Equation (3.15) is true. Such representations are not unique. For instance, based on Equation (3.4), we may choose $\hat{x}$ as the mathematical operation of multiplying an x-dependent function, usually called the *wave function* $\psi(x)$, with x and $\hat{p} = -i\hbar d/dx$ as a derivative.

In addition to the commutator, we need an adjointness relation, mapping an operator $\hat{A}$ to its adjoint $\hat{A}^{\dagger}$, in order to define the generalized complex conjugate used in Equation (3.7). Introducing this operation in a representation requires an *inner product* which assigns a complex number (ψ_1, ψ_2) to any pair of wave functions, ψ_1 and ψ_2. The *adjoint* $\hat{A}^{\dagger}$ of an operator $\hat{A}$ is then defined such that

$$(\psi_1, \hat{A}\psi_2) = (\hat{A}^\dagger\psi_1, \psi_2) \tag{3.16}$$

for any ψ_1 and ψ_2 on which $\hat{A}$ can be applied in the representation.

The standard inner product for x-dependent wave functions—that is, wave functions in the position representation—is given by

$$(\psi_1, \psi_2) = \int_{-\infty}^{\infty} \psi_1(x)^*\psi_2(x)\mathrm{d}x. \tag{3.17}$$

For the inner product to be meaningful, allowed wave functions should be square integrable, such that (ψ, ψ) is finite. We then normalize wave functions by requiring that these inner products are not just finite but equal to one:

$$(\psi, \psi) = \int_{-\infty}^{\infty} |\psi(x)|^2\mathrm{d}x = 1. \tag{3.18}$$

In the form $(\psi, \psi) = 1$, normalization is imposed in any representation, but the specific evaluation by an integral or a series depends on how the representation is defined.

The real number $|(\psi_1, \psi_2)|^2$ obtained from the inner product is related to a generalized angle between the two wave functions. It tells us how likely it is that we measure properties of a wave function ψ_2 if the system is actually described by wave function ψ_1. If the wave functions are orthogonal, that is $(\psi_1, \psi_2) = 0$, we do not measure any properties of ψ_2 in ψ_1. If the states are identical, $\psi_1 = \psi_2$, normalization implies that we are indeed able to measure all properties of ψ_2 in ψ_1 because $|(\psi_1, \psi_2)|^2 = 1$.

According to the definition of an adjoint in a representation by wave functions $\psi(x)$, any operator $\hat{f}$ that multiplies a wave function with some complex function $f(x)$ then has an adjoint given by simple complex conjugation, $\hat{f}^\dagger = \hat{f}^*$. A derivative operator such as $\hat{p}$ requires integration by parts in order to evaluate the adjointness condition:

$$\int_{-\infty}^{\infty} \psi_1(x)^*\hat{p}\psi_2(x)\mathrm{d}x = \int_{-\infty}^{\infty} (\hat{p}\psi_2(x))^*\psi_2(x)\mathrm{d}x. \tag{3.19}$$

Therefore, $\hat{p}^\dagger = \hat{p}$, and the momentum operator is real in the generalized sense of the adjoint. Alternatively, we may work with p-dependent functions on which $\hat{p}$ operates by multiplication with p (and is therefore obviously real), while $\hat{x} = i\hbar\mathrm{d}/\mathrm{d}p$ is a derivative.

These two choices of representations are equivalent and imply the same implementation of an algebra with basic commutator (3.15), which can be shown by relating them through Fourier transformation. However, there are inequivalent choices some of which have been used in certain approaches to quantum cosmology.

It turns out to be quite difficult to find inequivalent representations of Equation (3.15) because a mathematical result, the Stone–von Neumann theorem, asserts that any representation of the commutator

$$[\hat{x},\ \hat{U}_\epsilon] = \hbar\epsilon\hat{U}_\epsilon, \tag{3.20}$$

which is an exponentiated version of Equation (3.15) if $\hat{U}_\epsilon = \exp(-i\epsilon\hat{p})$, has a unique representation up to equivalence, *provided* the operator $\hat{U}_\epsilon$ is differentiable in ϵ. Evaluating the derivative of $\hat{U}_\epsilon$ by ϵ at $\epsilon = 0$ then allows us to derive $\hat{p}$ from $\hat{U}_\epsilon$. However, there are inequivalent representations of Equation (3.20) in which $\hat{U}_\epsilon$ is not differentiable at $\epsilon = 0$. For instance, we may define a set of infinitely many functions $f(p)$, each of which is a superposition of finitely many functions $e_x(p) = \exp(-ixp)$ for some real x. Defining

$$(f(p),\ g(p)) = \lim_{P\to\infty}\frac{1}{2P}\int_{-P}^{P}f(p)^*g(p)\mathrm{d}p, \tag{3.21}$$

we have $(e_x(p),\ e_x(p)) = 1$ (normalization) and $(e_{x_1}(p),\ e_{x_2}(p)) = 0$ for $x_1 \neq x_2$ (orthogonality). The functions $e_x(p)$ therefore form an infinite, uncountable orthonormal basis with respect to an inner product (3.21), which defines the so-called *Bohr representation*. This property is already different from the standard *Schrödinger representation* used in quantum mechanics, in which any orthonormal basis is infinite but countable, given for instance by the energy eigenfunctions in a potential with a discrete energy spectrum.

On the new basis $e_x(p)$, the commutator (3.20) can be represented by

$$\hat{x}e_x(p) = xe_x(p), \qquad \hat{U}_\epsilon e_x(p) = e_{x+\epsilon\hbar}(p) \tag{3.22}$$

because

$$\begin{aligned}
\left(\hat{x}\hat{U}_\epsilon - \hat{U}_\epsilon\hat{x}\right)e_x(p) &= (x + \epsilon\hbar)e_{x+\epsilon\hbar}(p) - xe_{x+\epsilon\hbar}(p) \\
&= \epsilon\hbar e_{x+\epsilon\hbar}(p) = \epsilon\hbar\hat{U}_\epsilon e_x(p).
\end{aligned} \tag{3.23}$$

However, $\hat{U}_\epsilon$ is not differentiable at $\epsilon = 0$, which follows from the orthogonality of $e_x(p)$ and $\hat{U}_\epsilon e_x(p) = e_{x+\epsilon\hbar}(p)$ for $\epsilon \neq 0$: If U_ϵ were differentiable, $(e_{x_1}(p),\ \hat{U}_\epsilon e_{x_2}(p))$ would be differentiable for any x_1 and x_2, in contradiction with

$$(e_x(p),\ \hat{U}_\epsilon e_x(p)) = (e_x(p),\ e_{x+\epsilon\hbar}(p)) = \begin{cases} 1 & \text{if } \epsilon = 0 \\ 0 & \text{if } \epsilon \neq 0 \end{cases}. \tag{3.24}$$

For systems that are more complicated than an x and a p both taking all real values—or combinations of several independent versions of this form for the three spatial dimensions or multiple particles—the task of finding and classifying inequivalent representations of basic commutators is usually very complicated. For instance, if x is replaced by the scale factor of a cosmological model, which is not allowed to take negative values, the representation theory is rather different. It is then useful to work, whenever possible, directly with algebraic relationships that follow from the commutator without referring to a specific representation of mathematical operations on functions $\psi(x)$ or $e_x(p)$.

3.1.4 States

Measurement results are numbers rather than operations. A physical interpretation of a non-commutative algebra therefore requires a mapping from the algebra, $\mathcal{A}$, to the real or complex numbers. The value of this mapping assigned to a certain mathematical operation is interpreted as the measurement result of a corresponding physical procedure. For instance measuring the position x of a particle is described by the mathematical operation of multiplying an x-dependent function with x. The momentum of a particle is independent of its position because, in classical mechanics, both values can be chosen freely at an initial time. In quantum mechanics, therefore, a measurement of the momentum p is described by an operation independent of multiplication with x. One candidate is the derivative by x, already used in Equation (3.1), and indeed the momentum of a wave is related to its wave number. In order to obtain the correct units, we need a constant, Planck's constant $\hbar$, with has units of the product xp, the units of an action. With this new fundamental parameter, we can relate a momentum measurement to the mathematical operation $-i\hbar \mathrm{d}/\mathrm{d}x$.

Measurement results depend not only on the apparatus but also on the state the measured system is in. The general concepts of measurement and state are sharply divided and completely independent in quantum mechanics, but their combination is needed to obtain a specific measurement result. A *measurement* is the mathematical abstraction of a (usually) complicated experimental device, which has been constructed before it is applied. It should therefore be described independently of the measurement results that it may give once it is used on a specific system. The *state* a system is in when a measurement is done does not depend on the measurement device but rather on other procedures that have been used to prepare it for a lab experiment, or in cosmology on some physical processes that led to an initial configuration which then evolved until it is observed. The independence of measurement and state implies that a state describes all the information contained in the system that could in principle be extracted by experiments or observations. This conclusion applies to both classical and quantum mechanics. However, it has much more far-reaching implications in quantum mechanics because the non-commuting behavior of different measurements implies strong restrictions on the available information.

These basic facts have several implications important for the common interpretation of quantum mechanics. First, the independence of measurements and measurement results is the reason why the mathematical description of the measurement process, by itself, need not be a number. It may be an operator, such as a derivative, provided there is a second mathematical ingredient, the description of the state, such that a suitable combination of operator and state uniquely determines a real or complex number. Since we already defined measurement operators as elements of an algebra $\mathcal{A}$, a state ω, in its most general form, is therefore a mapping from $\mathcal{A}$ to the complex numbers.

For quantum mechanics to work as we know it, such a mapping cannot be arbitrary. Two basic operations in an algebra are addition of its elements and multiplication with a number. These two operations (but not necessarily the non-commutative product of two algebra elements) should be respected by the mapping, which means that any ω should be linear. We can also normalize any ω by requiring that its value on the unit operation, which does not change the function it acts on, equals one. Finally, any allowed ω must obey certain inequalities, the most prominent of which is Heisenberg's uncertainty relation. As a condition on allowed states ω, this relation can be written as

$$
\begin{aligned}
&(\omega(\hat{x}^2) - \omega(\hat{x})^2)(\omega(\hat{p}^2) - \omega(\hat{p})^2) \\
&- \left(\frac{1}{2}\omega(\hat{x}\hat{p} + \hat{p}\hat{x}) - \omega(\hat{x})\omega(\hat{p})\right)^2 \geq \frac{\hbar^2}{4}
\end{aligned}
\tag{3.25}
$$

if $\hat{x}$ and $\hat{p}$ are the operations that describe measurements of position x and momentum p, respectively, or any pair of canonical operators.

Even though $\omega(\hat{x})$ and $\omega(\hat{p})$ are, for a given state, numbers that could be interpreted as the unique outcome of a single measurement of x and p, the reality of quantum mechanics is more complicated because of the non-commuting nature of $\hat{x}$ and $\hat{p}$. Since the operations do not commute, it cannot be possible to obtain unique outcomes for both x and p at the same time. We would have to choose whether we measure x first or p first, but then the results should depend on the choice, in contrast to what two unique numbers $\omega(\hat{x})$ and $\omega(\hat{p})$ would suggest. Moreover, the state changes after a measurement because we gain information about the system. This feature, which often leads to counter-intuitive implications, is the second important consequence of our conclusion that a state, being independent of a measurement, must contain all the information that can in principle be extracted from a given system by experiments. If we have already made one experiment, our information changes, and the state must be adjusted to a new "reality" after the measurement.

In order to overcome the difficulties of non-commuting measurements, $\omega(\hat{x})$ and $\omega(\hat{p})$ are interpreted as averages of large numbers of measurements of x or p, performed on a system which has meticulously been prepared such that it is in the same state, described by ω, each time a measurement has been performed. In statistical language, $\omega(\hat{x})$ and $\omega(\hat{p})$ are the expectation values of the corresponding measurements in a state ω. Quantities such as $\omega(\hat{x}^2) - \omega(\hat{x})^2 = (\Delta_\omega x)^2$ that appear in the uncertainty relation (3.25) are then the statistical variances, and $\frac{1}{2}\omega(\hat{x}\hat{p} + \hat{p}\hat{x}) - \omega(\hat{x})\omega(\hat{p}) = C_{\omega,xp}$ is the covariance. In these statistical terms, Heisenberg's uncertainty relation is written as

$$
(\Delta_\omega x)^2(\Delta_\omega p)^2 - C_{\omega,xp}^2 \geq \frac{\hbar^2}{4}.
\tag{3.26}
$$

In a large number of relevant examples, the state ω is closely related to the wave function ψ which we used when we introduced a correspondence between mathematical operations and physical measurements, or a representation of the algebra.

Given an operation $\hat{O}$, such as $\hat{x}\psi(x) = x\psi(x)$ or $\hat{p}\psi(x) = -i\hbar d\psi(x)/dx$, we can extract a number by defining a state

$$\omega_\psi(\hat{O}) = \int_{-\infty}^{\infty} \psi(x)^* \hat{O}\psi(x)dx \qquad (3.27)$$

with the complex conjugate $\psi(x)^*$ of $\psi(x)$. More generally, if (ψ_1, ψ_2) denotes the inner product of wave functions provided by a representation of the operator algebra, the expectation value of $\hat{O}$ in a state ω_ψ is given by

$$\omega_\psi(\hat{O}) = (\psi, \hat{O}\psi). \qquad (3.28)$$

Normalization of ψ, given by $(\psi, \psi) = 1$, then implies normalization of ω_ψ: $\omega_\psi(\hat{I}) = 1$ where $\hat{I}$ is the identity operator that does not change the state it acts on. (In the operator algebra, $\hat{I}\hat{A} = \hat{A}\hat{I} = \hat{A}$ for any operator $\hat{A}$.)

If $\hat{O} = \hat{x}$, the expression

$$\omega_\psi(\hat{x}) = \int_{-\infty}^{\infty} x|\psi(x)|^2 dx \qquad (3.29)$$

in the position representation takes the form of a statistical expectation value of x-measurements with probability density $|\psi(x)|^2$. If the wave function is normalized such that $\int_{-\infty}^{\infty} |\psi(x)|^2 dx = 1$, the state ω_ψ is normalized. If, for a given state ω, there is a wave function ψ such that $\omega = \omega_\psi$, ω is called a *pure state*.

Another example of defining states is given by *density matrices*, $\rho(x_1, x_2)$, depending on two copies of the position. We normalize a density matrix by requiring that

$$\int_{-\infty}^{\infty} \rho(x, x)dx = 1. \qquad (3.30)$$

Because a density matrix depends on two position values, it can be interpreted as an operator acting on wave functions by

$$\hat{\rho}\psi(x) = \int_{-\infty}^{\infty} \rho(x, y)\psi(y)dy. \qquad (3.31)$$

For any normalized ρ, we obtain a normalized state ω_ρ such that

$$\omega_\rho(\hat{O}) = \int_{-\infty}^{\infty} \left(\hat{O}_x\rho(x, y) \right)|_{y=x} dx \qquad (3.32)$$

where $\hat{O}_x$ indicates that $\hat{O}$ acts only on the x-dependence of $\rho(x, y)$. In a generic representation of the operator algebra on a space of wave functions with basis ψ_I, such that any wave function ψ in the representation can be written as a normalized sum

$$\psi = \sum_I c_I\psi_I \qquad (3.33)$$

with suitable complex coefficients c_I, the expectation value of $\hat{O}$ in a state ω_ρ is given by

$$\omega_\rho(\hat{O}) = \sum_I (\psi_I, \hat{O}\hat{\rho}\psi_I). \tag{3.34}$$

In this generic form, a density matrix $\hat{\rho}$ is an operator acting on wave functions in the representation such that

$$\sum_I (\psi_I, \hat{\rho}\psi_I) = 1. \tag{3.35}$$

A normalized density matrix ρ can be constructed from a normalized wave function ψ by defining $\rho(x, y) = \psi(x)\psi(y)^*$, but density matrices exist which are not of this form. The concept of a density matrix is therefore more general than that of a wave function. Any state ω that can be obtained from a density matrix but not from a wave function is called a *mixed state* (as opposed to a pure state).

3.2 Measurements

In the preceding section we concluded that states describe the complete information available about a system. A state ω determines an infinite set of numbers relative to a basic operator such as $\hat{x}$, given by $\omega(\hat{x}^n)$ for all positive integers $n > 0$. These numbers are interpreted as the moments of a probability distribution for measurements of the position x. If we know the probability distribution $\rho(x)$, its moments are defined as

$$\omega(\hat{x}^n) = \int_{-\infty}^{\infty} x^n \rho(x) \mathrm{d}x. \tag{3.36}$$

3.2.1 Moments of a State

Because a probability density must always be positive, $\rho(x) \geqslant 0$, the moments are not arbitrary but must obey certain inequalities, in addition to normalization, $\int_{-\infty}^{\infty} \rho(x)\mathrm{d}x = 1$. Provided these inequalities are satisfied, the moments determine a unique probability density which can, for instance, be constructed from the moments by first computing the Fourier transform $\tilde{\rho}(\lambda)$ of $\rho(x)$ as

$$\tilde{\rho}(\lambda) = \int_{-\infty}^{\infty} \rho(x)\exp(i\lambda x)\mathrm{d}x = \sum_{n=0}^{\infty} \frac{(i\lambda)^n}{n!}\omega(\hat{x}^n). \tag{3.37}$$

Inverse Fourier transformation then determines $\rho(x)$.

If we include all moments of a canonical pair of basic operators, $\hat{x}$ and $\hat{p}$, we can reconstruct a density matrix from the $\hat{x}$-$\hat{p}$ moments. (Additional inequalities then have to be imposed on the moments, including Heisenberg's uncertainty relation (3.25) and higher-order versions.) First, reconstruction using Fourier transformation allows us to determine the diagonal $\rho(x, x)$ of the density matrix through

$$\omega(\hat{x}^n) = \int_{-\infty}^{\infty} x^n \rho(x, x)\mathrm{d}x, \tag{3.38}$$

just as in the case of a probability density. The full density matrix, including $\rho(x, y)$ for $x \neq y$, can be obtained from a Taylor expansion around the diagonal. We first define the coefficient functions

$$\rho_j(x) = \frac{\partial^j \rho(y, x)}{\partial y^j}\bigg|_{y=x} \tag{3.39}$$

which appear in the Taylor expansion

$$\rho(x + d, x) = \sum_{j=0}^{\infty} \frac{d^j}{j!}\rho_j(x) \tag{3.40}$$

and are related to moments through

$$\omega(\hat{x}^n \hat{p}^j) = \left(\frac{\hbar}{i}\right)^j \int_{-\infty}^{\infty} x^n \rho_j(x)\mathrm{d}x. \tag{3.41}$$

We can therefore reconstruct all $\rho_j(x)$ from the moments, and then insert them in the Taylor series for $\rho(x + d, x)$.

If the state is known to be pure, we need only moments of the form $\omega(\hat{x}^n \hat{p})$ with a single $\hat{p}$ in order to reconstruct the phase $\alpha(x)$ of a wave function $\psi(x) = |\psi(x)|\exp(i\alpha(x))$, in addition to the probability density $|\psi(x)|^2$. Using

$$\mathrm{Re}\ \omega(\hat{x}^n \hat{p}) = \hbar \int_{-\infty}^{\infty} x^n |\psi(x)|^2 \frac{\mathrm{d}\alpha}{\mathrm{d}x}\mathrm{d}x, \tag{3.42}$$

the phase is obtained by integrating $\hbar^{-1}f(x)/|\psi(x)|^2$, where $f(x)$ is the reconstructed density that appears in the integrand of Equation (3.42).

3.2.2 Eigenvalues

If a wave function is available, quantum mechanics is able to make statements about the outcome of a single measurement, not just about averaged outcomes $\omega(\hat{O})$ of multiple measurements on a system prepared in the same way for each experiment. The outcome of a single measurement remains probabilistic, but it is often possible to rule out some outcomes (in the presence of quantum discreteness) and to determine the probabilities of the allowed ones. The outcome of a single measurement of an observable O is one of the eigenvalues λ of the corresponding operator $\hat{O}$, defined such that

$$\hat{O}\psi_\lambda = \lambda\psi_\lambda \tag{3.43}$$

for some wave function ψ_λ, the eigenfunction corresponding to the eigenvalue λ.

If the operator $\hat{O}$ is self-adjoint, such that $\hat{O}^\dagger = \hat{O}$ or

$$\int_{-\infty}^{\infty} \psi^*(x)\hat{O}\psi(x)\mathrm{d}x = \int_{-\infty}^{\infty} (\hat{O}\psi(x))^*\psi(x)\mathrm{d}x, \qquad (3.44)$$

all its eigenvalues are real: We have

$$\lambda = \int_{-\infty}^{\infty} \psi_\lambda^*(x)\hat{O}\psi_\lambda(x)\mathrm{d}x = \int_{-\infty}^{\infty} (\hat{O}\psi_\lambda(x))^*\psi_\lambda(x)\mathrm{d}x = \lambda^*. \qquad (3.45)$$

It is therefore meaningful to identify them with possible measurement outcomes. Moreover, for such an operator the eigenfunctions corresponding to distinct eigenvalues, $\lambda_1 \neq \lambda_2$, are orthogonal because

$$\begin{aligned}
\lambda_1 \int_{-\infty}^{\infty} \psi_{\lambda_2}^*(x)\psi_{\lambda_1}(x)\mathrm{d}x &= \int_{-\infty}^{\infty} \psi_{\lambda_2}^*(x)\hat{O}\psi_{\lambda_1}(x)\mathrm{d}x \\
&= \int_{-\infty}^{\infty} (\hat{O}\psi_{\lambda_2}(x))^*\psi_{\lambda_1}(x)\mathrm{d}x \\
&= \lambda_2 \int_{-\infty}^{\infty} \psi_{\lambda_2}^*(x)\psi_{\lambda_1}(x)\mathrm{d}x
\end{aligned} \qquad (3.46)$$

is then possible for $\lambda_2 \neq \lambda_1$ only if $\int_{-\infty}^{\infty} \psi_{\lambda_1}(x)^*\psi_{\lambda_2}(x)\mathrm{d}x = 0$.

It also follows from general results about operator representations that the eigenfunctions of a self-adjoint operator are complete, in the sense that any wave function ψ can uniquely be written as a superposition of the ψ_λ:

$$\psi = \sum_\lambda c_\lambda \psi_\lambda \qquad (3.47)$$

with complex coefficients c_λ. (The λ-sum is obtained in the case of discrete eigenvalues and should be replaced by an integral over λ if there is a continuous range of eigenvalues.) If a system is in the state described by a wave function ψ as in Equation (3.47), the probability of measuring a specific eigenvalue λ in a single measurement is given by $|c_\lambda|^2$. This statement is consistent with the original definition of probabilistic outcomes of multiple measurements: The expectation value of multiple measurement results λ, each with probability $|c_\lambda|^2$, agrees with the expectation value of the operator $\hat{O}$ in the state ω_ψ:

$$\begin{aligned}
\omega_\psi(\hat{O}) &= \int_{-\infty}^{\infty} \psi(x)^*\hat{O}\psi\,\mathrm{d}x \\
&= \sum_{\lambda_1,\lambda_2} c_{\lambda_1}^* c_{\lambda_2} \int_{-\infty}^{\infty} \psi_{\lambda_1}(x)^*\hat{O}\psi_{\lambda_2}(x)\mathrm{d}x = \sum_\lambda \lambda|c_\lambda|^2.
\end{aligned} \qquad (3.48)$$

For this statement, both the completeness and orthogonality properties of eigenfunctions are important, as well as their normalization.

Such statements about outcomes of individual measurements are possible only if we have a pure state described by a wave function because states ω or density matrices do not obey a superposition principle such as Equation (3.47). The wave function then describes all the information available about the system, after it has been prepared in this initial state. Since the information changes after a single measurement, which has a definite outcome after it has been made, the wave function must be adjusted. If a single measurement of O has led to the (eigen)value λ of $\hat{O}$, the state right after the measurement is the corresponding eigenfunction ψ_λ. (This statement assumes that the eigenvalue is not degenerate, that is, there is no independent eigenfunction $\psi_{\lambda'}$ with eigenvalue $\lambda' = \lambda$. If this assumption is not true, the state after measuring the value λ of $\hat{O}$ is given by

$$\psi_\lambda = \frac{\sum_{i=1}^{n} c_{\lambda_i} \psi_{\lambda_i}}{\sqrt{\sum_{i=1}^{n} |c_{\lambda_i}|^2}} \tag{3.49}$$

if ψ_{λ_i} are all the degenerate eigenstates with the same eigenvalue $\lambda_1 = \cdots = \lambda_n = \lambda$.)

3.2.3 The Divine Dividing Line

These implications rely on a clear dividing line between the system (described by the state) and an observer who performs the measurement and adjusts the state according to the information available. Since this information changes after the measurement, the state must change just by virtue of new information that needs to be taken into account because the state, by definition, describes *all* the information that can in principle be available. In addition, there is as usual dynamical evolution that the state may be subject to if we let some time pass by. However, in contrast to dynamical evolution, the adjustment of the state after a measurement must be done instantaneously. In order to emphasize the abrupt nature of this process, it is often referred to as the *collapse of the wave function.*

A measurement, performed by an observer, therefore has a much stronger influence on the system than nature, which is bound to its laws unfolding more patiently during the passage of time. This distinction is meaningful in laboratory experiments, in which the observer is assumed to have full control over "creating" and discharging a state. It requires a strict separation between the measured system and the observer, in particular of their states. Such a divine role of an observer completely outside the system to be measured is unavailable in cosmology. The role and interpretation of single measurements is therefore problematic in this context. The problem can be evaded by foregoing individual measurements and working instead only with expectation values and moments of a state, or a state ω instead of a wave function.

We take the same view in this book, but note that, conceptually, it does not completely solve the problem of how to make sense of a fundamentally probabilistic theory of cosmology: When we consider an expectation value of a cosmological observable, such as $\omega(\hat{V})$ where V is the volume of an expanding region in space, we need an ensemble of independent realizations of the same state, each giving rise to a different volume measurement such that the average is equal to $\omega(\hat{V})$ and the variance and higher moments of the distribution of measurement results are given by the $\hat{V}$-moments of ω. Do we then need to prepare multiple independent universes in order to have different realizations of the same state? This option is clearly not available to us, but it can be avoided by considering multiple independent regions in space (contained in our past light cone) which, for practical purposes, initially occupy the same state. Large-scale homogeneity at least in the late universe shows that such regions are readily available. At least in principle, we can then measure all their volumes at some later time, and the average can be compared with an $\omega(\hat{V})$ computed using some model of quantum cosmology. There is therefore a well-defined procedure to test fundamentally probabilistic theories in cosmology, provided the measurement does not require all of space at a given time to be included in the measurement. The latter requirement may be problematic in the early universe close to the Big Bang singularity, where quantum cosmology is expected to be most relevant. We will therefore return to this discussion once we have developed sufficiently many ingredients of early-universe quantum cosmology.

3.2.4 Decoherence

The appearance of a measurement result shown on the display of some apparatus is usually regarded as the "emergence" of a classical property from the quantum realm: Unlike the objects of modern physics, an experimental apparatus is a macroscopic object which does not quantum fluctuate at a significant level. It is constructed in such a way that it shows a definite measurement result for a certain period of time which allows the experimenter (or a computer) to register it. Because it is macroscopic, the apparatus consists of many degrees of freedom which interact with the measured system in a specific way so as to facilitate the measurement.

The feature of providing many degrees of freedom that interact with a measured system is not restricted to specially designed measurement devices. Even if we do not design an apparatus with the specific purpose of performing a measurement, microscopic systems are never completely isolated. They always interact with a large number of degrees of freedom, for instance the molecules of air in a lab experiment or photons of the cosmic microwave background radiation in the universe. These interactions may be weak, but they happen without cessation and may have an effect on the microscopic state. Instead of a sharp transition from some state to an eigenstate of an observable, as postulated in a collapse of the wave function which hurriedly keeps track of the information available to us, the measurement process can be viewed as a dynamical process that involves the system we are interested in as well as a large number of largely unrecorded "environmental" degrees of freedom. The proposal that this process can explain how measurements with classical results happen in quantum mechanics is called *decoherence* (Zeh 1970; Giulini et al. 2003; Schlosshauer 2007). It has been applied several times to cosmology (Polarski & Starobinsky 1996; Kiefer et al. 1998; Burgess et al. 2008; Kiefer et al. 2007, 2000; Barvinsky et al. 1999; Falciano et al. 2007; Peter & Pinto-Neto 2008; Pinto-Neto 2009; Kiefer & Kraemer 2012).

The measurement process can be expressed as a transition between states, *including* those of the apparatus. An apparatus is designed such that it can always be reset to a neutral initial state, described by a wave function $\phi_\varnothing$. Performing a measurement then implies that the apparatus changes to a recognizably different state, or some wave function ϕ_λ if the value λ is shown on the device. A good device is correlated with the system it measures, such that it should reach a state ϕ_λ only if the measured system is in a state described by the eigenfunction ψ_λ of the observable to be measured. Formally, we write the combined system-and-apparatus states as (tensor) products of their wave functions because the probabilities of finding the system and the apparatus in a specific state are products of the individual probabilities. (This operation $\otimes$ is distinct from pointwise products of wave functions using multiplication of complex numbers.) If the system is initially in an eigenstate, the measurement is therefore a transition from a combined initial wave function $\psi_\lambda \otimes \phi_\varnothing$ to a final wave function $\psi_\lambda \otimes \phi_\lambda$, the last factor indicating that the measurement result has been shown.

Generically, we cannot be sure that the system is initially in an eigenstate of the measured observable, but thanks to completeness it is always a superposition of eigenfunctions, $\psi = \sum_\lambda c_\lambda \psi_\lambda$. If the interaction process is linear on wave functions (which it is, according to the following section), during a measurement such a superposition should change from $\psi \otimes \phi_\varnothing$ to $\sum_\lambda c_\lambda \psi_\lambda \otimes \phi_\lambda$. However, the last state describes a superposition of measurement states registering the possible outcomes, given by the eigenvalues λ. It does not show a definite result with a single apparatus state ψ_λ. Without additional steps in the theoretical description of a measurement, the outcome of a specific eigenvalue, as postulated by the axioms of quantum mechanics, can only be described by an ad hoc assumption that the state changes drastically (or "collapses") to a specific eigenstate (von Neumann 1996).

Decoherence is such an additional step. It inserts a third factor into the products of wave functions, corresponding to the environmental degrees of freedom which interact with the system but are not registered by the apparatus. Because the environment interacts with the system, although weakly, its state too changes according to the state of the system. If it is initially in some state $\xi_\varnothing$, the final state after all interactions is given by $\sum_\lambda c_\lambda \psi_\lambda \otimes \phi_\lambda \otimes \xi_\lambda$.

The apparatus is still in a superposition. However, the pure product state is not the final outcome of the measurement because we do not register the state of the environment. Because we lack this information, we should replace the superposition of product wave functions for the combined probabilities of all degrees of freedom with a conditional probability that takes into account our ignorance of the environment. Partial ignorance means that the final state of the measurement should not be pure but mixed, described by a density matrix rather than a wave function. A pure state such as $\sum_\lambda c_\lambda \psi_\lambda \otimes \phi_\lambda \otimes \xi_\lambda$, described by a density matrix, is given by

$$\rho = \sum_{\lambda_1, \lambda_2} c_{\lambda_1}^* c_{\lambda_2} \psi_{\lambda_2} \otimes \phi_{\lambda_2} \otimes \xi_{\lambda_2} (\psi_{\lambda_1} \otimes \phi_{\lambda_1} \otimes \xi_{\lambda_1})^*. \tag{3.50}$$

Any reference to environment states is then eliminated by computing the reduced density matrix

$$\rho_{\text{reduced}} = \sum_I (\xi_I, \hat{\rho}\xi_I) \tag{3.51}$$

where ξ_I with I in some suitable index set describes a complete set of environment states (such that any environment state ξ can be written uniquely as $\xi = \sum_I c_I \xi_I$ with complex numbers c_I). The label I is therefore distinct from the label λ, which latter describes a complete set of system states according to the eigenvalues of an observable. Because the environment in its usual role has many more degrees of freedom than the system, the index set of all possible I is much larger than the index set given by all eigenvalues, λ, of $\hat{O}$.

The reduced density matrix, in general, does not show classical features because it contains interference terms in which $\lambda_1 \neq \lambda_2$. The diagonal ones in which $\lambda_1 = \lambda_2$, by contrast, have coefficients given by the probabilities $|c_\lambda|^2$ of individual measurement outcomes. We can therefore say that a classical probability distribution $|c_\lambda|^2$ emerges if the off-diagonal interference terms ($\lambda_1 \neq \lambda_2$) are much smaller than the diagonal probability terms.

For a classical measurement result λ to emerge for a given observable $\hat{O}$ with eigenvalues λ, the interaction of the system with the environment must be such that, to a good approximation, a pure density matrix $\xi_{\lambda_2}\xi_{\lambda_1}^*$ of the environment evolves a short time after the measurement has been performed into a state with reduction

$$\sum_I \xi_I^* \left(\xi_{\lambda_2}\xi_{\lambda_1}^* \right) \xi_I \propto \delta_{\lambda_1,\lambda_2}. \tag{3.52}$$

If this condition is fulfilled for a given observable with eigenvalues λ, the reduced density matrix in which we eliminate environment degrees of freedom after they have done their job of interacting with system and device, is of the form

$$\rho = \sum_\lambda |c_\lambda|^2 \psi_\lambda \otimes \phi_\lambda (\psi_\lambda \otimes \phi_\lambda)^* \tag{3.53}$$

which is free of interference terms.

In condition (3.52) we refer to both index sets, I and λ, which describe the degrees of freedom of the environment and the system, respectively. The condition can therefore be met only if the environment and the system interact in a suitable way. A controlled experiment couples the system state ψ to a device state ϕ, such that both interact with the environment state ξ so as to make a classical measurement λ of the desired observable $\hat{O}$ emerge after some time of interaction.

In less controlled situations, as in cosmology, we have to accept interactions as nature provides them. Cosmological observations tell us that classical properties have already emerged a long time ago out of a putative highly quantum state in the very early universe. In this case, we can no longer influence the transition to classical behavior. Conversely to the situation of a lab experiment, the condition (3.52) then tells us which observables should be considered classical during a given epoch of the universe, such as the time right after inflation. Theories of quantum cosmology, provided they include a sufficiently large number of degrees of freedom so as to model an environment, can then be tested by computing whether they imply sufficiently fast classicalization of quantities such as the density variations measured via the cosmic microwave background radiation.

3.3 Dynamics

Since quantum mechanics replaces the classical values of x and p with mathematical operations, the same is the case for the energy. For a conservative mechanical system with potential $V(x)$, the energy

$$E = \frac{p^2}{2m} + V(x) \tag{3.54}$$

is replaced by the energy operator (or Hamiltonian)

$$\hat{H} = \frac{\hat{p}^2}{2m} + V(\hat{x}). \tag{3.55}$$

3.3.1 Heisenberg's and Ehrenfest's equations

In classical mechanics, the Hamiltonian $H(x, p)$ determines the dynamics of a system through Hamilton's equations of motion,

$$\frac{\mathrm{d}x}{\mathrm{d}t} = \frac{\partial H}{\partial p}, \quad \frac{\mathrm{d}p}{\mathrm{d}t} = -\frac{\partial H}{\partial x}. \tag{3.56}$$

In quantum mechanics, the commutator (3.4) with basic operators has properties that are very similar to the partial derivatives on the right-hand side of Hamilton's equations. For instance, the commutator

$$\begin{aligned}
[\hat{p}, V(\hat{x})] &= -i\hbar \left[\frac{\mathrm{d}}{\mathrm{d}x}, V(x) \right] \\
&= -i\hbar \left(\frac{\mathrm{d}}{\mathrm{d}x} V(x) - V(x) \frac{\mathrm{d}}{\mathrm{d}x} \right) = -i\hbar \frac{\mathrm{d}V(x)}{\mathrm{d}x}
\end{aligned} \tag{3.57}$$

(acting as in Equation (3.4) on a generic function) behaves like a derivative with respect to x. Moreover,

$$[\hat{x}, \hat{p}^2] = -\hbar^2 \left(x \frac{\mathrm{d}^2}{\mathrm{d}x^2} - \frac{\mathrm{d}^2}{\mathrm{d}x^2} x \right) = 2\hbar^2 \frac{\mathrm{d}}{\mathrm{d}x} = 2i\hbar\hat{p} \tag{3.58}$$

is related to the partial derivative by p. Moreover, the signs on the right-hand sides of Equations (3.57) and (3.58) are opposite, as required for Hamilton's equations. We just divide by $i\hbar$ and obtain a quantum version of Hamilton's equations,

$$\frac{\mathrm{d}\hat{x}}{\mathrm{d}t} = \frac{[\hat{x}, \hat{H}]}{i\hbar}, \quad \frac{\mathrm{d}\hat{p}}{\mathrm{d}t} = \frac{[\hat{p}, \hat{H}]}{i\hbar}, \tag{3.59}$$

called *Heisenberg's equations of motion*.

Taking expectation values of Heisenberg's equations, generalized to any operator $\hat{O}$, these differential equations for operators determine how a state changes in time, according to

$$\frac{d\omega(\hat{O})}{dt} = \omega(d\hat{O}/dt) = \frac{\omega([\hat{O}, \hat{H}])}{i\hbar}. \qquad (3.60)$$

(If an operator $\hat{O}(t)$ has explicit time dependence, we should add the partial derivative $\omega(\partial\hat{O}/\partial t)$ to this equation.) Since we can measure only expectation values $\omega(\hat{O})$ but not the operator $\hat{O}$ or the state ω individually, two mathematical options can be used to express time evolution. The first one, already introduced in Equation (3.59), interprets evolution as a property of measurements, and therefore assigns time dependence to operators. Alternatively, one could keep operators, representing measurement procedures, constant and assign time dependence to the state ω, or to a wave function ψ or density matrix ρ.

For a pure state $\omega = \omega_\psi$, the same evolution (3.60) of expectation values $\omega_\psi(\hat{O})$ is obtained if ψ solves the *Schrödinger equation*

$$i\hbar\frac{\partial\psi}{\partial t} = \hat{H}\psi. \qquad (3.61)$$

If this equation is fulfilled, we indeed have

$$\begin{aligned}
\frac{d\omega_\psi(\hat{O})}{dt} &= \frac{d}{dt}\int_{-\infty}^{\infty} \psi(t, x)^* \hat{O}\psi(t, x)dx \\
&= \int_{-\infty}^{\infty} \frac{\partial\psi(t, x)^*}{\partial t}\hat{O}\psi(t, x)dx + \int_{-\infty}^{\infty} \psi(t, x)^*\hat{O}\frac{\partial\psi(t, x)}{\partial t}dx \\
&= \frac{i}{\hbar}\left(\int_{-\infty}^{\infty} (\hat{H}\psi)(x)\hat{O}\psi(x)dx - \int_{-\infty}^{\infty} \psi(t, x)^*\hat{O}\hat{H}\psi(x)dx\right) \\
&= \frac{\omega_\psi([\hat{O}, \hat{H}])}{i\hbar}.
\end{aligned} \qquad (3.62)$$

A pure state $\omega = \omega_\psi$ given by a wave function ψ can equally be expressed in terms of the density matrix $\rho_\psi(x, y) = \psi(x)\psi(y)^*$. The evolution Equation (3.63) then implies the *von Neumann equation*

$$i\hbar\frac{\partial\rho}{\partial t} = [\hat{H}, \hat{\rho}], \qquad (3.63)$$

writing a partial time derivative because ρ may also depend on other variables, such as x in a position representation. Compared with Equation (3.59), the opposite commutator is used. The resulting sign change implements the new picture of having an evolving state rather than evolving measurements.

In order to derive the von Neumann equation, recall that a density matrix ρ, defining a state ω_ρ, can be interpreted as an operator via

$$\hat{\rho}\psi(x) = \int_{-\infty}^{\infty} \rho(x, y)\psi(y)\mathrm{d}y. \tag{3.64}$$

Equation (3.63) is then required by the Schrödinger equation upon equating the product rule

$$i\hbar\frac{\partial\rho_\psi(x, y)}{\partial t} = i\hbar\left(\frac{\partial\psi(x)}{\partial t}\psi(y)^* + \psi(x)\frac{\partial\psi(y)^*}{\partial t}\right) \tag{3.65}$$

with the commutator ,

$$\begin{aligned}
[\hat{H}, \hat{\rho}_\psi](x, y) &= \int_{-\infty}^{\infty} H(x, z)\rho_\psi(z, y)\mathrm{d}z \\
&\quad - \int_{-\infty}^{\infty} \rho_\psi(x, z)H(z, y)\mathrm{d}z \\
&= \int_{-\infty}^{\infty} H(x, z)\psi(z)\mathrm{d}z\,\psi(y)^* \\
&\quad - \psi(x)\int_{-\infty}^{\infty} (\psi(z)H(z, y))^*\mathrm{d}z \\
&= (\hat{H}\psi)(x)\psi(y)^* - \psi(x)(\hat{H}\psi)(y)^*.
\end{aligned} \tag{3.66}$$

In this calculation, we write operator products as continuous-variable matrix products of the functions $\rho(x, y)$ and $H(x, y)$. The latter represents $\hat{H}$ in the position representation by inverting the transformation (3.64) between operators and functions of two position variables. Moreover, we have used the reality of $H(x, y)$, just as we used the self-adjointness of $\hat{H}$ in Equation (3.62).

With these definitions, the basic expectation values, $\omega(\hat{x})$ and $\omega(\hat{p})$, obey equations of motion that resemble the classical equations:

$$\frac{\mathrm{d}\omega(\hat{x})}{\mathrm{d}t} = \frac{\omega(\hat{p})}{m}, \quad \frac{\mathrm{d}\omega(\hat{p})}{\mathrm{d}t} = -\omega\left(\frac{\mathrm{d}V(x)}{\mathrm{d}x}\Big|_{x=\hat{x}}\right). \tag{3.67}$$

According to these equations, referred to as *Ehrenfest's theorem*, the time derivative of the x-expectation value is related to the p-expectation value exactly in the classical manner. However, the time derivative of the p-expectation value differs from the classical result unless

$$\omega\left(\frac{\mathrm{d}V(x)}{\mathrm{d}x}\Big|_{x=\hat{x}}\right) = \frac{\mathrm{d}V(x)}{\mathrm{d}x}\Big|_{x=\omega(\hat{x})}. \tag{3.68}$$

This relation is true only if $V(x)$ is at most quadratic, which is not generic. If $V(x)$ is not quadratic, $\mathrm{d}V(x)/\mathrm{d}x$ is non-linear and there is a difference between evaluating ω at $(\mathrm{d}V(x)/\mathrm{d}x)|_{x=\hat{x}}$ and inserting $\omega(\hat{x})$ in $\mathrm{d}V(x)/\mathrm{d}x$: Any function that is non-linear in x, such as x^n with $n > 1$, generically implies a non-zero difference $\omega(\hat{x}^n) - \omega(\hat{x})^n$. For instance, if $n = 2$, we have $\omega(\hat{x}^2) - \omega(\hat{x})^2 = 0$ only if the x-variance is zero.

3.3.2 Harmonic Oscillator

The harmonic oscillator with Hamiltonian operator

$$\hat{H} = \frac{\hat{p}^2}{2m} + \frac{1}{2}m\nu^2\hat{x}^2 \tag{3.69}$$

presents a simple example because it implies decoupled equations of motion for moments of a fixed order. For expectation values, Ehrenfest's theorem shows that the quantum equations

$$\frac{d\omega(\hat{x})}{dt} = \frac{\omega(\hat{p})}{m}, \quad \frac{d\omega(\hat{p})}{dt} = -m\nu^2\omega(\hat{x}) \tag{3.70}$$

are identical with the classical ones, solved by

$$\begin{aligned}\omega(\hat{x})(t) &= A\sin(\nu t) + B\cos(\nu t),\\ \omega(\hat{p}) &= m\nu(A\cos(\nu t) - B\sin(\nu t))\end{aligned} \tag{3.71}$$

with constants A and B.

For second-order moments, we can use the product rule of the commutator in order to derive their equations of motion:

$$\frac{d\omega(\hat{x}^2)}{dt} = \frac{\omega([\hat{x}^2, \hat{H}])}{i\hbar} = \frac{\omega(\hat{x}[\hat{x}, \hat{H}] + [\hat{x}, \hat{H}]\hat{x})}{i\hbar} = \frac{1}{m}\omega(\hat{x}\hat{p} + \hat{p}\hat{x}). \tag{3.72}$$

The position variance, $(\Delta_\omega x)^2 = \omega(\hat{x}^2) - \omega(\hat{x})^2$, therefore has a time derivative

$$\frac{d(\Delta_\omega x)^2}{dt} = \frac{2}{m}\Delta_\omega(xp) \tag{3.73}$$

proportional to the covariance

$$\Delta_\omega(xp) = \frac{1}{2}\omega(\hat{x}\hat{p} + \hat{p}\hat{x}) - \omega(\hat{x})\omega(\hat{p}) \tag{3.74}$$

which is another second-order moment.

In general, the covariance is time dependent and has an equation of motion which can again be derived from commutators. We obtain

$$\begin{aligned}\frac{d\Delta_\omega(xp)}{dt} &= \frac{\omega([\hat{x}\hat{p} + \hat{p}\hat{x}, \hat{H}])}{2i\hbar} - \frac{d\omega(\hat{x})}{dt}\omega(\hat{p}) - \omega(\hat{x})\frac{d\omega(\hat{p})}{dt}\\ &= \frac{(\Delta_\omega p)^2}{m} - m\nu^2(\Delta_\omega x)^2.\end{aligned} \tag{3.75}$$

Here we encounter the momentum variance, which has time derivative

$$\frac{d(\Delta_\omega p)^2}{dt} = \frac{\omega([\hat{p}^2, \hat{H}])}{i\hbar} - 2\frac{d\omega(\hat{p})}{dt}\omega(\hat{p}) = -2m\nu^2\Delta_\omega(xp). \tag{3.76}$$

We now have a set of three coupled, linear differential equations, which is solved by

$$(\Delta_\omega x(t))^2 = C\sin(2\nu t) + D + E\cos(2\nu t)$$
$$\Delta_\omega(xp)(t) = m\nu(C\cos(2\nu t) - E\sin(2\nu t)) \qquad (3.77)$$
$$(\Delta_\omega p(t))^2 = -m^2\nu^2(C\sin(2\nu t) - D + E\cos(2\nu t))$$

with constants C, D and E. The combination

$$(\Delta_\omega x(t))^2(\Delta_\omega p(t))^2 - \Delta_\omega(xp)(t)^2 = m^2\nu^2(-C^2 + D^2 - E^2) \qquad (3.78)$$

is time-independent and obeys Heisenberg's uncertainty relation 3.26; provided

$$D^2 \geqslant \frac{\hbar^2}{4m^2\nu^2} + C^2 + E^2. \qquad (3.79)$$

For $C = E = 0$, the covariance vanishes at all times while the two variances are constant and non-zero. This behavior is realized for dynamical coherent states with constant variances, shown in Figure 3.1, while $C \neq 0$ or $E \neq 0$ imply the behavior of squeezed coherent states in which all second-order moments oscillate, see Figure 3.2.

Note, however, that our solutions apply to the second-order moments of any state of the harmonic oscillator. In any such state, fluctuations imply an energy contribution

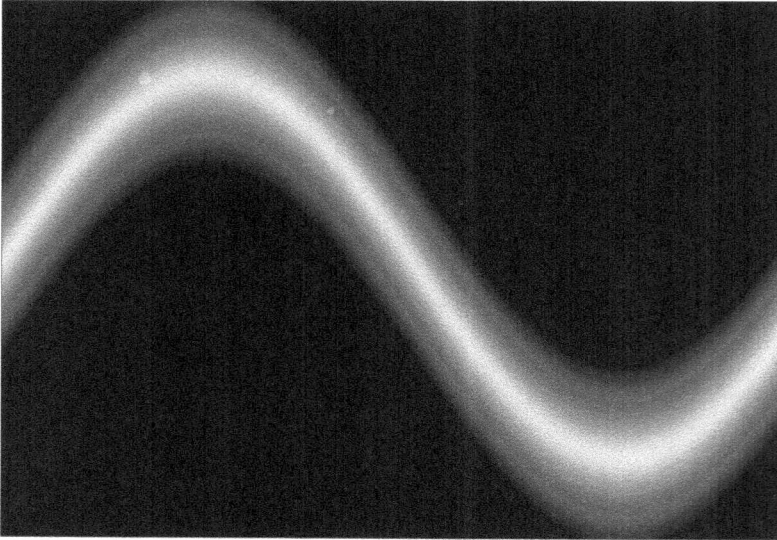

Figure 3.1. Illustration of a dynamical coherent state of the harmonic oscillator, time changing along the horizontal axis while the vertical axis marks the position. Colors indicate the height of the Gaussian wave function in arbitrary units, computed using the evolving expectation values and second-order moments. Notice that the position variance is constant in such a state. It only looks larger around the local extrema because the expectation value is changing more slowly in these regions. (The position variance should be measured in the vertical direction, not transversally to the curve followed by the state. However, in our intuitive interpretation of such a diagram we follow the latter method.)

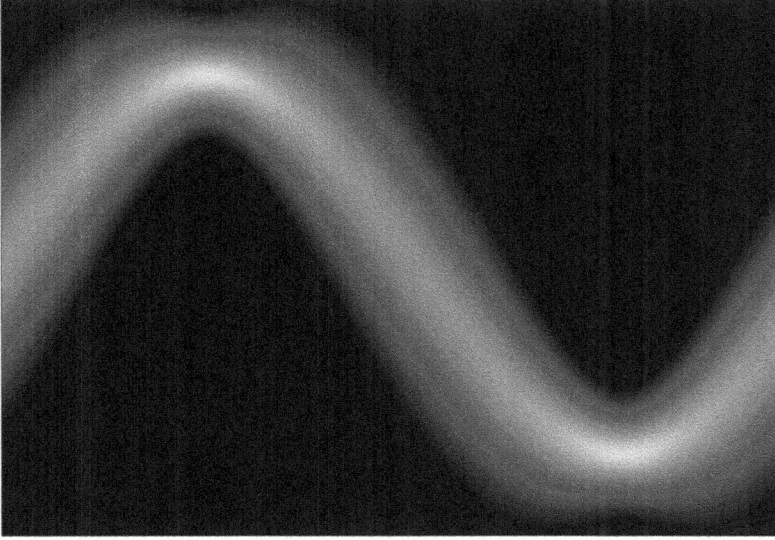

Figure 3.2. Dynamics of a squeezed coherent state of the harmonic oscillator, illustrated as described in Figure 3.1. A comparison with Figure 3.1 shows that the time-dependent position variance in this example is increased between local extrema, while it is suppressed around local extrema.

$$E_{\text{fluct}} = \frac{(\Delta_\omega p)^2}{2m} + \frac{1}{2}m\nu^2(\Delta_\omega x)^2 \tag{3.80}$$

in addition to the classical-type energy

$$E_{\text{class}} = \frac{\omega(\hat{p})^2}{2m} + \frac{1}{2}m\nu^2\omega(\hat{x})^2, \tag{3.81}$$

obtained by splitting the expectation value

$$\omega(\hat{H}) = \frac{\omega(\hat{p}^2)}{2m} + \frac{1}{2}m\nu^2\omega(\hat{x}^2) \tag{3.82}$$

into expectation-value and fluctuation terms From our general solutions we derive a time-independent

$$E_{\text{fluct}} = m\nu^2 D. \tag{3.83}$$

Its minimum value, given by $C = E = 0$ and $D = \frac{1}{2}m^{-1}\nu^{-1}\hbar$ based on Equation (3.79), equals the zero-point energy

$$E_{\text{fluct,min}} = \frac{1}{2}\hbar\nu \tag{3.84}$$

of the harmonic-oscillator ground state.

3.3.3 Effective Potentials

If we are interested in specific classes of states, it is possible to solve the moment equations in different ways. For instance, some questions require stationary states in which all moments are time-independent. The second-order equations of motion then imply that $\Delta_\omega(xp) \stackrel{.}{=} 0$, which in turn implies $\Delta_\omega p = m\nu\Delta_\omega x$ when evaluated for the harmonic oscillator. The fluctuation energy of any stationary state in this system is therefore given by

$$E_{\text{fluct}} = m\nu^2(\Delta_\omega x)^2. \tag{3.85}$$

In order to compute the ground-state energy, we can use a variational principle which is based on the fact that the expectation value $\hat{H}$ is minimized if and only if it is computed in the ground state. In its usual application, the variational method is applied to a suitable set of trial wave functions ψ to determine the expectation values $\omega_\psi(H)$ to be varied, because the full set of all possible wave functions would be too large and unwieldy. There is therefore always a question of how best to parameterize a sufficiently large class of trial functions which allows us to get close to the ground state. If the ground state is not already known, as in most examples of interest, it is very difficult to tell whether we do get close in a given set of trial states.

In the harmonic example, the fluctuation energy depends on a single parameter $(\Delta_\omega x)^2$, which can be derived from any state. The only condition is that this moment actually comes from a state, which means that it must obey Heisenberg's uncertainty relation 3.26. Since this inequality is linear in $(\Delta_\omega x)^2$, it is minimized at the boundary of allowed values, attained for states saturating the uncertainty relation. The resulting minimum, $(\Delta_\omega x_{\min})^2 = \frac{1}{2}m^{-1}\nu^{-1}\hbar$, implies a fluctuation energy equal to the ground-state energy $E_0 = \frac{1}{2}\hbar\nu$. Of course, standard textbooks provide the ground state of the harmonic oscillator by a Gaussian wave function, and it would have been easy to apply the variational method using Gaussian trial states. In more general cases, however, there is an advantage to a parameterization of states by moments, provided suitable approximations allow us to include a certain finite number of them.

Canonical Variables
Minimization and other calculations required for an application of the variational method in a parameterization by moments can be made more efficient by expressing the moments through canonical degrees of freedom. That is, the relevant variables are written as functions of three different types of coordinates: a certain number of configuration variables s_i, the same number of momentum variables p_i, and (if necessary) a certain number of constants of motion C_I which have vanishing time derivatives. Configuration and momentum variables (or canonical variables) are characterized by the existence of a Hamiltonian $H(s_i, p_{s_i}, C_I)$ such that Hamilton's equations are realized,

$$\frac{ds_i}{dt} = \frac{\partial H}{\partial p_{s_i}}, \quad \frac{dp_{s_i}}{dt} = -\frac{\partial H}{\partial s_i}, \quad \frac{dC_I}{dt} = 0. \tag{3.86}$$

Even though these equations look classical, they describe quantum states because now the canonical variables parameterize moments. A suitable Hamiltonian $H(s_i, p_{s_i}, C_I)$ should therefore be determined from the expectation value $\omega(\hat{H})$ of the Hamilton operator in a generic state.

According to Ehrenfest's equations, the expectation values $\omega(\hat{x})$ and $\omega(\hat{p})$ of any classical canonical pair x and p are canonical. However, moments do not come immediately in the form of canonical variables. For instance, there are three independent second-order moments, $(\Delta_\omega x)^2$, $\Delta_\omega(xp)$, and $(\Delta_\omega p)^2$, which, in general, could be parameterized by canonical variables in two possible ways: using either one s, one p and one C, or three C_I. The equations of motion (3.73), (3.75), and (3.76) for the harmonic oscillator do not directly indicate how such a parameterization should be performed.

With hindsight (Jackiw & Kerman 1979; Jalabert & Pastawski 2001; Prezhdo 2006) or systematic methods based on Poisson geometry (Baytaş et al. 2019, 2020), one can see that a good parameterization is given by $s = \Delta_\omega x$ with momentum

$$p_s = \frac{\Delta_\omega(xp)}{\Delta_\omega x}. \tag{3.87}$$

Equation (3.73) then has the required form, $\dot{s} = p_s/m$. Moreover, it is possible to show that the uncertainty product

$$U = (\Delta_\omega x)^2(\Delta_\omega p)^2 - \Delta_\omega(xp)^2 \tag{3.88}$$

is time-independent for any Hamiltonian that depends on second-order moments and basic expectation values. We can therefore identify U with the third canonical parameter, C. Inverting these relations, the moments parameterized through canonical variables are given by

$$(\Delta_\omega x)^2 = s^2, \quad \Delta_\omega(xp) = sp_s, \quad (\Delta_\omega p)^2 = p_s^2 + \frac{U}{s^2}. \tag{3.89}$$

In this parameterization, the quantum harmonic oscillator has the expected Hamiltonian

$$\begin{aligned}
\omega(\hat{H}) &= \frac{\omega(\hat{p}^2)}{2m} + \frac{1}{2}m\nu^2\omega(\hat{x}^2) \\
&= \frac{\omega(\hat{p})^2 + (\Delta_\omega p)^2}{2m} + \frac{1}{2}m\nu^2(\omega(\hat{x})^2 + (\Delta_\omega x)^2) \\
&= \frac{\omega(\hat{p})^2}{2m} + \frac{p_s^2}{2m} + \frac{U}{2ms^2} + \frac{1}{2}m\nu^2\omega(\hat{x})^2 + \frac{1}{2}m\nu^2s^2.
\end{aligned} \tag{3.90}$$

We can now minimize this canonical Hamiltonian with respect to the moment variables s and p_s. Minimization with respect to p_s requires $p_s = 0$, which implies that the ground state is stationary with respect to the second-order moments. The variable s appears in a combination of two terms, $U/(2ms^2) + \frac{1}{2}m\nu^2s^2$, which is minimized by

$$s^4 = \frac{U}{m^2 v^2}. \tag{3.91}$$

The constant U has a range limited by the lower bound $\hbar^2/4$ of the uncertainty relation applied to Equation (3.88). Minimization of all three canonical variables respecting this boundary implies

$$s^2 = \frac{\hbar}{2mv} \tag{3.92}$$

as in our previous calculation.

Semiclassical Expansion
As an approximation, the same canonical variables can be used for any potential if we perform a Taylor expansion around the expectation value of $\hat{x}$,

$$\begin{aligned} V(\hat{x}) = V(\omega(\hat{x})) + V'(\omega(\hat{x}))(\hat{x} - \omega(\hat{x})) \\ + \frac{1}{2} V''(\omega(\hat{x}))(\hat{x} - \omega(\hat{x}))^2 + \cdots. \end{aligned} \tag{3.93}$$

At this stage, the expansion is formal because we have not determined in which sense we consider $\hat{x} - \omega(\hat{x})$ to be small, which is not very clear if an operator $\hat{x}$ is involved. However, in an effective Hamiltonian we only need the expectation value of the potential. Calculating

$$\omega(V(\hat{x})) = V(\omega(\hat{x})) + \frac{1}{2} V''(\omega(\hat{x}))\omega((\hat{x} - \omega(\hat{x}))^2) + \cdots \tag{3.94}$$

by inserting the expansion 3.93; eliminating the linear term because $\omega(\hat{x} - \omega(\hat{x})) = 0$ for any normalized state, we replace operators by moments. The leading quantum term depends on the variance $\omega((\hat{x} - \omega(\hat{x}))^2) = \omega(\hat{x}^2) - \omega(\hat{x})^2$, which should be sufficiently small relative to $V(\omega(\hat{x}))/V''(\omega(\hat{x}))$ for the approximation to be reliable.

The expansion is therefore meaningful provided the values of higher-order moments that would result from higher-order terms in the Taylor expansion decrease sufficiently quickly with increasing order. A large class of states to which the approximation can be applied is such that the moments of order n decrease according to $(\hbar/h_0)^{n/2}$, where h_0 is a suitable classical parameter, depending on the specific system, with the units of $\hbar$ to provide a unitless number $\hbar/h_0$. For instance, the second-order moments, $n = 2$, of an unsqueezed state close to saturation of Heisenberg's uncertainty relation obey $(\Delta_\omega x)^2(\Delta_\omega p)^2 \approx \hbar^2/4$, such that each of them is of the order $\hbar = \hbar^{n/2}$. Any state with moments obeying this relationship of the order with $\hbar$ will be considered semiclassical.

Expansions of expectation values by moments, even with a semiclassicality assumption, may still have complicated convergence properties. It is known that such expansions usually do not converge but rather form asymptotic series. Nevertheless, the semiclassicality assumption improves the behavior because for generic states one could not even be sure that expectation values are analytic in $\hbar$; see for instance (Heller 1976). An exception is given by polynomial potentials, in which case the expansion (3.94) of $\omega(V(\hat{x}))$ around $\omega(\hat{x})$ is a finite series, and merely rewrites bare moments $\omega(\hat{x}^n)$ in a direct application of ω to the monomials of $V(\hat{x})$ in terms of central moments, $\omega((\hat{x} - \omega(\hat{x}))^n)$. But even then, solutions of equations of motion generated by the quantum potential $\omega(V(\hat{x}))$ are usually asymptotic series in $\hbar$.

In order to derive equations of motion, we use the expanded $\omega(V(\hat{x}))$ in an expected Hamiltonian,

$$\omega(\hat{H}) = \frac{\omega(\hat{p})^2}{2m} + \frac{(\Delta_\omega p)^2}{2m} + V(\omega(\hat{x})) + \frac{1}{2}V''(\omega(\hat{x}))(\Delta_\omega x)^2 + \cdots. \tag{3.95}$$

In terms of canonical variables s and p_s, in addition to $\omega(\hat{x})$ and $\omega(\hat{p})$, we have

$$\omega(\hat{H}) = \frac{\omega(\hat{p})^2}{2m} + \frac{p_s^2}{2m} + \frac{U}{2ms^2} + V(\omega(\hat{x})) + \frac{1}{2}V''(\omega(\hat{x}))s^2 + \cdots \tag{3.96}$$

which is identical with the Hamiltonian of a classical system of two degrees of freedom, both with mass m, interacting according to the effective potential

$$V_{\text{eff}}(\omega(\hat{x}), s) = \frac{U}{2ms^2} + V(\omega(\hat{x})) + \frac{1}{2}V''(\omega(\hat{x}))s^2 + \cdots. \tag{3.97}$$

To this order, there are two new, s-dependent terms compared with the classical potential. The last one, $\frac{1}{2}V''(\omega(\hat{x}))s^2$ has implications for tunneling because it is always negative around a local maximum of the classical potential, where $V''(\omega(\hat{x})) < 0$. The quantum correction therefore lowers the classical potential; see Figure 3.3 for an example. For sufficiently large s, corresponding to a state which has spread out by splitting into tunneled and reflected wave packets, the effective potential may be small enough to let the expectation value $\omega(\hat{x})$ bypass the classical barrier without violating energy conservation.

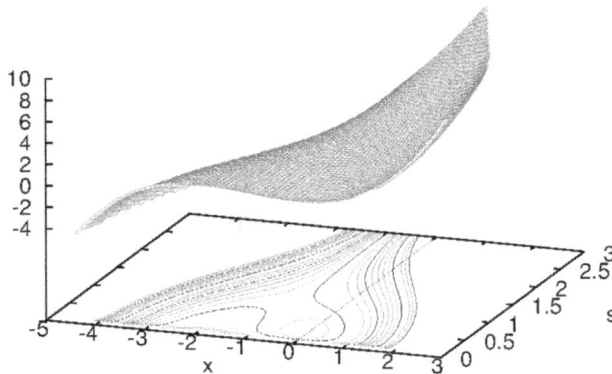

Figure 3.3. Effective potential (3.97) for a cubic classical potential, shown in Figure 3.4. As shown by the contour lines, in the vicinity of the local minimum of the classical potential, a path around the barrier opens up in the s-direction.

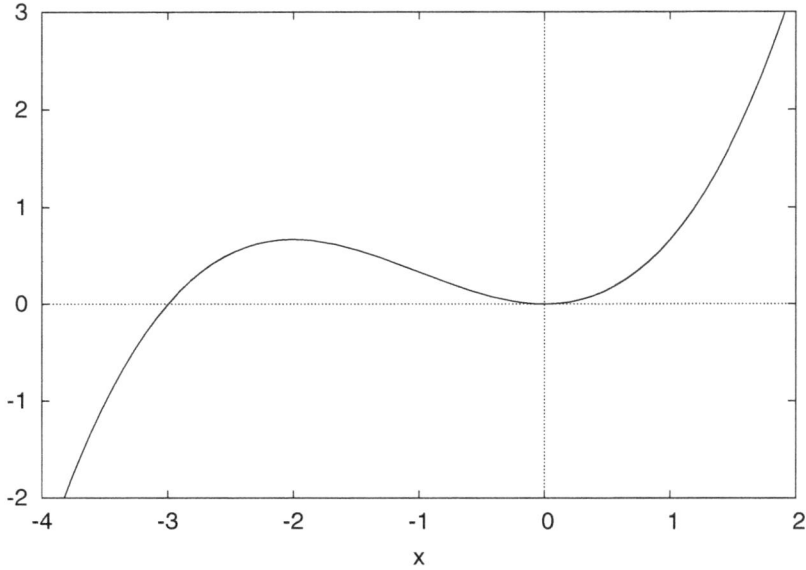

Figure 3.4. Cubic classical potential that gives rise to the effective potential shown in Figure 3.3.

The first new term in the effective potential, $U/(2ms^2)$, appears universally in all systems as it does not depend on the classical potential. (As derived, it originates from the expected kinetic energy.) It can be interpreted intuitively by viewing the combination

$$\frac{p_s^2}{2m} + \frac{U}{2ms^2} = \frac{p_X^2 + p_Y^2}{2m} \tag{3.98}$$

as the standard kinetic energy of two Cartesian coordinates, X and Y, written in polar coordinates on the plane. Formally, the two versions of the kinetic energy are mapped into each other by the canonical transformation

$$s = \sqrt{X^2 + Y^2}, \quad p_s = \frac{Xp_X + Yp_Y}{\sqrt{X^2 + Y^2}}. \tag{3.99}$$

The doubling of degrees of freedom means that we have introduced an auxiliary variable in addition to s, which can be identified as the polar angle

$$\phi = \arctan\frac{Y}{X} \tag{3.100}$$

with momentum

$$p_\phi = Xp_Y - Yp_X. \tag{3.101}$$

However, this auxiliary variable is not dynamical because it is constrained to take the constant value $p_\phi = \sqrt{U}$ in order to match the general two-dimensional kinetic energy

$$\frac{p_X^2 + p_Y^2}{2m} = \frac{p_s^2}{2m} + \frac{p_\phi^2}{2ms^2}.$$ (3.102)

with Equation (3.98). This constraint is consistent because (i) $U \geqslant \hbar^2/4$ is always positive, and (ii) the auxiliary degree of freedom ϕ does not appear in the effective potential, such that its momentum, just like U, is conserved.

Since the constrained $p_\phi > 0$ cannot be zero, there is always rotation in the auxiliary plane. The resulting centrifugal barrier, related to the original potential contribution $U/(2ms^2)$, enforces Heisenberg's uncertainty relation in a classical-type description.

Higher Moments and Cross-correlations
As the order of moments increases, they describe more and more degrees of freedom. We have $n + 1$ moments at order n, which sum up to

$$\sum_{n=2}^{N}(n + 1) = \frac{1}{2}(N^2 + 3N - 4)$$ (3.103)

moments up to order N. For multiple classical degrees of freedom, the number of moments is even larger because there are cross moments such as $\omega(\hat{x}_1 \hat{x}_2) - \omega(\hat{x}_1)\omega(\hat{x}_2)$ for two degrees of freedom, x_1 and x_2, in addition to the individual moments of each degree of freedom. These moments illustrate the non-local nature of quantum mechanics: If the individual classical degrees of freedom correspond to the positions of particles lined up in one dimension, even distant ones that may not directly interact with each other generically can have a non-zero correlation $\omega(\hat{x}_1 \hat{x}_2) - \omega(\hat{x}_1)\omega(\hat{x}_2)$ with non-trivial dynamics. There are $N(2N + 1)$ second-order moments for N degrees of freedom.

The ten second-order moments of two degrees of freedom have a canonical parameterization by two fluctuation variables, s_1 and s_2 with momenta p_{s_1} and p_{s_2}, respectively, two angles, α and β with momenta p_α and p_β, respectively, as well as two conserved quantities, C_1 and C_2 (Baytaş et al. 2019, 2020). The two individual variances are then

$$(\Delta_\omega x_1)^2 = s_1^2, \quad (\Delta_\omega x_2)^2 = s_2^2,$$ (3.104)

while the two individual covariances are

$$\Delta_\omega(x_1 p_1) = s_1 p_{s_1}, \quad \Delta_\omega(x_2 p_2) = s_2 p_{s_2}.$$ (3.105)

The two angles and their momenta appear in cross-correlations, such as

$$\Delta_\omega(x_1 x_2) = s_1 s_2 \cos \beta$$ (3.106)

and

$$\Delta_\omega(x_1 p_2) = s_1 p_{s_2} \cos \beta - \frac{s_1}{s_2}(p_\alpha + p_\beta)\sin \beta, \tag{3.107}$$

as well as the momentum variances such as

$$\Delta_\omega\!\left(p_1{}^2\right) = p_{s_1}^2 + \frac{U_1}{s_1^2} \tag{3.108}$$

where

$$U_1 = (p_\alpha - p_\beta)^2$$
$$+ \frac{1}{2\sin^2\beta}\!\left(\left(C_1 - 4p_\alpha^2\right) - \sqrt{C_2 - C_1^2 + \left(C_1 - 4p_\alpha^2\right)^2}\, \sin(\alpha + \beta)\right). \tag{3.109}$$

The equations simplify for states such that $\alpha = 0$, $p_\alpha = 0$ and $C_2 = 0$, in which case

$$\Delta_\omega\!\left(p_1{}^2\right) = p_{s_1}^2 + \frac{p_\beta^2}{s_1^2} + \frac{C_1}{2s_1^2\sin^2\beta} \tag{3.110}$$

has the form of a three-dimensional momentum squared written in spherical coordinates,

$$p_X^2 + p_Y^2 + p_Z^2 = p_{r,1}^2 + \frac{p_{\vartheta,1}^2}{r_1^2} + \frac{p_{\varphi,1}^2}{r_1^2\sin^2\vartheta_1} \tag{3.111}$$

where

$$r_1 = s_1 = \sqrt{X^2 + Y^2 + Z^2}, \quad \vartheta_1 = \beta = \arctan\left(\sqrt{X^2 + Y^2}/Z\right) \tag{3.112}$$

while the angle φ_1 is now auxiliary, with constrained momentum $p_{\varphi,1} = \sqrt{C_1/2}$.

3.3.4 Low-energy Effective Potential

The number of moments increases rather rapidly with the number of classical degrees of freedom. Moreover, the inclusion of moments in quantum dynamics leads to non-local interactions through cross-correlations: In a discrete model of a one-dimensional quantum field theory, for instance, we may line up a number N of degrees of freedom x_i, $i = 1, \ldots, N$, which classically interact through some next-neighbor local potential $V(x_i, x_{i+1})$. A second-order expansion of the potential therefore contains the cross-correlation $\Delta_\omega(x_i x_{i+1})$ in addition to the usual fluctuations $(\Delta_\omega x_i)^2$. Since

$$\{\Delta_\omega(x_{i+1} p_i), \Delta_\omega(x_i x_{i+1})\} = -(\Delta_\omega x_{i+1})^2, \tag{3.113}$$

derived from basic commutators as in Hamilton's equations for moments such as Equation (3.72), is non-zero for generic states, the time derivative of $\Delta_\omega(x_{i+1}p_i)$ is non-zero. Similarly,

$$\{\Delta_\omega(x_{i-1}p_i), \Delta_\omega(x_{i-1}x_i)\} = -(\Delta_\omega x_{i-1})^2 \tag{3.114}$$

implies that $\Delta_\omega(x_{i-1}p_i)$ generically has a non-zero time derivative.

Once a state has a non-zero $\Delta_\omega(x_{i\pm1}p_i)$, the time derivative of $\Delta_\omega(x_i x_{i\pm1})$ is non-zero because

$$\{\Delta_\omega(x_i x_{i\pm1}), (\Delta_\omega p_i)^2\} = 2\Delta_\omega(x_{i\pm1}p_i) \tag{3.115}$$

contributes to the time derivative through the kinetic term of the Hamiltonian. A state therefore generically develops next-neighbor cross-correlations.

Since this argument applies to any i, after some time we have non-zero time derivative of $\Delta_\omega(x_{i+2}p_i)$, using

$$\{\Delta_\omega(x_{i+2}p_i), \Delta_\omega(x_i x_{i+1})\} = -\Delta_\omega(x_{i+1}x_{i+2}). \tag{3.116}$$

Once this moment is non-zero, we have a non-zero time derivative of $\Delta_\omega(x_i x_{i+2})$, using the kinetic term of the Hamiltonian. Iteration then shows that eventually all cross-correlations, even over long distances, may become non-zero—unless the moments of a specific state, such as the ground state, are such that some of the time derivatives used in our arguments vanish due to cancellations based on specific coefficients of the next-neighbor moments in an effective potential. Generically, therefore, quantum theories imply non-local behavior.

The description simplifies if we are interested in states near the ground state (assuming that it exists) of a system with small, perturbative classical interactions. To zeroth order in perturbation theory, the ground state is determined by the product of the ground-state wave functions of all individual degrees of freedom. Such a product is free of cross-correlations, which eliminates non-locality from the quantum description. When interactions are included in perturbation theory, cross-correlations may be generated dynamically. But being perturbative, they are much smaller than the variances of individual ground states, which have lower bounds thanks to the uncertainty relation irrespective of the strength of perturbative interactions. To low orders in perturbation theory, cross-correlations therefore remain approximately zero compared with individual moments of the degrees of freedom.

As a more specific example, our discrete system of N degrees of freedom of unit mass may be viewed as a discretized version of the Hamiltonian

$$H = \int_{-L/2}^{L/2} \left(\frac{1}{2}\left(c^2 p_\phi(x)^2 + \left(\frac{\partial\phi}{\partial x}\right)^2 \right) + V(\phi(x)) \right) dx, \tag{3.117}$$

for a self-interacting scalar field ϕ on Minkowski spacetime with one finite spatial dimension of length L. (See the parenthesis following Equation (1.83) for the origin of the factor of c^2 in the kinetic term.) Assuming spatial translation invariance, we have used the same potential, $V(\phi(x))$, at each point x. General solutions for $\phi(x)$ may nevertheless be inhomogeneous, but the ground state of a translation-invariant system should respect this symmetry. We can use the ground-state value, ϕ_0 such that $V(\phi_0)$ minimizes the potential at each point, as a constant background for an expansion, $\phi(x) = \phi_0 + \delta\phi(x)$. Similarly, $p_\phi(x) = \delta p_\phi(x)$ because $p_\phi = 0$ in the static ground state. The Hamiltonian is then expanded as

$$H = \int_{-L/2}^{L/2} \left(\frac{1}{2}\left(c^2\delta p_\phi(x)^2 + \left(\frac{\partial\delta\phi}{\partial x}\right)^2 \right) + V(\phi_0) + \frac{1}{2}V''(\phi_0)\delta\phi(x)^2 \right)dx \qquad (3.118)$$

to second order in $\delta\phi$. Although the potential is now quadratic in $\delta\phi(x)$, the field values at different points are still coupled through local interactions, owing to the spatial derivative term. We can decouple the modes by using Fourier transformation,

$$\delta\phi(x) = \frac{1}{2\pi} \sum_{n=-\infty}^{\infty} \delta\tilde{\phi}_{k_n} e^{ik_n x} \qquad (3.119)$$

and

$$\delta p_\phi(x) = \frac{1}{2\pi} \sum_{n=-\infty}^{\infty} \delta\tilde{p}_{\phi,k_n} e^{-ik_n x}, \qquad (3.120)$$

where $k_n = 2\pi n/L$. The Hamiltonian then equals

$$H = \frac{1}{2\pi} \sum_{n=-\infty}^{\infty} \left(\frac{1}{2}c^2\delta\tilde{p}_{\phi,k_n}^2 + \frac{1}{2}k_n^2\delta\tilde{\phi}_{k_n}^2 + \frac{1}{2}V''(\phi_0)\delta\tilde{\phi}_{k_n}^2 \right) + LV(\phi_0). \qquad (3.121)$$

(Reality of $\delta\phi(x)$ and $\delta p_\phi(x)$ implies that $\delta\tilde{\phi}_{k_n} = \delta\tilde{\phi}_{-k_n}$ and $\delta\tilde{p}_{\phi,k_n} = \delta\tilde{p}_{\phi,-k_n}$.) The potential for the modes is therefore

$$V(\delta\tilde{\phi}_k) = V(\phi_0) + \frac{1}{4\pi L} \sum_{n=-\infty}^{\infty} \left(V''(\phi_0) + k_n^2 \right)\delta\tilde{\phi}_{k_n}^2. \qquad (3.122)$$

To second order in moments with zero cross-correlations, justified by the proximity to the ground state, we can use a canonical parameterization of fluctuations by variables s_{k_n} and p_{s,k_n} for each mode independently. The Hamiltonian then receives an additional kinetic contribution $\frac{1}{2}c^2 p_{s,k_n}^2$ as well as an effective potential

$$V_{\text{eff}}(\phi_0, s_{k_n}) = V(\phi_0) + \frac{1}{4\pi L} \sum_{n=-\infty}^{\infty} \left(\frac{c^2 U_{k_n}}{s_{k_n}^2} + \left(V''(\phi_0) + k_n^2 \right)s_{k_n}^2 \right) + \cdots. \qquad (3.123)$$

There are independent fluctuation and uncertainty variables, s_{k_n} and U_{k_n}, for each mode, k_n. In order to be close to the ground state, we minimize the effective potential with respect to s_{k_n} and U_{k_n} (subject to $U_{k_n} \geqslant \hbar^2/4$). The result,

$$s_{k_n} = \sqrt[4]{\frac{c^2 U_{k_n}}{m(V''(\phi_0) + k_n^2)}} \qquad (3.124)$$

with $U_{k_n} = \hbar^2/4$, leads to the *low-energy effective potential*

$$V_{\text{low-energy}}(\phi_0) = V(\phi_0) + \frac{c\hbar}{4\pi L} \sum_{n=-\infty}^{\infty} \left(\sqrt{V''(\phi_0) + k_n^2} - |k_n|\right). \qquad (3.125)$$

We have subtracted $|k_n|$ from the integrand in order to obtain $V_{\text{low-energy}}(\phi_0) = V(\phi_0)$ for a constant potential. (Subtracting $\frac{1}{2}\hbar|k_n|$ from the effective potential removes the zero-point contributions (3.84) to the energy implied by infinitely many mode oscillators.) In the limit $L \to \infty$, the series is turned into an integral,

$$V_{\text{low-energy}}(\phi_0) = V(\phi_0) + \frac{c\hbar}{4\pi} \int_{-\infty}^{\infty} \left(\sqrt{V''(\phi_0) + k^2} - |k|\right) \mathrm{d}k. \qquad (3.126)$$

For a classical potential $V(\phi(x)) = \lambda\phi(x)^4$, for instance, we have the low-energy effective potential

$$V_{\text{low-energy}}(\phi_0) = \lambda\phi_0^4 + \frac{c\hbar}{4\pi} \int_{-\infty}^{\infty} \left(\sqrt{12\lambda\phi_0^2 + k^2} - |k|\right) \mathrm{d}k. \qquad (3.127)$$

This potential can be written with a Fourier version of spacetime integration,

$$V_{\text{low-energy}}(\phi_0) = \lambda\phi_0^4 + \frac{ic\hbar}{4\pi} \int_{-\infty}^{\infty} \int_{-\infty}^{\infty} \log\left(1 + \frac{12\lambda\phi_0^2}{k^2 - k_0^2}\right) \mathrm{d}k\,\mathrm{d}k_0. \qquad (3.128)$$

In this form, the low-energy effective potential was originally derived by (Coleman & Weinberg 1973; in four spacetime dimensions). The moment derivation presented here is adapted from Bojowald & Brahma (2014).

3.3.5 States at Non-zero Temperature

A large class of mixed states is given by systems in thermodynamic equilibrium at non-zero temperature T. The constituent degrees of freedom, such as molecules in a container of gas, do not occupy the same state because they keep interacting with one another, exchanging energy. However, in equilibrium the average energy and other quantities have fixed values depending on the temperature. If the degrees of freedom are quantum, there are two different averages, one over the ensemble of degrees of freedom, and one in the quantum-mechanical sense of an expectation value in the microscopic state of a single constituent.

A useful tool to derive thermodynamical averages in classical statistical mechanics is the *partition function*

$$Z = \frac{1}{2\pi\hbar} \int_{-\infty}^{\infty} \int_{-\infty}^{\infty} \exp(-\beta E(x, p)) \mathrm{d}x\,\mathrm{d}p \qquad (3.129)$$

with the inverse temperature, $\beta = 1/(k_B T)$ in terms of *Boltzmann's constant k_B*. The energy function $E(x, p)$ determines the statistical weight, or probability $0 \leqslant \exp(-\beta E(x, p)) \leqslant 1$, by which a region in phase space contributes, making such a contribution less likely for energies much larger than $k_B T$. Even in classical statistics, Planck's constant appears as a normalization factor of the partition function, attributing a fundamental area of $2\pi\hbar$ to phase space. The ensemble average of an observable O, described classically by a phase-space function $O(x, p)$, is then given by

$$\langle O \rangle_E = \frac{1}{2\pi\hbar Z} \int_{-\infty}^{\infty} \int_{-\infty}^{\infty} O(x, p)\exp(-\beta E(x, p))\mathrm{d}x\mathrm{d}p. \qquad (3.130)$$

We divide by Z in order to normalize the probability distribution $\exp(-\beta E(x, p))$.

In quantum statistical physics, the partition function is computed using similar weights, $\exp(-\beta \hat{H})$ with the Hamilton operator $\hat{H}$. The phase-space integral is replaced by a sum over all independent pure states of the system, usefully parameterized by the choice of a suitable basis of states, ψ_n:

$$Z = \frac{1}{2\pi\hbar} \sum_n \omega_{\psi_n}(\exp(-\beta \hat{H})). \qquad (3.131)$$

This function of the temperature is the trace of the operator $\exp(-\beta\hat{H})/(2\pi\hbar)$ that appears in the definition of the density matrix

$$\hat{\rho}_\beta = \frac{1}{2\pi\hbar Z} \exp(-\beta \hat{H}) \qquad (3.132)$$

belonging to a mixed state of the statistical ensemble.

If this matrix is expressed in the basis ψ_n of a representation, its diagonal components are determined by the statistical weights

$$\rho_{\beta,nn} = \frac{1}{2\pi\hbar Z}\omega_{\psi_n}(\exp(-\beta \hat{H})). \qquad (3.133)$$

The quantum-mechanical expectation value of an observable represented by an operator $\hat{O}$ is given by the mixed-state expectation value

$$\langle \hat{O} \rangle_{\hat{\rho}_\beta} = \frac{1}{2\pi\hbar Z} \sum_n \omega_{\psi_n}(\hat{O} \exp(-\beta \hat{H})) = \sum_{n,m} O_{nm}\rho_{\beta,mn} \qquad (3.134)$$

with the matrix elements

$$\rho_{\beta,mn} = \int_{-\infty}^{\infty} \psi_n(x)^*\hat{\rho}_\beta\psi_m(x)\mathrm{d}x \qquad (3.135)$$

of $\hat{\rho}_\beta$ in the representation with basis ψ_n. This expectation value is different from a classical-type expectation value with weights 3.133; which would include only contributions with $n = m$.

Since we have a phase-space description of semiclassical moments in terms of canonical variables, given by Equation (3.89), we can, following (Baytaş et al. 2019), replace the quantum partition function with a semiclassical version,

$$Z = \frac{1}{(2\pi\hbar)^2} \int_{-\infty}^{\infty} \int_{-\infty}^{\infty} \int_{0}^{\infty} \int_{-\infty}^{\infty} \int_{\hbar^2/4}^{\infty}$$
$$\times \exp(-\beta E(x, p, s, p_s, U)) dx dp ds dp_s dU \qquad (3.136)$$

where integration ranges take into account the inequalities $s \geq 0$ and $U \geq \hbar^2/4$. The normalization factor has been squared because we now have two canonical pairs. The lone conserved quantity U does not contribute to the phase-space area.

For a harmonic oscillator, the moment variables provide an additional factor

$$Z = \frac{1}{2\pi\hbar} \int_{0}^{\infty} \int_{-\infty}^{\infty} \int_{\hbar^2/4}^{\infty} ds\, dp_s\, dU$$
$$\times \exp\left(-\beta\left(\frac{1}{2m}p_s^2 + \frac{U}{2\,ms^2} + \frac{1}{2}m\nu^2 s^2\right)\right) \qquad (3.137)$$
$$= \frac{1}{2\hbar} \frac{2 + \beta\hbar\nu}{\nu^3\beta^3} \exp\left(-\frac{1}{2}\beta\hbar\nu\right)$$

to the classical partition function (3.129). It allows us to compute the ensemble average of the quantum variance, defined as

$$\langle s^2 \rangle_E = \frac{1}{(2\pi\hbar)^2} \int_{-\infty}^{\infty} \int_{-\infty}^{\infty} \int_{0}^{\infty} \int_{-\infty}^{\infty} \int_{\hbar^2/4}^{\infty} s^2$$
$$\times \exp(-\beta E(x, p, s, p_s, U)) dx dp ds dp_s dU, \qquad (3.138)$$

via a partial derivative of Z by ν:

$$\langle s^2 \rangle_E = -\frac{1}{m\nu\beta} \frac{\partial \log Z}{\partial \nu} = 3\frac{k_B T}{m\nu^2} + \frac{1}{2m}\frac{\hbar^2}{2k_B T + \hbar\nu}. \qquad (3.139)$$

The fluctuation momentum, p_s, has zero average, $\langle p_s \rangle_E = 0$, because

$$\int_{-\infty}^{\infty} p_s \exp\left(-\frac{1}{2}\beta p_s^2/m\right) dp_s = 0. \qquad (3.140)$$

These quantities combine the quantum mechanical average in a microscopic state, used to define the meaning of s and p_s as fluctuation parameters, with the thermodynamical ensemble average over many copies of the microscopic system at a given temperature. As an example, the microscopic state may be viewed as belonging to a single diatomic molecule with a bond that could be modeled as a one-dimensional harmonic oscillator. Many such molecules at a given temperature then form the ensemble. Each microscopic state in general has a quantum fluctuation of the bond length (x) that differs from other molecules in the ensemble. The precise individual values are not determined by the temperature, but the average is given by Equation (3.139).

The ensemble average of the uncertainty parameter U can be derived if we generalize the partition function by inserting a multiplier λ,

$$
\begin{aligned}
Z_\lambda &= \frac{1}{2\pi\hbar} \int_0^\infty \int_{-\infty}^\infty \int_{\hbar^2/4}^\infty \, \mathrm{d}s \, \mathrm{d}p_s \, \mathrm{d}U \\
&\quad \times \exp\left(-\beta\left(\frac{1}{2m}p_s^2 + \frac{\lambda U}{2\,ms^2} + \frac{1}{2}m\nu^2 s^2\right)\right) \\
&= \frac{1}{2\hbar} \frac{2 + \beta\sqrt{\lambda}\,\hbar\nu}{\lambda\beta^3\nu^3} \exp\left(-\frac{1}{2}\beta\sqrt{\lambda}\,\hbar\nu\right).
\end{aligned}
\tag{3.141}
$$

This multiplier, unlike β, does not have a physical interpretation and must be set equal to one in all final expressions. By inserting it in the generalized partition function, we are able to compute the average uncertainty

$$
\langle U \rangle_{\mathrm{E}} = \frac{2}{\beta^2\nu} \frac{1}{Z} \frac{\partial^2 Z}{\partial\nu\partial\lambda}\bigg|_{\lambda=1} = \frac{\hbar^2}{4} + \frac{6}{\beta^2\nu^2} + \frac{\hbar^2}{2 + \beta\hbar\nu},
\tag{3.142}
$$

shown in Figure 3.5.

The averages

$$
\langle p_s^2 \rangle_{\mathrm{E}} = \frac{\int_{-\infty}^\infty p_s^2 \exp\left(-\frac{1}{2}\beta p_s^2/m\right)}{\int_{-\infty}^\infty \exp\left(-\frac{1}{2}\beta p_s^2/m\right)} = \frac{m}{\beta}
\tag{3.143}
$$

and

$$
\langle U/s^2 \rangle_{\mathrm{E}} = -\frac{2m}{\beta} \frac{1}{Z} \frac{\partial Z}{\partial\lambda}\bigg|_{\lambda=1} = \frac{m}{2\beta} \frac{8 + 4\beta\hbar\nu + (\beta\hbar\nu)^2}{2 + \beta\hbar\nu}
\tag{3.144}
$$

help us to derive the average moments

$$
(\Delta_{\mathrm{E}}x)^2 = \langle s^2 \rangle_{\mathrm{E}} = 3\frac{k_{\mathrm{B}}T}{m\nu^2} + \frac{1}{2m}\frac{\hbar^2}{2k_{\mathrm{B}}T + \hbar\nu}
\tag{3.145}
$$

$$
\Delta_{\mathrm{E}}(xp) = \langle sp_s \rangle_{\mathrm{E}} = 0
\tag{3.146}
$$

$$
(\Delta_{\mathrm{E}}p)^2 = \left\langle p_s^2 + \frac{U}{s^2} \right\rangle_{\mathrm{E}} = \frac{m}{2}\frac{12(k_{\mathrm{B}}T)^2 + 10\hbar\nu k_{\mathrm{B}}T + \hbar^2\nu^2}{2k_{\mathrm{B}}T + \hbar\nu}
\tag{3.147}
$$

of a mixed state in thermodynamical equilibrium; see Figure 3.6. For $T \to 0$, these values approach the limit given by the harmonic-oscillator ground state.

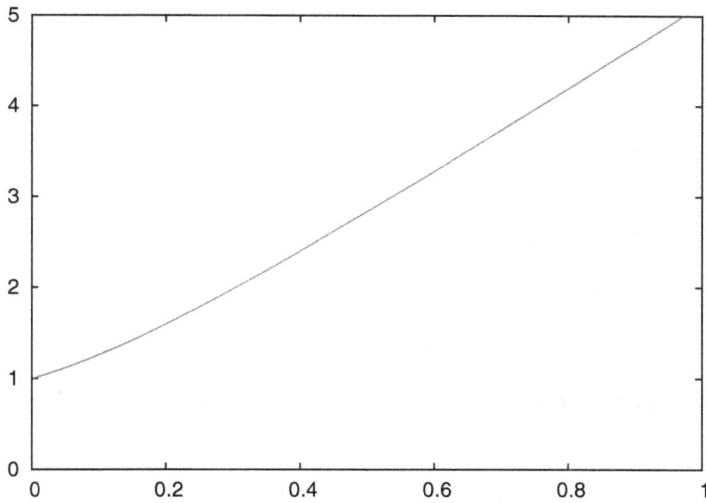

Figure 3.5. Ensemble average of the dimensionless uncertainty product, $2\sqrt{U}/\hbar$ from Equation (3.142), as a function of the dimensionless $k_BT/(\hbar\nu)$.

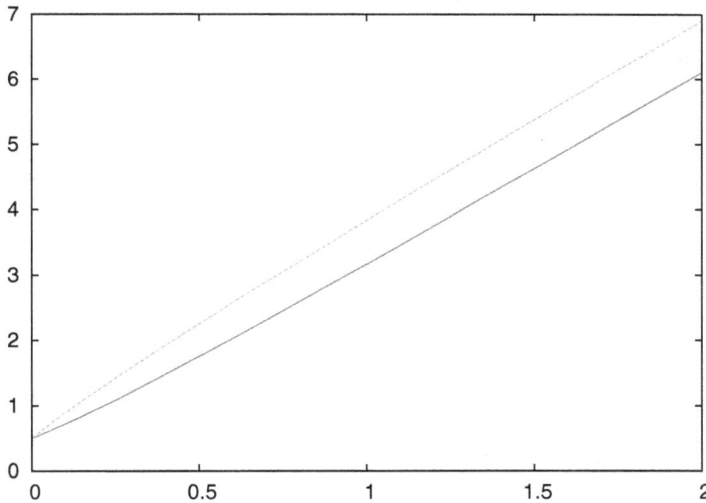

Figure 3.6. Position and momentum variance in thermal states of the harmonic oscillator as functions of the dimensionless temperature, $k_BT/(\hbar\nu)$. The red curve shows the dimensionless position variance $m\nu(\Delta_E x)^2/\hbar$ from Equation (3.145), while the green curve shows the dimensionless momentum variance $(\Delta_E p)^2/(m\hbar\nu)$ from Equation (3.147).

3.3.6 Time-correlation Functions and the Fluctuation–Dissipation Theorem

It is often of interest to compute expectation values of products of observables evaluated at different times. For instance, transition probabilities in tunneling or ionization applications are of this form. A useful result, the *fluctuation–dissipation theorem*, allows one to derive properties of the energy spectrum from such expectation values. If we use the position as an observable, its time-correlation function

$$C_{x,\omega}(t) = \omega(\hat{x}(0)\hat{x}(t)) \tag{3.148}$$

in a state ω can be computed exactly for the harmonic oscillator.

In order to do so, we first rewrite $C_{x,\omega}(t)$ in the form

$$\omega(\hat{x}(0)\hat{x}(t)) = \omega(\hat{x}\exp(i\hat{H}t/\hbar)\hat{x}\exp(-i\hat{H}t/\hbar)), \tag{3.149}$$

using the formal solution

$$\hat{O}(t) = \exp(i\hat{H}t/\hbar)\hat{O}\exp(-i\hat{H}t/\hbar) \tag{3.150}$$

of Heisenberg's equation of motion (3.59) for any observable $\hat{O}$. Upon expanding the exponential functions in a Taylor series, Equation (3.149) equals

$$\omega(\hat{x}(0)\hat{x}(t)) = \sum_{n=0}^{\infty} \frac{1}{n!}\left(\frac{it}{\hbar}\right)^n \omega(\hat{x}[[\hat{H}, \hat{x}]]_n) \tag{3.151}$$

$$= \omega(\hat{x}^2) + \frac{it}{\hbar}\omega(\hat{x}[\hat{H}, \hat{x}]) - \frac{t^2}{2\hbar^2}\omega(\hat{x}[\hat{H}, [\hat{H}, \hat{x}]]) + \mathcal{O}(t^3). \tag{3.152}$$

The coefficients are expressed in terms of the iterated commutators $[[\hat{H}, \hat{x}]]_n$, defined recursively by $[[\hat{H}, \hat{x}]]_0 = \hat{x}$ and $[[\hat{H}, \hat{x}]]_{n+1} = [[[\hat{H}, \hat{x}]]_n, \hat{x}]$.

For the harmonic-oscillator Hamiltonian, the iterated commutators with $\hat{x}$ have the simple form

$$[[\hat{H}, \hat{x}]]_{2n+1} = -i\nu^{2n}\hbar^{2n+1}\frac{\hat{p}}{m} \tag{3.153}$$

$$[[\hat{H}, \hat{x}]]_{2n} = \nu^{2n}\hbar^{2n}\hat{x}, \tag{3.154}$$

which can be shown by induction. Splitting the series into contributions from odd and even n, respectively, we therefore have

$$\omega(\hat{x}(0)\hat{x}(t)) = \sum_{n=0}^{\infty} \frac{1}{(2n + 1)!}\frac{(-1)^n\nu^{2n}}{m}t^{2n+1}\omega(\hat{x}\hat{p})$$

$$+ \sum_{n=0}^{\infty} \frac{1}{(2n)!}(-1)^n\nu^{2n}t^{2n}\omega(\hat{x}^2). \tag{3.155}$$

The remaining equal-time expectation values can be related to moments we already computed for the ground state of the harmonic oscillator. First, the required bare moments are related to their corresponding central moments by $\omega(\hat{x}\hat{p}) = \Delta_\omega(xp) + \frac{1}{2}i\hbar$ with $\Delta_\omega(xp) = 0$ for any stationary state, and $\omega(\hat{x}^2) = (\Delta_\omega x)^2$ for any state with $\omega(\hat{x}) = 0$. For the ground state, we have already computed $(\Delta_\omega x)^2 = \hbar/(2m\nu)$, while it turns out that $(\Delta_\omega x)^2 = (\ell + 1)\hbar/(2m\nu)$ for the excited states with $\ell \geqslant 1$. For all these states, we can sum up the series of the time-correlation function,

$$\omega(\hat{x}(0)\hat{x}(t)) = \frac{\hbar}{2m\nu}\left(\sum_{n=0}^{\infty}\frac{1}{n!}(i\nu t)^n + \ell\sum_{n=0}^{\infty}\frac{(-1)^n}{(2n)!}(\nu t)^{2n}\right)$$

$$= \frac{\hbar}{2m\nu}(\exp(i\nu t) + \ell\cos(\nu t)). \tag{3.156}$$

Another example is given by a mixed state at temperature T, for which the moments (3.145) give us

$$\omega_E(\hat{x}(0)\hat{x}(t)) = \sum_{n=0}^{\infty}\frac{1}{(2n+1)!}\frac{(-1)^n\nu^{2n}}{m}t^{2n+1}\frac{1}{2}i\hbar$$

$$+ \sum_{n=0}^{\infty}\frac{1}{(2n)!}(-1)^n\nu^{2n}t^{2n}\left(3\frac{k_BT}{\nu^2} + \frac{1}{2}\frac{\hbar^2}{2k_BT + \hbar\nu}\right) \tag{3.157}$$

$$= \frac{\hbar}{2m\nu}\left(\exp(-i\nu t) + 8\frac{k_BT}{\hbar\nu}\frac{3k_BT + \hbar\nu}{2k_BT + \hbar\nu}\cos(\nu t)\right).$$

Time-correlation functions are closely related to energy spectra. For a general observable $\hat{O}$ and Hamilton operator $\hat{H}$, as well as a state ω_{ψ_m} described by an eigenfunction ψ_m of $\hat{H}$ such that $\hat{H}\psi_m = E_m\psi_m$, we compute

$$\omega_{\psi_m}(\hat{O}(0)\hat{O}(t)) = (\psi_m, \hat{O}\exp(-it\hat{H}/\hbar)\hat{O}\exp(it\hat{H}/\hbar)\psi_m)$$

$$= \sum_n(\psi_m, \hat{O}\psi_n)(\psi_n, \exp(-it\hat{H}/\hbar)\hat{O}\exp(it\hat{H}/\hbar)\psi_m)$$

$$= \sum_n(\psi_m, \hat{O}\psi_n)e^{it(E_m-E_n)/\hbar}(\psi_n, \hat{O}\psi_m) \tag{3.158}$$

$$= \sum_n e^{it(E_m-E_n)/\hbar}|O_{mn}|^2$$

where we used completeness of the eigenfunctions ψ_n of $\hat{H}$ with eigenvalues E_n. The coefficients

$$O_{mn} = (\psi_m, \hat{O}\psi_n) = \int_{-\infty}^{\infty}\psi_m(x)^*\hat{O}\psi_n(x)dx \tag{3.159}$$

are defined as matrix elements of $\hat{O}$ in the eigenbasis of $\hat{H}$. The final result (3.158) shows that the Fourier frequencies of the time-correlation functions $\omega_{\psi_m}(\hat{O}(0)\hat{O}(t))$, multiplied by $\hbar$, determine the allowed energy differences $\Delta E_{mn} = E_m - E_n$ in the energy spectrum.

In a complicated system we may not be able to prepare an initial state that corresponds to a specific eigenstate ω_{ψ_n}. However, if the state is not described by an eigenfunction of $\hat{H}$, we can always expand it as a superposition of these eigenfunctions as long as it is pure,

$$\psi = \sum_m c_m \psi_m \tag{3.160}$$

with suitable coefficients c_m. In this case,

$$
\begin{aligned}
\omega_\psi(\hat{O}(0)\hat{O}(t)) &= \sum_m c_m(\psi,\ \hat{O}\exp(-it\hat{H}/\hbar)\hat{O}\exp(it\hat{H}/\hbar)\psi_m) \\
&= \sum_{m,n} c_m(\psi,\ \hat{O}\psi_n)(\psi_n,\ \exp(-it\hat{H}/\hbar)\hat{O}\exp(it\hat{H}/\hbar)\psi_m) \\
&= \sum_{m,n} c_m(\psi,\ \hat{O}\psi_n)e^{it(E_m-E_n)/\hbar}(\psi_n,\ \hat{O}\psi_m) \\
&= \sum_{m,n} e^{it(E_m-E_n)/\hbar}C_{m,n}
\end{aligned}
\tag{3.161}
$$

with coefficients

$$C_{m,n} = c_m(\psi,\ \hat{O}\psi_n)(\psi_n,\ \hat{O}\psi_m). \tag{3.162}$$

Although the coefficients are now more complicated combinations of matrix elements of the observable $\hat{O}$, the Fourier spectrum of the time-correlation function again determines allowed energy differences in the spectrum.

In quantum field theory, time-correlation functions are often expressed in an alternative form, using the product of two operators at different times, t_1 and t_2, in *time ordering*,

$$\omega(T\hat{x}(t_1)\hat{x}(t_2)) = \begin{cases} \omega(\hat{x}(t_1)\hat{x}(t_2)) & \text{if } t_1 < t_2 \\ \omega(\hat{x}(t_2)\hat{x}(t_1)) & \text{if } t_1 > t_2 \end{cases}. \tag{3.163}$$

Thus, for an energy eigenstate of the harmonic oscillator,

$$\omega(T\hat{x}(t_1)\hat{x}(t_2)) = \frac{\hbar}{2m\upsilon}(\exp(-i\upsilon|t_1 - t_2|) + \ell\cos(\upsilon|t_1 - t_2|)). \tag{3.164}$$

3.4 Transition Amplitude and Path Integrals

Probabilities of transitions between different states, ϕ_1 and ϕ_2, can often be described in terms of the transition amplitude (ϕ_1, ϕ_2). The norm squared of this inner product is interpreted as the probability of finding a system in the state ϕ_1 if it is actually in the state ϕ_2. The question thus posed is perhaps more interesting, and certainly more dynamical, if the two states describe the system at different times, such as $(\phi_1(0), \phi_2(t))$. The norm squared of this value then tells us how likely it is that a system prepared in the state ϕ_1 at time $t = 0$ evolves into a state ϕ_2 some time t later.

3.4.1 Transition Amplitudes

Using a complete set of normalized and orthogonal wave functions, ψ_n, it is possible to derive any transition amplitude from the amplitudes $(\psi_m(0), \psi_n(t))$:

$$(\phi_1(0), \phi_2(t)) = \sum_{m,n}(\phi_1(0), \psi_m(0)) \tag{3.165}$$
$$\cdot (\psi_m(0), \psi_n(t)) \cdot (\psi_n(t), \phi_2(t)).$$

The coefficients $(\phi_1(0), \psi_m(0))$ and $(\psi_n(t), \phi_2(t))$ express the two states as superpositions of the complete wave functions at fixed times 0 and t, respectively. For instance, $\psi_1(0) = \sum_n c_n(0)\psi_n(0)$ requires that $c_n(0) = (\psi_n(0), \psi_1(0))$ for an orthonormal set of wave functions ψ_n. The two factors $(\phi_1(0), \psi_m(0))$ and $(\psi_n(t), \phi_2(t))$ in Equation (3.165) are therefore unrelated to transitions in time, but they help us to describe any transition amplitude in terms of the basic amplitudes $(\psi_m(0), \psi_n(t))$ in a complete, orthonormal set of states.

Position Eigenstates

A common case of complete states in this example is given by position eigenstates $\psi_{x_0}(x) = \delta(x - x_0)$, with a continuous range of the state label x_0. The state label in this case is the eigenvalue of the position operator in such a state because $x\delta(x - x_0) = x_0\delta(x - x_0)$ follows simply from the condition that $\delta(x - x_0) = 0$ if $x \neq x_0$. Unfortunately, delta functions, and therefore position eigenstates, are not normalizable wave functions. We will therefore have to stretch the formal validity of our calculations at several places in this section. The final results will nevertheless be useful.

A time-independent Hamiltonian $\hat{H}$ determines the transition operator

$$\hat{T}(0, t) = \exp(-it\hat{H}/\hbar), \tag{3.166}$$

defined such that

$$\psi(t) = \hat{T}(0, t)\psi(0) \tag{3.167}$$

for any wave function ψ that solved the Schrödinger Equation (3.61). The position-dependent transition amplitude is then given by the matrix elements

$$(x_1(0), x_2(t)) = (\psi_{x_1}|\hat{T}(0, t)|\psi_{x_2}) \tag{3.168}$$

of the transition operator in position eigenstates.

For $x_2 \approx x_1$, which is likely realized for small t, we can write the matrix elements as approximate expectation values,

$$(x_1(0), x_2(t)) \approx \wp_{x_1}(\exp(-it\hat{H}/\hbar)), \tag{3.169}$$

where we use the notation $\wp$ to indicate a "non-state," or a mapping from the algebra of observables to the complex numbers that is not normalized and not even normalizable. Most of the following calculations will be done with wave functions, which, thanks to the theory of distributions such as the delta "function," are often more forgiving when it comes to violations of normalizability.

Free Particle

For the free particle, $\hat{H} = \hat{p}^2/(2m)$, we can compute the transition amplitudes in different ways. For instance, we can write the position eigenstate

$$\psi_{x_1}(x) = \delta(x - x_1) = \frac{1}{2\pi} \int_{-\infty}^{\infty} e^{ik(x-x_1)}dk \tag{3.170}$$

as a superposition of plane waves $\psi_k(x) = e^{ikx}$ (which, like position eigenstates, are not normalizable). Plane waves are eigenfunctions of $\hat{H}$ with eigenvalues $E_k = \hbar^2 k^2/(2m)$, which implies that

$$\hat{T}(0, t)\psi_k = e^{-iE_k t/\hbar}\psi_k. \tag{3.171}$$

Therefore,

$$\begin{aligned}(x_1(0), x_2(t)) &= \frac{1}{2\pi} \int_{-\infty}^{\infty} e^{ik(x_1-x_2)-iE_k t/\hbar}dk \\ &= \frac{1}{2\pi} \int_{-\infty}^{\infty} e^{ik(x_1-x_2)-\frac{1}{2}i\hbar k^2 t/m}dk.\end{aligned} \tag{3.172}$$

The Gaussian integral in this expression does not converge absolutely (when the integrand is replaced by its absolute value), which is perhaps not surprising because the eigenfunctions ψ_x are not normalizable. For $x_1 = x_2$ and $t = 0$ we therefore should obtain an infinite result, in agreement with Equation (3.172). For non-zero t, the integral is nevertheless (conditionally) convergent, as we will show in Section 3.4.2. However, this result cannot easily be obtained by applying standard Gaussian integrations directly to Equation (3.172).

As a shortcut, we obtain a meaningful (and even useful) result if we compute the integral in "Euclidean time" defined as $t_E = it$. With respect to this parameter,

$$\begin{aligned}(x_1(0), x_2(-it_E)) &= \frac{1}{2\pi} \int_{-\infty}^{\infty} e^{ik(x_2-x_1)-E_k t_E/\hbar}dk \\ &= \frac{1}{2\pi} \int_{-\infty}^{\infty} e^{ik(x_1-x_2)-\frac{1}{2}\hbar k^2 t_E/m}dk \\ &= \sqrt{\frac{m}{2\pi\hbar t_E}} \exp\left(-\frac{1}{2}\frac{m}{\hbar t_E}(x_1 - x_2)^2\right)\end{aligned} \tag{3.173}$$

is finite as long as $t_E \neq 0$. Formally, we can then replace $t_E = it$ in the last result and obtain the transition amplitude

$$(x_1(0), x_2(t)) = \sqrt{\frac{m}{2\pi i\hbar t}} \exp\left(\frac{1}{2}i\frac{m}{\hbar t}(x_1 - x_2)^2\right) \tag{3.174}$$

in time t.

Momentum Representation

Alternatively, we can solve the Schrödinger equation in the momentum representation,

$$i\hbar\frac{\partial\psi(p,\,t)}{\partial t} = \frac{p^2}{2m}\psi(p,\,t) \tag{3.175}$$

with some initial wave function $\psi(p,\,0) = \psi_0(p)$. Here, the $\hat{p}$-operator just multiplies a wave function by p. The result,

$$\psi(p,\,t) = \psi_0(p)\exp\left(-\frac{1}{2}i\frac{p^2}{m\hbar}t\right)$$

$$= \int_{-\infty}^{\infty}\psi_0(\bar{p})\exp\left(-\frac{1}{2}i\frac{\bar{p}^2}{m\hbar}t\right)\delta(p-\bar{p})\mathrm{d}\bar{p} \tag{3.176}$$

$$= \int_{-\infty}^{\infty}(\bar{p}(0),\,p(t))\psi_0(\bar{p})\mathrm{d}\bar{p},$$

determines the momentum-dependent transition amplitude

$$(p_1(0),\,p_2(t)) = \exp\left(-\frac{1}{2}i\frac{p_1^2}{m\hbar}t\right)\delta(p_2-p_1). \tag{3.177}$$

This result is well-defined without an intermediate step using Euclidean time. However, the double Fourier transformation

$$(x_1(0),\,x_2(t)) = \frac{1}{2\pi\hbar}\int_{-\infty}^{\infty}\int_{-\infty}^{\infty}(p_1(0),\,p_2(t))e^{ip_2 x_2/\hbar}e^{-ip_1 x_1/\hbar}\mathrm{d}p_1\,\mathrm{d}p_2 \tag{3.178}$$

converges absolutely only if we replace t with Euclidean time. The result is the same as Equation (3.173).

3.4.2 Picard–Lefshetz Theory

The introduction of Euclidean time is a formal device that can be made more precise with a slight change of perspective on the required integrations. The original problem to be addressed is the integration over p required in

$$(x_1(0),\,x_2(t)) = \frac{1}{2\pi\hbar}\int_{-\infty}^{\infty}\exp\left(-\frac{1}{2}i\frac{p^2}{m\hbar}t\right)e^{ip(x_2-x_1)/\hbar}\mathrm{d}p, \tag{3.179}$$

using the delta function in Equation (3.177). The integrand oscillates and does not drop off as p approaches plus or minus infinity. The integral is therefore not absolutely convergent because we integrate a constant over an infinite range if we replace the integrand with its absolute value. The integral nevertheless has a finite value and is therefore conditionally convergent. This claim will be demonstrated by the following derivations using complex analysis.

If we make the integration variable p complex, rather than t for Euclidean time, we can appeal to Cauchy's theorem and try to change the integration contour in the complex plane such that imaginary contributions to p imply a drop-off of the integrand. Since the integrand is holomorphic (that is, it depends only on p but not on p^*) without poles, given by an exponentiated polynomial in p, deforming the contour does not change the value of the integral, but it may improve its convergence properties.

Picard–Lefshetz theory provides a systematic procedure for finding a suitable contour in this situation. In general, we might be looking at an integral of the form $\mathcal{I} = \int_{-\infty}^{\infty} \exp(S(p))\mathrm{d}p$ with some function $S(p)$. In the first step, we look for extrema of the exponent $S(p)$, defined by the condition $\mathrm{d}S/\mathrm{d}p = 0$. Around an extremum, therefore, the exponent changes very slowly such that the integrand does not oscillate much. Since we are evaluating this condition in the complex plane of p, in a second step we can look for a path through a given extremum along which oscillations are suppressed compared with the original integration path given by the real line.

Holomorphic Functions
Any local extremum of the imaginary part of a holomorphic function is a saddle point: The holomorphic nature,

$$\frac{\partial S}{\partial p^*} = 0 \tag{3.180}$$

can be written as two partial differential equations for the real and imaginary parts of S if we transform from independent complex coordinates (p, p^*) to independent real coordinates (R, I) defined as the real and imaginary parts of p: $p = R + iI$ and $p^* = R - iI$.

Inverting these equations,

$$R = \frac{1}{2}(p + p^*), \quad I = \frac{1}{2i}(p - p^*). \tag{3.181}$$

The chain rule therefore turns Equation (3.180) into

$$\begin{aligned}
0 &= \frac{\partial S}{\partial p^*} = \frac{\partial R}{\partial p^*}\frac{\partial S}{\partial R} + \frac{\partial I}{\partial p^*}\frac{\partial S}{\partial I} = \frac{1}{2}\frac{\partial S}{\partial R} + \frac{i}{2}\frac{\partial S}{\partial I} \\
&= \frac{1}{2}\left(\frac{\partial \operatorname{Re} S}{\partial R} - \frac{\partial \operatorname{Im} S}{\partial I}\right) + \frac{1}{2}i\left(\frac{\partial \operatorname{Im} S}{\partial R} + \frac{\partial \operatorname{Re} S}{\partial I}\right).
\end{aligned} \tag{3.182}$$

This result implies two independent equations from its real and imaginary parts, the Chauchy–Riemann equations

$$\frac{\partial \operatorname{Re} S}{\partial R} = \frac{\partial \operatorname{Im} S}{\partial I} \tag{3.183}$$

$$\frac{\partial \operatorname{Im} S}{\partial R} = -\frac{\partial \operatorname{Re} S}{\partial I}. \tag{3.184}$$

The Cauchy–Riemann equations simplify the Hessian

$$\begin{pmatrix} \partial^2 \operatorname{Im} S/\partial R^2 & \partial^2 \operatorname{Im} S/\partial R \partial I \\ \partial^2 \operatorname{Im} S/\partial I \partial R & \partial^2 \operatorname{Im} S/\partial I^2 \end{pmatrix} \tag{3.185}$$

of the imaginary part of a holomorphic function, which determines whether a local extremum is a saddle point (unless the second-order derivatives vanish at the extremum). In particular, a local extremum of a function in two dimensions is a saddle point if the Hessian has eigenvalues of opposite signs. Using Cauchy–Riemann equations, Equation (3.185) equals

$$\begin{pmatrix} \partial^2 \operatorname{Im} S/\partial R^2 & \partial^2 \operatorname{Re} S/\partial R^2 \\ \partial^2 \operatorname{Re} S/\partial R^2 & -\partial^2 \operatorname{Im} S/\partial R^2 \end{pmatrix}, \tag{3.186}$$

a matrix of the form

$$\begin{pmatrix} A & B \\ B & -A \end{pmatrix} \tag{3.187}$$

which has eigenvalues $\lambda_{\pm} = \pm\sqrt{A^2 + B^2}$. As long as $A \neq 0$ or $B \neq 0$, the local extremum is therefore a saddle point of the imaginary part of S.

Since a local extremum of the imaginary part of a holomorphic S is a saddle point, it is crossed by curves in the complex plane along which the imaginary part of S does not change. Integrating over these curves eliminates all oscillations in the original integral and therefore improves its convergence.

Free Particle
In our specific example (3.179), $dS/dp = 0$ is solved by $p_0 = m(x_2 - x_1)/t$, equal to the classical momentum of a free particle moving from x_1 to x_2 in time t. (At this place, we have to assume $t \neq 0$ for a finite saddle point.) The imaginary part of $S(p_0)$ is equal to

$$\operatorname{Im} S(p_0) = \frac{m}{2t\hbar}(x_2 - x_1)^2. \tag{3.188}$$

Curves of constant imaginary part are therefore determined by the condition

$$\operatorname{Im} S(R + iI) = -\frac{(R^2 - I^2)t}{2m\hbar} + \frac{R}{\hbar}(x_2 - x_1) \tag{3.189}$$
$$= \operatorname{Im} S(p_0)$$

for a complex momentum $p = R + iI$. We can solve this quadratic equation for I as a function of R, such that

$$I(R) = \pm \left| R - \frac{m}{t}(x_2 - x_1) \right| \tag{3.190}$$

describes a cross through p_0 (for fixed t).

We now insert the imaginary part $I(R)$ of p, evaluated along our curves of constant Im S, in the exponent

$$
\begin{aligned}
S(R + iI) &= \frac{t}{m\hbar} I\left(R - \frac{m(x_2 - x_1)}{t} \right) \\
&+ \frac{it}{2m\hbar}\left(I^2 - R\left(R - \frac{2m(x_2 - x_1)}{t} \right) \right).
\end{aligned}
\tag{3.191}
$$

As desired,

$$\text{Im } S(R + iI(R)) = \frac{m(x_2 - x_1)^2}{2\hbar t}. \tag{3.192}$$

is independent of R. The real part of S shows that the choice of contour given by

$$I(R) = -\left(R - \frac{m(x_2 - x_1)}{t} \right) \text{ if } t > 0 \tag{3.193}$$

$$I(R) = R - \frac{m(x_2 - x_1)}{t} \text{ if } t < 0 \tag{3.194}$$

leads to Gaussian suppression at the asymptotic ends of the contour ($R \to \pm\infty$): For this choice, we have

$$\text{Re } S(R + iI(R)) = -\frac{|t|}{m\hbar}\left(R - \frac{m(x_2 - x_1)}{t} \right)^2. \tag{3.195}$$

Cauchy's theorem guarantees that the integral over our new contour $I(R)$ equals the original integral over the real line provided the combined contour can be closed off at infinity with vanishing contributions to the integral. The resulting closed contours are illustrated in Figure 3.7 for the two cases, $t > 0$ and $t < 0$. In order to see that the circular contours do not contribute to the integral provided their radii are moved to infinity, we may parameterize them in polar coordinates (ρ, φ) on the plane: $R(\rho, \varphi) = \rho \cos\varphi$ and $I(\rho, \varphi) = \rho \sin\varphi$, or $p = \rho e^{i\varphi}$. The contribution to $S(p)$ of highest polynomial order in ρ, obtained from the quadratic term $-\frac{1}{2}ip^2 t/(m\hbar)$ in Equation (3.179), then equals

$$S(\rho e^{i\varphi}) \sim \frac{t\rho^2}{2m\hbar}(\sin(2\varphi) - i\cos(2\varphi)). \tag{3.196}$$

For $\rho \to \infty$, the integrals of S over the required circular contours do not contribute provided $\text{sgn}(t \sin(2\varphi)) = -1$. As shown in Figure 3.7, this condition is fulfilled for both signs of t.

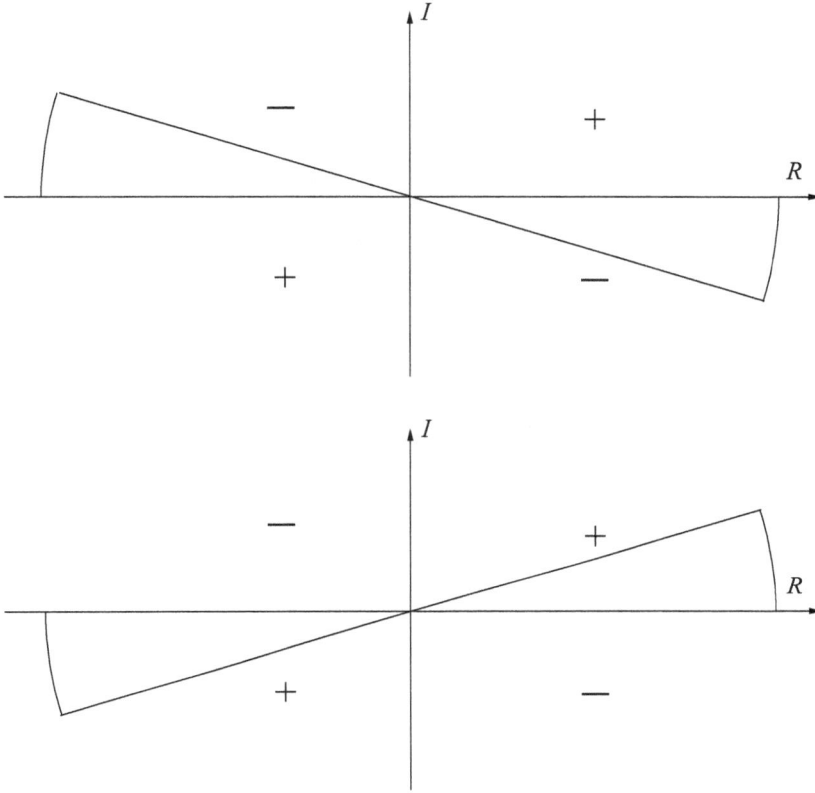

Figure 3.7. Closed integration contours used for an application of Cauchy's theorem to the transition amplitude of a free particle. The integration over the real line can be deformed into an integration over a suitable straight line at an agle determined by an application of Picard–Lefschetz theory. The circular contours vanish when they are moved out to infinity. The two distinct cases of $t > 0$ (top) and $t < 0$ (bottom) require different contours. The $+$ and $-$ signs indicate the sign of $\sin(2\varphi)$ with the polar angle φ in the plane, used in a parameterization of the circular contours. The contours are such that $\operatorname{sign}(t \sin(2\varphi)) < 0$ for both signs of t, implying exponential fall-off as the radii of circular segments approach infinity.

We can therefore replace the integral (3.179) over real p with an integral over complex $p = R + iI(R)$ with $I(R)$ given in Equation (3.193). Since the integration path is now given in terms of R, we should also transform the integration measure from $\mathrm{d}p$ to $\mathrm{d}R$. These two measures are not equal because an increment of R contributes to an increment of p through both the real and imaginary parts. The required transformation can be derived by first considering the two-dimensional measure $\mathrm{d}p\mathrm{d}p^*$ on the complex plane, which is then transformed to $\mathrm{d}R\mathrm{d}I$. Using $p^* = R - iI$, the measures are related by

$$\mathrm{d}p\mathrm{d}p^* = \begin{vmatrix} \partial p/\partial R & \partial p/\partial I \\ \partial p^*/\partial R & \partial p^*/\partial I \end{vmatrix} \mathrm{d}R\mathrm{d}I = 2\mathrm{d}R\mathrm{d}I. \tag{3.197}$$

Therefore, $dp = \sqrt{2}\,dR$ because R and I contribute symmetrically to p and p^*.
The result is a standard integral,

$$\sqrt{2}\int_{-\infty}^{\infty} \exp(S(R + iI(R)))dR$$

$$= \sqrt{2}\,e^{i\,\mathrm{Im}\,S(R+iI(R))} \int_{-\infty}^{\infty} e^{\mathrm{Re}\,S(R+iI(R))}dR$$

$$= \sqrt{2}\,\exp\left(i\frac{m(x_2 - x_1)^2}{2\hbar t}\right)\int_{-\infty}^{\infty} \exp\left(-\frac{|t|}{m\hbar}\left(R - \frac{m(x_2 - x_1)}{t}\right)^2\right)dR \qquad (3.198)$$

$$= \sqrt{\frac{2\pi m\hbar}{|t|}}\,\exp\left(\frac{im}{2t\hbar}(x_2 - x_1)^2\right)$$

using the R-independent (3.192) and integrating the Gaussian implied by Equation (3.195). The transition amplitude

$$(x_1(0),\, x_2(t)) = \frac{1}{2\pi\hbar}\int_{-\infty}^{\infty} \exp(S(R + iI(R)))dR$$

$$= \sqrt{\frac{m}{2\pi\hbar|t|}}\,\exp\left(\frac{im}{2t\hbar}(x_2 - x_1)^2\right) \qquad (3.199)$$

therefore agrees with the less well-defined Equation (3.174).

3.4.3 Path Integrals

Yet another way of computing transition amplitudes is given by the path integral. This method is important not only as a new calculational procedure, but also as a possible way to introduce quantization independently of the canonical method which starts with an algebra of observables. Several approaches to quantum gravity and cosmology rely crucially on the path-integral formulation of quantum mechanics.

Transition amplitudes can be formulated as path integrals by noting that the exponential function $\exp(-it\hat{H}/\hbar)$ in Equation (3.169) can be simplified by a Taylor expansion if t is "small." Defining the smallness of an expression with units is always tricky. In the present case, the argument of the exponential function is free of units and could be used instead of t, but it is an operator rather than a number that could be declared small. Physically, t should be much smaller than some relevant characteristic timescale of the system, but such a scale usually depends on the state the system is in. The state used in Equation (3.169), however, is a non-oscillating position eigenstate (or, rather, a non-state $\varnothing$), and does not determine a clear characteristic timescale.

In order to make sure that the argument of the exponential function is sufficiently small for a Taylor expansion, we take a limit in which t approaches zero. Even then, we may not be completely justified when we use a Taylor expansion because there may be states, and in particular non-states, in which the Hamiltonian does not have a finite expectation value, $\varnothing(\hat{H})$. Taking the limit of $t \to 0$ could then lead to a non-zero value $\lim_{t\to 0}(\wp(\hat{H})t/\hbar) \neq 0$ in the argument of the exponential function, and perhaps even to a value greater than one.

While the limit may be mathematically ill-defined for some states, physically, we are usually not interested in states in which the expected energy is infinite and could lead to a non-zero $\wp(\hat{H})t$ even for $t \to 0$. We can then proceed with a Taylor expansion. However, if we insert the Taylor expansion in Equation (3.169),

$$(x_1(0), x_2(t)) = \sum_{n=0}^{\infty} \frac{(-i/\hbar)^n}{n!} \wp_{x_1}(\hat{H}^n)t^n, \qquad (3.200)$$

we notice that we not only need a small expected energy, but small moments $\wp(\hat{H}^n)$ as well. The energy, therefore, cannot be strongly fluctuating or otherwise lead to large moments. In particular in quantum cosmology of the early universe, we may well be led to states in which the expected energy is very large or even infinite and fluctuates very much. For now, we will merely remember this potential problem and ignore its consequences.

Assuming that δt is sufficiently small for a first-order Taylor expansion to be applicable, we obtain

$$(x_1(0), x_2(\delta t)) \approx \wp_{x_1}(\hat{I}) - i\frac{\wp_{x_1}(\hat{H})\delta t}{\hbar}. \qquad (3.201)$$

Since the position eigenstates are not normalizable, $\wp_{x_1}(\hat{I})$ is infinite where $\hat{I}$ is the identity operator. We will write this infinity as $\wp_{x_1}(\hat{I}) = \lim_{x_2 \to x_1}\delta(x_2 - x_1)$, and later make it finite by working with $x_2 \neq x_1$ for non-zero (but still small) δt. For a mechanics Hamiltonian of the form $\hat{H} = \hat{p}^2/(2m) + V(\hat{x})$, we compute $\wp_{x_1}(V(\hat{x})) = V(x_1)\lim_{x_2 \to x_1}\delta(x_2 - x_1)$ because all second-order and higher position moments of $\hat{x}$ vanish in an eigenstate of $\hat{x}$. The expectation value of $\hat{p}^2$, by contrast, is not even well-defined because the momentum variance $(\Delta_{\wp_{x_1}}p)^2$ is infinite, as a consequence of the uncertainty relation with zero position variance in a position eigenstate.

We compute the contribution from $\wp_{x_1}(\hat{p}^2)$ by inserting the complete momentum eigenfunctions:

$$(\psi_{x_1}, \hat{p}^2\psi_{x_2}) = \int_{-\infty}^{\infty} (\psi_{x_1}|\hat{p}^2 \phi_p) \cdot (\phi_p, \psi_{x_2})dp \qquad (3.202)$$

where

$$\phi_p(x) = \frac{1}{2\pi\hbar} \exp(ipx/\hbar) \qquad (3.203)$$

are the momentum eigenfunctions, and therefore

$$(\psi_x, \phi_p) = \frac{1}{2\pi\hbar} \exp(ipx/\hbar). \qquad (3.204)$$

Using the eigenfunction property, we have

$$\hat{p}^2 \phi_p = p^2 \phi_p, \qquad (3.205)$$

such that

$$(\psi_{x_1}, \hat{p}^2 \psi_{x_2}) = \frac{1}{2\pi\hbar} \int_{-\infty}^{\infty} p^2 \exp(ip(x_1 - x_2)/\hbar)dp \qquad (3.206)$$

is given by an integral which is not absolutely convergent, but let us proceed ignoring this problem. (This convergence problem is related to those encountered in our first two derivations of the transition amplitude.)

With these contributions, we arrive at

$$(x_1(0), x_2(\delta t)) \approx \delta(x_2 - x_1) - i\frac{\wp_{x_1}(\hat{H})\delta t}{\hbar}$$

$$= \delta(x_2 - x_1) - \frac{i\delta t}{4\pi\hbar^2 m} \int_{-\infty}^{\infty} p^2 \exp(ip(x_1 - x_2)/\hbar)dp \qquad (3.207)$$

$$- i\frac{V(x_1)\delta t}{\hbar}\delta(x_2 - x_1).$$

All three terms can be combined in an integral if we use

$$\delta(x_2 - x_1) = \frac{1}{2\pi\hbar} \int_{-\infty}^{\infty} \exp(ip(x_2 - x_1)/\hbar)dp, \qquad (3.208)$$

such that

$$(x_1(0), x_2(\delta t)) \approx \frac{1}{2\pi\hbar} \int_{-\infty}^{\infty} \left(1 - \frac{i\delta t}{\hbar}\left(\frac{p^2}{2m} + V(x_1) \right) \right)$$

$$\times \exp(ip(x_2 - x_1)/\hbar)dp. \qquad (3.209)$$

A transition amplitude for times t that are not small can be written as a combination of small-time transition amplitudes, using the composition formula

$$(x_1(0), x_2(t)) = \int_{-\infty}^{\infty} (x_1(0), x(t')) \cdot (x(t'), x_2(t))dx(t') \qquad (3.210)$$

based on the completeness of position eigenstates, where $0 < t' < t$ is an arbitrary time between 0 and t. Applying this formula N times, we define $\delta t = t/(N + 1)$ and obtain

$$(x_1(0), x_2(t)) = \int_{-\infty}^{\infty} \cdots \int_{-\infty}^{\infty} (x_1(0), x^{(1)}(\delta t)) \cdot (x^{(1)}(\delta t), x^{(2)}(2\delta t))$$

$$\cdots (x^{(N-1)}((N - 1)\delta t), x^{(N)}(N\delta t)) \qquad (3.211)$$

$$\cdot (x^{(N)}(N\delta t), x_2(t))dx^{(1)}(\delta t)\cdots dx^{(N)}(N\delta t).$$

In the limit of $N \to \infty$, $\delta t \to 0$, the resulting infinite number of integrations is a path integral because we can view the auxiliary integration variables, $x^{(n)}(n\delta t)$ with

$n = 1, \dots, N$, as actual positions taken by a particle at intermediate times, holding the initial and final positions, $x_1(0)$ and $x_2(t)$, fixed. Integrating the auxiliary positions from negative infinity to infinity means that the quantum particle probes all possible paths from $x_1(0)$ to $x_2(t)$ (even discontinuous ones). The final transition amplitude is given by the superposition of all transition amplitudes taken over the paths.

Still in the limit of $N \to \infty$, $\delta t \to 0$, we can replace each infinitesimal transition amplitude, $(x^{(n-1)}((n-1)\delta t), x^{(n)}(n\delta t))$ for $1 \leqslant n \leqslant N$, with the small-$t$ expression (3.209):

$$(x_1(0), x_2(t))$$

$$= \int_{-\infty}^{\infty} \int_{-\infty}^{\infty} \cdots \int_{-\infty}^{\infty} \int_{-\infty}^{\infty} \left(1 - \frac{i\delta t}{\hbar}\left(\frac{(p^{(0)})^2}{2m} + V(x_1(0))\right)\right) e^{ip^{(0)}(x^{(1)} - x_1(0))/\hbar}$$

$$\cdot \left(1 - \frac{i\delta t}{\hbar}\left(\frac{(p^{(1)})^2}{2m} + V(x^{(1)})\right)\right) e^{ip^{(1)}(x^{(2)} - x^{(1)})/\hbar}$$

$$\cdots \left(1 - \frac{i\delta t}{\hbar}\left(\frac{(p^{(N-1)})^2}{2m} + V(x^{(N-1)})\right)\right) e^{ip^{(N-1)}(x^{(N)} - x^{(N-1)})/\hbar} \quad (3.212)$$

$$\cdot \left(1 - \frac{i\delta t}{\hbar}\left(\frac{(p^{(N)})^2}{2m} + V(x^{(N)})\right)\right) e^{ip^{(N)}(x_2(t) - x^{(N)})/\hbar}$$

$$\frac{dp^{(0)} dx^{(1)} dp^{(1)} \cdots dx^{(N)} dp^{(N)}}{(2\pi\hbar)^{N+1}}.$$

All auxiliary variables, $x^{(n)}$ and $p^{(n)}$ as well as the additional $p^{(0)}$ for the initial transition amplitude, depend on the discrete times $n\delta t$, which we have suppressed in order to shorten the equation.

In order to shorten it further, we multiply the exponential functions,

$$e^{ip^{(0)}(x^{(1)} - x_1(0))/\hbar} \cdots e^{ip^{(N)}(x_2(t) - x^{(N)})/\hbar} = \exp\left(\frac{i}{\hbar}\sum_{n=0}^{N} p^{(n)}(x^{(n+1)} - x^{(n)})\right) \quad (3.213)$$

where we define $x^{(N+1)} = x_2(t)$ and $x^{(0)} = x_1(0)$. Moreover, for small δt, we write $x^{(n+1)} - x^{(n)} \approx \dot{x}^{(n)}\delta t$, such that

$$e^{ip^{(0)}(x^{(1)} - x_1(0))/\hbar} \cdots e^{ip^{(N)}(x_2(t) - x^{(N)})/\hbar} \approx \exp\left(\frac{i}{\hbar}\sum_{n=0}^{N} p^{(n)}\dot{x}^{(n)}\delta t\right)$$

$$\approx \exp\left(\frac{i}{\hbar}\int_0^t p(t')\dot{x}(t')dt'\right) \quad (3.214)$$

takes the form of an exponentiated integral. Similarly, we write the product of potential-dependent factors,

$$\left(1 - \frac{i\delta t}{\hbar}\left(\frac{(p^{(0)})^2}{2m} + V(x_1(0))\right)\right)\cdots\left(1 - \frac{i\delta t}{\hbar}\left(\frac{(p^{(N)})^2}{2m} + V(x^{(N)})\right)\right)$$

$$\approx \exp\left(-\frac{i\delta t}{\hbar}\sum_{n=0}^{N}\left(\frac{(p^{(n)})^2}{2m} + V(x^{(n)})\right)\right) \qquad (3.215)$$

$$\approx \exp\left(-\frac{i}{\hbar}\int_0^t\left(\frac{p(t')^2}{2m} + V(x(t'))\right)dt'\right).$$

Therefore,

$$(x_1(0),\ x_2(t))$$

$$= \int \mathcal{D}_0x\,\mathcal{D}p\,\exp\left(\frac{i}{\hbar}\int_0^t\left(p(t')\dot{x}(t') - \frac{p(t')^2}{2m} - V(x(t'))\right)dt'\right) \qquad (3.216)$$

with the path-integral measure $\mathcal{D}_0x\,\mathcal{D}p$, in the limit of $N \to \infty$, indicating the infinite number integrations over $x(t')$ and $p(t')$ in Equation (3.212). The subscript zero in $\mathcal{D}_0x$ indicates the presence of a boundary condition, keeping $x(0) = x_1(0)$ and $x(t) = x_2(t)$ fixed. There is no such boundary condition on p.

The final exponent, $\int_0^t (p(t')\dot{x}(t') - \frac{1}{2}p(t')^2/m - V(x(t')))dt'$, equals the classical action provided we can view $p(t')$ as a function of $\dot{x}(t')$ such that $p(t') = m\dot{x}(t')$. This relationship is indeed implied by the Gaussian integrations over $p(t')$:

$$\int_{-\infty}^{\infty} \exp\left(\frac{i}{\hbar}\left(p\dot{x} - \frac{p^2}{2m}\right)\delta t\right)\frac{\mathrm{d}p}{2\pi\hbar} = \int_{-\infty}^{\infty} \exp\left(\frac{1}{\hbar}\left(p\dot{x} - \frac{p^2}{2m}\right)\delta t_\mathrm{E}\right)\frac{\mathrm{d}p}{2\pi\hbar}$$

$$= \int_{-\infty}^{\infty} \exp\left(-\frac{(p - m\dot{x})^2}{2m\hbar}\delta t_\mathrm{E} + \frac{m\dot{x}^2}{2\hbar}\delta t_\mathrm{E}\right)$$

$$\frac{\mathrm{d}p}{2\pi\hbar}$$

$$= \sqrt{\frac{m}{2\pi\hbar\delta t_\mathrm{E}}}\,\exp\left(\frac{m\dot{x}^2}{2\hbar}\delta t_\mathrm{E}\right) \qquad (3.217)$$

where we used Euclidean time $t_\mathrm{E} = it$ as in Section 3.4.1. Performing all momentum integrations in Equation (3.216) and transforming back to Lorentzian time, we obtain

$$(x_1(0), x_2(t)) = \int \mathcal{D}_0x\,\exp\left(\frac{i}{\hbar}\int_0^t\left(\frac{1}{2}m\dot{x}(t')^2 - V(x(t'))\right)dt'\right)$$

$$= \int \mathcal{D}_0x\,\exp\left(\frac{iS[x(t)]}{\hbar}\right). \qquad (3.218)$$

(Factors of $\sqrt{m/(2\pi\hbar\delta t_\mathrm{E})}$ can be absorbed in the path-integral measure.)

3.4.4 Relevance

Introducing the path integral seems like a lot of effort to compute a function $(x_1(0), x_2(t))$ which is not directly measurable because an initial state can rarely be prepared as a precise position eigenstate. However, even though the transition amplitude itself may not be of primary interest in physical applications, it can be used to derive the evolution of a wave function,

$$\psi(x_2, t) = \int_{-\infty}^{\infty} (x_1(0), x_2(t)) \cdot \psi(x_1, 0) \mathrm{d}x_1, \tag{3.219}$$

which is well-defined and finite, for instance using Equation (3.174) for the free particle. The integral of the wave function is absolutely convergent because the initial state $\psi(x_1, 0)$, unlike a position eigenstate ψ_{x_1}, is normalized. By discussing methods to compute the (initially) ill-defined expression $(x_1(0), x_2(t))$ based on non-normalizable position eigenstates, independently of the evolved wave function (3.219), we can split a complicated calculation into smaller chunks. The introduction of Euclidean time only facilitates such a piecemeal calculation; it does not introduce a new physical situation in which four-dimensional Euclidean space would actually replace Minkowski spacetime.

The explicit calculation of path integrals is possible for several potentials, but in most cases it requires more effort than a canonical derivation of evolution based on energy spectra. However, transition amplitudes based on path integrals can be computed efficiently for harmonic oscillators and perturbations around them, including large numbers of coupled harmonic oscillators. The path integral then appears as a systematic method to decouple such systems, in which way it has been used in particle physics and many-body systems. The widespread application in particle physics relies on the fact that non-interacting field theories on Minkowski spacetime are related to systems of infinitely many coupled harmonic oscillators. Weak interactions can be described by perturbation theory, which is also done efficiently on the basis of path integrals. Finally, questions relevant in particle physics can often be stated in terms of scattering amplitudes, which are closely related to the transition amplitudes used in quantum mechanics.

Moreover, knowing the transition amplitude, once it has been made well-defined, can help us to understand quantum evolution per se, independently of the specific form of an initial state. This property is especially relevant in quantum gravity or cosmology, in which we do not know much about suitable states. For instance, gravity does not have a Hamiltonian bounded from below, see Equation (1.18),—it does not even have a Hamiltonian but rather a Hamiltonian constraint—and therefore there is no ground state that could be used as a distinguished state of a system. By replacing states with transition amplitudes, computed through a path integral, one can therefore evade the problem of what a "natural" state should be for gravity or cosmology.

The last step in the derivation of path integrals is also relevant in quantum gravity: Although we started with a canonical formulation in which x and p are used as independent phase-space variables, we were able to transform the expression to a

Lagrangian formulation in which x and $\dot{x}$ appear. In a generally covariant system, the action is manifestly covariant, while the covariance of a canonical formulation can be determined only by rather complicated calculations using the hypersurface-deformation structure we explored in Section 2.3. Several approaches to quantum gravity and cosmology make use of this advantage by trying to introduce quantum gravity through the definition of a path integral.

Then again, the path integral in quantum gravity is mainly of a formal advantage, not only because relevant questions in strong-field gravity, compared with particle physics, are usually unrelated to transition amplitudes in systems of perturbed coupled harmonic oscillators. In addition, the problems of states and covariance reappear in quantum gravity, even if it is based on path integrals, once we start asking questions about possible predictions of such theories. In order to interpret transition amplitudes, we should apply them to an initial state and see how they describe evolution to a final state. The problem of states, or the question of what a natural state of quantum gravity should be, therefore reappears.

Moreover, a definition of path integrals in quantum gravity replaces the path-integral measure $\mathcal{D}x$ with some integration over spacetime metrics, $\mathcal{D}g_{\alpha\beta}$. The metric tensor is not invariant, and therefore the problem of covariance appears in a technical form: We have to show that the path-integral measure is generally covariant. In quantum gravity and cosmology, the path integral often hides the crucial problems that should be addressed, and which are uncovered only by careful canonical considerations.

We will discuss different approaches to quantum gravity in Chapter 5. In some of them, path-integral methods as well as Euclidean time are made use off in crucial ways, while others focus on canonical constructions. But first, in Chapter 4, we will see how lessons from quantum mechanics can be implemented in homogeneous cosmological models that do not require a full approach to quantum gravity. Even in these restricted models, the question of general covariance will be found to be important.

References

Barvinsky, A. O., Kamenshchik, A., Yu,, & Kiefer, C. 1999, NuPhB, 552, 420

Baytaş, B., Bojowald, M., & Crowe, S. 2020, AnPhy, 420, 168247

Baytaş, B., Bojowald, M., & Crowe, S. 2019, PhRvA, 99, 042114

Bojowald, M., & Brahma, S. 2014, arXiv:1411.3636

Burgess, C. P., Holman, R., & Hoover, D. 2008, PhRvD, 77, 063534

Coleman, S., & Weinberg, E. 1973, PhRvD, 7, 1888

Falciano, F. T., Pinto-Neto, N., & Santini, E. S. 2007, PhRvD, 76, 083521

Giulini, D., Kiefer, C., Joos, E., et al. 2003, Decoherence and the Appearance of a Classical World in Quantum Theory (Berlin: Springer)

Heller, E. J. 1976, JChPh, 65, 1289

Jackiw, R., & Kerman, A. 1979, PhLA, 71, 158

Jalabert, R. A., & Pastawski, H. M. 2001, PhRvL, 86, 2490

Kiefer, C., & Kraemer, M. 2012, PhRvL, 108, 021301

Kiefer, C., Lohmar, I., Polarski, D., & Starobinsky, A. A. 2007, CQGra, 24, 1699

Kiefer, C., Polarski, D., & Starobinsky, A. A. 1998, IJMPD, 7, 455

Kiefer, C., Polarski, D., & Starobinsky, A. A. 2000, PhRvD, 62, 043518

Peter, P., & Pinto-Neto, N. 2008, PhRvD, 78, 063506

Pinto-Neto, N. 2009, PhRvD, 79, 083514

Polarski, D., & Starobinsky, A. A. 1996, CQGra, 13, 377

Prezhdo, O. 2006, Theoretical Chemistry Accounts, 116, 206

Schlosshauer, M. 2007, Decoherence and the Quantum-to-classical Transition (Berlin: Springer)

von Neumann, J. 1996, Mathematical Foundations of Quantum Mechanics (Princeton, NJ: Princeton University Press)

Zeh, H. D. 1970, FoPh, 1, 69 Reprinted in Wheeler, J. A. & Zurek W. H. (ed) 1983, Quantum Theory and Measurement (Princeton, NJ: Princeton University Press)

Chapter 4

Minisuperspace Models

If the strong symmetry assumption of exact spatial homogeneity is used, quantum cosmology in so-called *minisuperspace models* can be developed in much the same way as in single-particle quantum mechanics. The term *superspace* here refers to the space of metrics as the configuration space of gravitational theories (and not to supersymmetries), which is distinct from the actual space used to determine the positions of particles. Nevertheless, related canonical or path-integral methods of quantization can be used. The position and momentum of a particle in one dimension are then replaced by the volume and the Hubble parameter of an isotropic model, implying models related to one-dimensional quantum mechanics. Anisotropic cosmological models with three independent scale factors (or the volume and two anisotropy parameters) as well as their rates of change have the same number of degrees of freedom as a particle in three dimensions. There are, however, two crucial differences in quantum cosmology as a consequence of general covariance.

First, the volume of a homogeneous space is not uniquely defined without specifying the region whose size it measures. If space is compact, as in an isotropic model with positive spatial curvature, one may take the full spatial volume as a somewhat preferred choice. However, if the geometry is exactly homogeneous, its expansion can be described equally well by the changing volume of any region smaller than all of space. Since space may well be infinitely large in isotropic models with non-positive curvature, any finite region is, in general, smaller than all of space.

Moreover, for cosmological observations made at a specific time in the history of the universe, one should restrict attention to the finite region determined by the maximum distance from which messengers at the speed of light can reach us from a distant source, such as the neutralization event that released the cosmic microwave background after the initial primordial plasma had cooled down enough. Given this limitation, we may never know whether space is compact, even if it "is." In order to

doi:10.1088/2514-3433/ab9c98ch4

avoid theoretical bias, cosmological models should therefore be able to describe models with finite or infinite full volume.

This condition then requires us to implement models based on different choices of finite regions within the same space. Since the volume of a region can be changed not only by a choice of region but also by a transformation of spatial coordinates, this condition is related to general covariance. It turns out that the classical invariance of Friedmann's equation with respect to rescaling the volume V to λV with some $\lambda > 0$, observed in Chapter 1, is *not* realized after quantization. As a quick argument as to why this is the case, recall that there are some quantum effects, such as the Casimir force, which depend on the size of the region in which quantum degrees of freedom are implemented.

Second, the dynamics of a cosmological minisuperspace model does not unfold in an absolute time. As another remnant of general covariance in the restricted setting of exact homogeneity, we are able to reparameterize the dynamics by replacing a time coordinate t with some $f(t)$, for any monotonic function f. Any pair $(V(t), H(t))$ of solutions of the two phase-space variables of an isotropic model is mapped to another pair of solutions, $(V(f(t)), H(f(t)))$. As a consequence, the energy of the system is uniquely defined only if it is zero because the energy is canonically conjugate to time, $\{t, E\} = 1$. After time reparameterization, the quantity canonically conjugate to the new time $f(t)$ is no longer E but $E/f'(t)$, which, for non-zero E, is time dependent unless f is linear. The quantity canonically conjugate to time therefore cannot be considered a valid observable for the conserved energy of the system, unless we restrict time reparameterizations to linear transformations.

In a generally covariant theory, in which any monotonic $f(t)$ must be possible as a valid new time coordinate, the only solution of this dilemma is to require that the total energy of any closed gravitational system vanish. The usual Hamiltonian is then replaced by a Hamiltonian constraint, such as Equation (1.17), according to which energy contributions from gravity and matter must be completely balanced. Minisuperspace models of cosmology are always constrained systems, which go beyond standard quantum mechanics but can be treated in an extended version. They present a new difficulty of how to realize quantum evolution in a constrained system, subject to time reparameterization invariance and lacking a unique non-zero Hamiltonian.

While minisuperspace models with their exact spatial homogeneity are too restricted for reliable description of cosmological observations, they can help us to understand questions of general covariance in a quantum setting.

4.1 Volume Scaling

The reason why quantum effects break the classical invariance under rescaling the volume, $V \mapsto \lambda V$, can easily be seen in a formal setting of path integrals for isotropic minisuperspace models. The transition amplitude for the volume at two times, $(V_1(0), V_2(\tau))$, is then written as

$$(V_1(0),\ V_2(\tau)) = \int \mathcal{D}_0 V \exp\left(\frac{iS[V(\tau)]}{\hbar}\right) \tag{4.1}$$

according to Equation (3.218). (At this time we use proper time τ, postponing a discussion of time reparameterization until Section 4.2.) However, neither the volume V in $\mathcal{D}_0 V$ nor the action

$$S[V(\tau)] = \int \mathrm{d}\tau \mathrm{d}^3 x \sqrt{\det(q)}\, \mathcal{L}(V,\ \dot{V}) = V_0 \int \mathrm{d}\tau \mathcal{L}(V,\ \dot{V}) \tag{4.2}$$

is scaling invariant, where we integrate the homogeneous Lagrangian density $\mathcal{L}$ over a finite region with coordinate volume $V_0 = \int \mathrm{d}^3 x \sqrt{\det(q)}$ using the spatial metric q_{ij}. The transition amplitude therefore cannot be scaling invariant in general.

However, we know that the classical limit of the theory is scaling invariant. The consistency of this limit with quantum non-invariance can be seen by a formal argument in which we include Planck's constant $\hbar$ in our scaling transformation. To this end, questions of units make it easier to present the argument using the Hamiltonian version of the path integral, (3.216), which integrates over all phase-space variables, V and $p_V = -c^2 H/(4\pi G)$ in our case; see Equation (1.16). The measure, $\mathcal{D}_0 V \mathcal{D} p_V$ has the same units as $\hbar$, and it can be normalized by using a result from quantum statistics that assigns an elementary area of $2\pi\hbar$ to any two-dimensional phase space. The normalized Hamiltonian path integral then takes the form

$$\begin{aligned}
(V_1(0),\ V_2(\tau)) &= \int \frac{\mathcal{D}_0 V \mathcal{D} p_V}{2\pi\hbar} \exp\left(\frac{iS[V(\tau),\ p_V(\tau)]}{\hbar}\right) \\
&= \int \frac{\mathcal{D}_0 V \mathcal{D} H}{-8\pi^2 c \ell_\mathrm{P}^2} \exp\left(\frac{iS[V(\tau),\ H(\tau)]}{\hbar}\right)
\end{aligned} \tag{4.3}$$

with the Planck length ℓ_P.

The path integral is formally scaling invariant if we rescale both V and $\hbar$ (or ℓ_P^2) by the same amount, $V \mapsto \lambda V$ and $\hbar \mapsto \lambda\hbar$. In the classical limit, $\hbar \to 0$ is invariant with respect to this transformation, and the only remaining transformation left from the full λ-invariance is the volume rescaling. In quantum mechanics, however, $\hbar$ is non-zero and has a specific value determined by observations that should not be changed. Therefore, any physical quantum model in which only V is subject to rescaling but not $\hbar$ breaks the classical scaling invariance.

4.1.1 Scalar Field

In order to understand the details and meaning of scaling transformations in quantum systems, it is useful to start with a scalar model. In addition, the scalar may also be considered a matter ingredient of a minisuperspace model. To this end, we assume that there is a scalar field ϕ on Minkowski spacetime, which is spatially homogeneous. The field Lagrangian

$$L = \int d^3x \left(-\frac{1}{2} \sum_{\alpha=0}^{3} \sum_{\beta=0}^{3} \eta^{\alpha\beta} \frac{\partial\phi}{\partial x^\alpha} \frac{\partial\phi}{\partial x^\beta} - W(\phi) \right) = \frac{1}{2c^2} V_0 \dot{\phi}^2 - V_0 W(\phi) \qquad (4.4)$$

therefore takes the form of a classical-mechanics system with kinetic and potential contributions. However, unlike standard mechanics examples, it is not invariant with respect to scaling transformations because it is proportional to V_0.

As a consequence, the momentum of ϕ,

$$p_\phi = \frac{\partial L}{\partial \dot{\phi}} = \frac{V_0}{c^2} \dot{\phi}, \qquad (4.5)$$

is not scaling invariant, nor is the Hamiltonian

$$H = p_\phi \dot{\phi} - L = \frac{c^2 p_\phi^2}{2V_0} + V_0 W(\phi). \qquad (4.6)$$

For zero scalar potential, $W(\phi) = 0$, the Hamiltonian resembles the mechanics Hamiltonian of a free particle with scaling-dependent mass, $m = V_0$. The path integral for this system, (3.217), depends on m only through the ratio $m/\hbar$, which is invariant under our formal rescaling of both V_0 and $\hbar$, but not under the physical transformation in which V_0 is rescaled but $\hbar$ is fixed to its observed value.

4.1.2 Quantum Corrections

A systematic derivation of quantum corrections and their scaling dependence is possible in a canonical formulation of quantum mechanics. We can easily adapt the effective Hamiltonian to first order in $\hbar$, (3.96), to the coefficients of the scalar Hamiltonian:

$$\omega(\hat{H}) = \frac{c^2 \omega(\hat{p}_\phi)^2}{2V_0} + \frac{c^2 p_s^2}{2V_0} + \frac{c^2 U}{2V_0 s^2} + V_0 W(\omega(\hat{\phi})) + \frac{1}{2} V_0 W''(\omega(\hat{\phi})) s^2 \qquad (4.7)$$

with the canonically conjugate quantum variables s and p_s related to the variances of ϕ and p_ϕ according to Equation (3.89), while $U \geqslant \hbar^2/4$ is a constant.

Again, we see that the effective Hamiltonian has a well-defined scaling behavior only if we transform both V_0 and $\hbar$, in which case $\omega(\hat{H})$ scales like V_0. (Like Equation (4.5), both quantum momenta $\omega(\hat{p}_\phi)$ and p_s scale like V_0, while U scales like $\hbar^2$.) However, the quantum correction $c^2 U/(2V_0 s^2)$ does not scale in this way if $\hbar$ is held fixed. Quantum corrections therefore break the classical scaling invariance. Since the term $c^2 U/(2V_0 s^2)$ represents a potential barrier at small s that enforces uncertainty relations, as seen in Section 3.3.3, violations of the classical scaling behavior are related to quantum uncertainty.

The relationship between scaling behaviors and uncertainty relations can be seen more directly if we explicitly compute quantum fluctuations. An effective classical potential depending only on $\omega(\hat{\phi})$, but not on s, is obtained if we minimize $\omega(\hat{H})$, given in Equation (4.7), with respect to the quantum variable $s = \Delta_\omega \phi$ at fixed $\omega(\hat{\phi}) = \bar{\phi}$. The result,

$$s^4 = \frac{c^2 U}{V_0^2 W''(\bar{\phi})}, \tag{4.8}$$

scales like $\omega(\hat{\phi})^4$ only if we scale both V_0 and $U \propto \hbar^2$. Therefore, the ratio $(\Delta_\omega \phi)/\omega(\hat{\phi})$ is not scaling independent if $\hbar$ is held fixed at its physical value, even though statistical intuition would suggest that both the volume fluctuation and the volume expectation value should scale in the same way if they are described by a scaling-invariant theory. This, however, cannot be true because $\Delta_\omega \phi$, unlike $\omega(\hat{\phi})$, is subject to an uncertainty relation with a fixed lower bound (for given momentum fluctuation).

Since the volume fluctuation does not scale like V_0, the low-energy effective Hamiltonian, obtained by evaluating Equation (4.7) at the s-minimum (4.8), as well as $p_s = 0$ which minimizes the kinetic energy of quantum variables, does not respect the classical scaling of Equation (4.6) like V_0: We obtain

$$H_{\text{low-energy}} = \frac{c^2 p_\phi^2}{2 V_0} + V_0 W_{\text{eff}}(\bar{\phi}) \tag{4.9}$$

with the effective scalar potential

$$W_{\text{eff}}(\bar{\phi}) = W(\bar{\phi}) + \frac{c \sqrt{U W''(\bar{\phi})}}{V_0} \tag{4.10}$$

whose quantum correction is not scaling invariant (unless $\hbar$ and therefore $\sqrt{U}$ is rescaled).

The same behavior can be seen for the volume variables of the gravitational contribution to the Hamiltonian (constraint). This contribution, proportional to VH^2 in spatially flat isotropic models, is a non-standard kinetic term from the point of view of classical mechanics. Nevertheless, an effective Hamiltonian can be derived with the same methods used for the scalar field, given by

$$\omega(\hat{V}\hat{H}^2)_{\text{symm}} = \omega(\hat{V})\omega(\hat{H})^2 + \omega(\hat{V})(\Delta_\omega H)^2 + \omega(\hat{H})\Delta_\omega(VH) + \cdots \tag{4.11}$$

$$= \omega(\hat{V})\omega(\hat{H})^2 + \omega(\hat{V})\left(p_h^2 + \frac{U_h}{h^2}\right) + \omega(\hat{H})h p_h \tag{4.12}$$

where h, p_h and U_h are gravitational quantum variables analogous to s, p_s and U in the scalar model. The presence of the term $\omega(\hat{V})U_h/h^2$ breaks the classical scaling behavior.

4.1.3 Infrared Renormalization

The dependence of homogeneous quantum cosmological models on the size of the region chosen to determine the expansion behavior requires a physical explanation if such models are to be viable. To this end, the scalar model is again useful because in this case we know the full low-energy effective potential, the Coleman–Weinberg example (3.128) (Coleman & Weinberg 1973) or the canonical version (3.126) (Bojowald & Brahma 2014); derived without using the homogeneity assumption. The quantum correction is more complicated in this case, given by an integral over modes k rather than an algebraic function of the scalar. Nevertheless, for small k, that is in the infrared, the integrand of the canonical version (3.126) resembles the minisuperspace correction in that both are proportional to the square root of the second derivative of the potential. (The factor of $\hbar/2$ in Equation (3.126) represents the limiting value of $\sqrt{U}$.)

The infrared regime of large wavelengths is precisely where we expect homogeneous models to provide a good approximation. However, delineating the infrared regime requires us to choose an infrared scale ℓ_{min} such that only large wave lengths $\ell \geqslant \ell_{min}$ are included in a truncated infrared theory. Large wave lengths imply small wave numbers, $k \leqslant k_{max} = 2\pi/\ell_{min}$. In this language, introducing a new length scale is the quantum ingredient that violates the classical scaling behavior.

We may use Equation (3.126) for our new scalar model, provided we identify the potential $V(\phi)$ in Equation (3.126) with $W(\phi)$ used here. As long as k_{max} is small compared with $\sqrt{W''(\bar{\phi})}$, the integral in Equation (3.126) can be approximated by the volume of infrared modes, $\frac{4}{3}\pi k_{max}^3$, times the integrand at $k = 0$,

$$\frac{c\hbar}{4\pi}\int_{-\infty}^{\infty}\left(\sqrt{W''(\bar{\phi}) + k^2} - |k|\right)\mathrm{d}k \approx \frac{c\hbar k_{max}^3}{3}\sqrt{W''(\bar{\phi})}. \tag{4.13}$$

The infrared region may be treated by a minisuperspace approximation because it contains long-wavelength modes which are close to being homogeneous. Therefore, k_{max} should be related to the volume V_0 of the averaging region, such that $2\pi/k_{max} \sim V_0^{1/3}$: A wave with wave number k_{max} or corresponding wave length $\ell_{min} = 2\pi/k_{max}$ then just fits into a diameter of a spatial sphere with volume V_0. Any wave of larger wave lengths has a basic period that extends over a longer distance, and can therefore be approximated by a constant amplitude. With this relationship, the infrared contribution to the effective potential is proportional to $c\hbar\sqrt{W''(\bar{\phi})}/V_0$, precisely of the form derived in the minisuperspace model.

With this key insight, a viable treatment of modes in a cosmological model of expanding space can be described as follows: Starting with some initial state at small curvature, where our universe is very nearly homogeneous on large scales, the minisuperspace approximation is well justified for wave lengths larger than some region of cosmological dimensions. Any modes with smaller wavelengths should be treated by a field theory with an infrared scale related to the large volume.

When we try to understand the high-curvature behavior, we may then evolve this state backwards in time, maintaining the infrared scale as long as the typical length of inhomogeneity modes does not change too much. However, as we approach large curvature in gravitational collapse, inhomogeneity grows even within co-moving volumes. (That is, the scale of inhomogeneity decreases faster than suggested just by the decreasing scale factor.) In order to maintain the approximation with an infrared scale separating a field theory from a minisuperspace model, we should reduce the size of the averaging region, increasing the infrared scale k_{max}. The approximation deteriorates because more modes are approximated by the integrand in Equation (4.13) evaluated at zero k, rather than being integrated. Adjusting the infrared scale, or the corresponding V_0, amounts to infrared renormalization. While V_0 is continually decreased in this process, quantum corrections become more and more significant because effective potentials or other quantum corrections depend on $1/V_0$.

At very large curvature, the BKL scenario suggests that homogeneous models again become relevant in a classical picture, but only in tiny spatial regions with microscopic V_0. Here, quantum corrections are essential and cannot be ignored. Moreover, once we reach this stage, the application of a minisuperspace model is no longer justified in a quantum discussion, in contrast to the classical situation, because an infrared scale has been pushed all the way into the ultraviolet. This problem will be discussed further in Section 6.3.

4.2 Problem of Time

A Hamiltonian generates Hamilton's equations of motion. In a generally covariant theory, in which we have a Hamiltonian constraint, the dynamics is still determined by Hamilton's equations, but the time coordinate is not unique. A large number of important conceptual questions in quantum cosmology are related to this basic feature of relativistic systems. For instance, how do we transform quantum states into each other, evolved using different time coordinates, making sure that predictions do not depend on which coordinate we choose? Or, if a time coordinate is used which is not globally valid but can be applied only in a certain neighborhood of an event, how can unitary quantum evolution, with an operator $\exp(-it\hat{H}/\hbar)$ defined for all real t, be consistent with having a finite range of time?

In full generality, the problem of time (Kuchař 1992; Isham 1993; Anderson 2012) remains unresolved, but several aspects can be illustrated well in model systems. One reason why the role of time is more obscure in relativistic quantum systems than in classical cosmology is that standard quantum mechanics provides an operator for the spatial position of a particle, but not for time. It is, however, easy to change this deficiency in a way that also demonstrates why we have to deal with a Hamiltonian constraint.

4.2.1 Hamiltonian Constraints

Given a canonical degree of freedom q with momentum p, we have Hamilton's equations generated by a Hamiltonian $H(q, p; t)$ which, as indicated, may depend explicitly on time. The Hamiltonian is also equal to the energy of the system, $E = H(q, p; t)$ (which is not conserved if the Hamiltonian does depend on time). A basic algebraic manipulation turns the energy equation into a constraint equation,

$$C_H(q, p; t, -E) = -E + H(q, p; t) = 0. \tag{4.14}$$

The dependence of C_H on two pairs of variables, (q, p) and (t, E), suggests to view E, or rather $p_t = -E$, as the momentum canonically conjugate to t. We can then formally derive Hamilton's equations for all four variables, generated by $C_H(q, p; t, p_t) = p_t + H(q, p; t)$ in some parameter λ:

$$\frac{dq}{d\lambda} = \frac{\partial C_H}{\partial p} = \frac{\partial H}{\partial p} \tag{4.15}$$

$$\frac{dp}{d\lambda} = -\frac{\partial C_H}{\partial q} = -\frac{\partial H}{\partial q} \tag{4.16}$$

$$\frac{dt}{d\lambda} = \frac{\partial C_H}{\partial p_t} = 1 \tag{4.17}$$

$$\frac{dp_t}{d\lambda} = -\frac{\partial C_H}{\partial t} = -\frac{\partial H}{\partial t}. \tag{4.18}$$

The third equation can easily be solved for any Hamiltonian, identifying the canonical variable $t = \lambda + t_0$ with the evolution parameter, up to choosing the origin of time. Substituting this solution in the remaining equations then reproduces the original Hamilton's equations of H,

$$\frac{dq}{dt} = \frac{\partial H}{\partial p} \tag{4.19}$$

$$\frac{dp}{dt} = -\frac{\partial H}{\partial q}, \tag{4.20}$$

as well as energy (non-)conservation

$$\frac{dE}{dt} = \frac{\partial H}{\partial t}. \tag{4.21}$$

The same system allows us to understand how the freedom of choosing different time coordinates is realized in relativistic systems. Since the Hamiltonian constraint is always zero on physical solutions, it does not have a unique expression for a given dynamics. Instead of C_H in Equation (4.14), we may as well use a constraint

$$C(q, p; t, p_t) = N(q, p; t, p_t)C_H(q, p; t, p_t) \tag{4.22}$$

with a phase-space function N that is never zero. While multiplying a constraint with such a function does not change the solution space of the constraint equation $C(q, p; t, p_t) = 0$, it modifies the rate of change measured by the parameter λ in Hamilton's equations generated by C:

$$\frac{dq}{d\lambda'} = \frac{\partial C}{\partial p} = N\frac{\partial H}{\partial p} + \frac{\partial N}{\partial p}C_H \tag{4.23}$$

$$\frac{dp}{d\lambda'} = -\frac{\partial C}{\partial q} = -N\frac{\partial H}{\partial q} - \frac{\partial N}{\partial q}C_H \tag{4.24}$$

$$\frac{dt}{d\lambda'} = \frac{\partial C}{\partial p_t} = N + \frac{\partial N}{\partial p_t}C_H \tag{4.25}$$

$$\frac{dp_t}{d\lambda'} = -\frac{\partial C}{\partial t} = -N\frac{\partial H}{\partial t} - \frac{\partial N}{\partial t}C_H. \tag{4.26}$$

These four λ'-equations for q, p, t and p_t have the same form, consisting of N times the standard term expected from Hamilton's equations plus a contribution proportional to C_H. Since $C_H = 0$ on all solutions (including those of $C = 0$), the second terms vanish. The third equation then still looks simple, but unless N is constant it can be solved only formally:

$$t(\lambda') = \int N\bigl(q(\lambda'),\, p(\lambda');\, t(\lambda'),\, p_t(\lambda')\bigr)d\lambda'. \tag{4.27}$$

The solution is implicit because the integral refers to the same function $t(\lambda')$ that we are trying to solve for. It therefore requires solutions $q(\lambda')$, $p(\lambda')$ and $p_t(\lambda')$ of the first two equations and the last one. Fortunately, the infinitesimal version $dt = Nd\lambda'$ is sufficient to see how different choices of N amount to different parameterizations of time.

We may then divide the three unsolved equations by N and use the algebraic solution $dt = Nd\lambda'$ of Equation (4.25) to see that

$$\frac{dq}{Nd\lambda'} = \frac{dq}{dt} = \frac{\partial H}{\partial p} \tag{4.28}$$

$$\frac{dp}{Nd\lambda'} = \frac{dp}{dt} = -\frac{\partial H}{\partial q} \tag{4.29}$$

$$\frac{dp_t}{Nd\lambda'} = \frac{dp_t}{dt} = -\frac{\partial H}{\partial t} \tag{4.30}$$

are still equivalent to the original Hamilton's equations, but expressed at an intermediate stage through a transformed parameter λ' such that $d\lambda = Nd\lambda'$ with respect to the original λ. In this way, different choices of the multiplier N of the Hamiltonian constraint correspond to different choices of time coordinates λ, which all describe the same physical relationships between phase-space degrees of freedom.

We had to make an assumption on N for this procedure to work: N cannot be zero anywhere on phase space because the solution spaces of $C_H = 0$ and $C = NC_H = 0$ would not necessarily agree if the condition were violated. Accordingly, $d\lambda'/d\lambda = N$ is never zero, and therefore the coordinate transformation of time is monotonic. The procedure can describe only global changes of coordinates. It does not allow us to implement the definition of time coordinates in local charts of a manifold, as required in general cases with non-trivial topology.

4.2.2 Deparameterization

The procedure described in the preceding subsection is called "parameterization" because physical time, t, is turned into a mere mathematical parameter, λ, that progresses along evolving solutions but does not necessarily keep track of the physical rate of change determined by t. Such a procedure is not necessary in non-relativistic systems in which an absolute time, t, exists which determines the rate of change for all observers. In such systems, parameterization provides a conceptual result, showing that it is possible to turn an absolute time into one of the phase-space variables, making it more similar to the spatial position q.

Dynamical systems described by a Hamiltonian constraint can be quantized using methods of constrained quantization (Bergmann 1949; Bergmann & Brunings 1949; Dirac 1950, 1958, 1969). Constrained quantization proceeds by first turning the basic phase-space variables, here q, p, t, and p_t, into operators such that canonical commutation relations are satisfied. In a second step, these operators are combined to form an operator $\hat{C}_{\mathrm{H}}$ that quantizes C_{H}. The classical constraint equation $C_{\mathrm{H}} = 0$ is then imposed as a state equation, $\omega(\hat{A}\hat{C}_{\mathrm{H}}) = 0$ for all operators $\hat{A}$ polynomial in $\hat{q}$, $\hat{p}$, $\hat{t}$, and $\hat{p}_t$. If $\omega = \omega_\psi$ is a pure state with wave function ψ, the quantum constraint equation reads $\hat{C}_{\mathrm{H}}\psi = 0$. It is easy to see that the standard representation of quantum mechanics, discussed in Section 3.1.3, turns the quantized (4.14) into Schrödingers equation on a pure state.

If one applies constrained quantization to our example, there are operators for both q and t, subject to a Hamiltonian constraint $\hat{C}_{\mathrm{H}}$ instead of quantum evolution via Schrödinger or Heisenberg equations with respect to an absolute parameter t. Constrained quantization therefore implies a more equal treatment of space (position) and time, except for the algebraic difference that the parameterized Hamiltonian constraint C_{H}, by construction, is linear in the momentum p_t of t but need not be linear in the momentum p of q.

In relativistic systems, one usually tries to invert this process because we start with a Hamiltonian constraint such as Equation (1.18) which provides Hamilton's equations of motion in some coordinate time parameter, corresponding to λ in the preceding subsection, which should be interpreted in terms of physical variables. There is no unique choice for a coordinate time because different observers experience the rate of change differently, depending on their state of motion. (In cosmology, proper time of co-moving observers may be preferred in many applications, but is not unique as a time coordinate.) There is no operator for a time coordinate λ, but if we can find an analog of the phase-space variable t encountered in parameterized systems, it would have an operator associated to it in a quantization and be conceptually more similar to spatial operators such as the volume in a quantum cosmological model. Instead of using time coordinates, which are not universally observable and therefore cannot be quantized, we would then describe evolution "relationally" by relations between different observables such as V and t (Dirac 1950; Bergmann 1961).

The choice of such a t may not be unique for a given system, but if at least one choice exists, relational evolution can be formulated. This procedure is therefore applied frequently in different approaches to quantum cosmology, even though several major questions about it remain. For instance, neither existence nor uniqueness of a valid choice of t is guaranteed. Existence is not obvious because the form in which t and its momentum p_t appear in a parameterized non-relativistic system is subject to certain assumptions which may be violated in relativistic systems. Relativistic systems usually have a quadratic dependence on the energy, or $-p_t$, in contrast to the linear dependence in C_H used so far. The procedure can be generalized to constraints quadratic in p_t, but then leads to further restrictions as shown in Bojowald & Tsobanjan (2019).

Moreover, while parameterization and relational evolution at first sight bring us closer to the physical way of keeping track of time, comparing a measured observable with the position of a physical system such as planetary orbits or the hands of a clock, the technical requirements of the procedure imply that t changes monotonically with respect to any coordinate time λ. It would have been inconsistent to allow the multiplier N in Equation (4.22) to go through zero because this would have changed the solution space of the constraint. In deparameterized quantum systems, only monotonically changing times can be allowed in order to ensure unitarity. Deparameterized relational evolution therefore does *not* describe the fundamental physical mechanisms underlying the way we keep track of time, because they refer to calibrated or precisely measured *periodic* systems. Nevertheless, deparameterization may help us to derive a qualitative understanding of relativistic quantum evolution. It can also be found as an approximation to more complicated unitary evolution with respect to fundamentally periodic degrees of freedom (Wendel et al. 2020). (For more results about local internal times, see for instance Bojowald et al. 2011a, 2011b; Höhn et al. 2012; Giacomini et al. 2019; Vanrietvelde et al. 2020, 2018).

4.2.3 Examples

In order to understand some of these difficulties, we should look at details of specific systems that can be deparameterized. In classical or quantum cosmology, we start with a set of phase-space variables quantized to basic operators for a quantum description, subject to a constraint C or its quantization, $\hat{C}$. Initially, none of the phase-space variables or basic operators are distinguished as candidates for time, and therefore the constraint, in general, is not of the form of C_H with a linear dependence on one of the momenta. However, it may be possible to find a suitable N such that $C = NC_H$ is of the required form, up to a multiplier N.

A common example in quantum cosmology is an isotropic model with dust as a matter ingredient; see for instance Acacio de Barros et al. (1998); Husain & Pawlowski (2011a, 2011b) or Brown & Kuchař (1995) for the general canonical formalism. The energy density of dust, because of dilution in an expanding universe, is given by a constant divided by the expanding volume, and therefore contributes just a constant to the matter energy in the Hamiltonian constraint (1.18). Any constant may be identified with the momentum of a variable that does not appear anywhere in the constraint, because Hamilton's equations are then consistent with a conservation of the momentum. We therefore call the constant dust energy $p_{T_1} = H_{dust}$, canonically conjugate to some parameter T_1 used as time. The constraint (1.18) then takes the desired form,

$$C_1 = C_{H,1} = p_{T_1} - \frac{6\pi G}{c^2} V p_V^2.$$

(4.31)

As a second example, we may use the cosmological constant, $\Lambda = p_{T_2}$, as a momentum of some time variable T_2. The energy *density* provided by a cosmological constant is constant, and therefore the energy contribution to the Hamiltonian constraint equals $H_\Lambda = \Lambda V$, implying

$$C_2 = p_{T_2} V - \frac{6\pi G}{c^2} V p_V^2 = N C_{H,2}$$

(4.32)

with a non-constant $N = V$ and

$$C_{H,2} = p_{T_2} - \frac{6\pi G}{c^2} p_V^2.$$

(4.33)

The Hamiltonian constraints (4.31) and (4.33) imply different equations of motion. For instance,

$$\frac{dV}{dT_1} = \frac{\partial C_{H,1}}{\partial p_V} = -\frac{12\pi G}{c^2} V p_V$$

(4.34)

while

$$\frac{dV}{dT_2} = \frac{\partial C_{H,2}}{\partial p_V} = -\frac{12\pi G}{c^2} p_V.$$

(4.35)

In order to compare these equations, we should transform them to proper time, using Hamilton's equations of motion for the full C_1 and C_2 before the factorization into NC_H. For dust, $C_1 = C_{H,1}$, such that Equation (4.34) is already of the correct form for proper-time evolution. For a cosmological constant, Hamilton's equation generated by C_2 for T_2 implies

$$\frac{dT_2}{d\tau} = \frac{\partial C_2}{\partial p_{T_2}} = V.$$

(4.36)

We then transform Equation (4.35) to proper time by multiplying it with V and referring to the chain rule of time derivatives. With respect to proper time, Equation (4.35) is therefore equivalent to Equation (4.34).

This part of Hamilton's equations does not depend much on the precise dynamics because it merely tells us how the Hubble parameter, proportional to p_V according to Equation (1.16), is related to the rate of change of the volume. The dynamical content of Hamilton's equations is given by the time derivatives implied for p_V, amounting to a second-order time derivatives of the volume, or its acceleration. These equations in the two matter models are

$$\frac{\mathrm{d}p_V}{\mathrm{d}T_1} = -\frac{\partial C_{H,1}}{\partial V} = \frac{6\pi G}{c^2}p_V^2 \tag{4.37}$$

$$\frac{\mathrm{d}p_V}{\mathrm{d}T_2} = -\frac{\partial C_{H,2}}{\partial V} = 0. \tag{4.38}$$

They are inequivalent when transformed to proper time.

This inequivalence is not problematic because dust and a cosmological constant represent different forms of energy, which drive the expansion of the universe in inequivalent ways. However, we should obtain equivalent dynamics if we combine the dust and Λ-contributions in a single constraint

$$C^{\mathrm{dust}-\Lambda} = p_{T_1} + Vp_{T_2} - \frac{6\pi G}{c^2}Vp_V^2. \tag{4.39}$$

We should then be able to choose either T_1 as time, with Hamiltonian constraint

$$C_{H,1}^{\mathrm{dust}-\Lambda} = C^{\mathrm{dust}-\Lambda} \tag{4.40}$$

or T_2, with Hamiltonian constraint

$$C_{H,2}^{\mathrm{dust}-\Lambda} = p_{T_2} + \frac{p_{T_1}}{V} - \frac{6\pi G}{c^2}p_V^2. \tag{4.41}$$

Classically, one can show that the two systems imply the same dynamics with respect to proper time. But they lead to inequivalent quantum dynamics, as can be seen from the presence of different quantum corrections in the effective Hamiltonian constraints

$$\omega\left(\hat{C}_{H,1}^{\mathrm{dust}-\Lambda}\right) = \omega(\hat{p}_{T_1}) + \omega(\hat{V})\omega(\hat{p}_{T_2}) + \Delta_\omega(Vp_{T_2})$$
$$- \frac{6\pi G}{c^2}\left(\omega(\hat{V})\omega(\hat{p}_V)^2 + \omega(\hat{V})\left(\Delta_\omega p_V^2\right) + \omega(\hat{p}_V)\Delta_\omega(Vp_V)\right) \tag{4.42}$$

$$\omega\left(\hat{C}_{H,2}^{\mathrm{dust}-\Lambda}\right) = \omega(\hat{p}_{T_2}) + \frac{\omega(\hat{p}_{T_1})}{\omega(\hat{V})} - \frac{1}{\omega(\hat{V})^2}\Delta_\omega(Vp_{T_1}) + \frac{\omega(\hat{p}_{T_1})}{\omega(\hat{V})^3}(\Delta_\omega V^2)$$
$$- \frac{6\pi G}{c^2}\left(\omega(\hat{p}_V)^2 + \left(\Delta_\omega p_V^2\right)\right), \tag{4.43}$$

defined analogously to the effective potential (3.94) and expanded up to first-order corrections in $\hbar$ through moment terms A detailed analysis of relationships between moments implied by these two effective constraints shows that the dynamics they imply are inequivalent (Bojowald & Halnon 2018); but here it suffices to see that the two effective constraints in Equation (4.42) cannot be transformed into each other by a simply factor as in Equations (4.31) and (4.33). Further examples, using different methods, have been constructed and analyzed in Malkiewicz (2015); Malkiewicz et al. (2020); Gielen & Menéndez-Pidal (2020). Without further conditions, quantization therefore does not preserve time reparameterization invariance.

4.3 Algebras

The role of different approaches to quantum gravity in cosmological models can be seen in the choice of basic phase-space variables and their corresponding operators. So far, we have worked with a canonical pair, (V, p_V), to describe the expansion of an isotropic space. The classical volume V cannot be negative, and therefore the range of eigenvalues of the operator $\hat{V}$, which determine possible outcomes of single measurements, should be restricted to the same range.

4.3.1 Positive Volume

However, the standard commutation relation $[\hat{V}, \hat{p}_V] = i\hbar$ for a canonical pair implies that both $\hat{V}$ and $\hat{p}_V$ have the full range of real numbers as eigenvalues. It is not possible to restrict the range of eigenvalues of $\hat{V}$ in any representation that also contains a self-adjoint $\hat{p}_V$ because $\exp(i\delta\hat{p}_V/\hbar)$, by virtue of the commutation relation, acts as a translation operator on wave functions $\psi(V)$:

$$\exp(i\delta\hat{p}_V/\hbar)\psi(V) = \sum_{n=0}^{n}\left(\frac{i\delta}{\hbar}\right)^n \hat{p}_V^n \psi(V) = \sum_{n=0}^{n} \delta^n \frac{\partial^n \psi(V)}{\partial V^n} = \psi(V + \delta), \qquad (4.44)$$

using the Taylor series of the exponential function and then a Taylor expansion of ψ around V. Therefore, if the representation contains an eigenstate ψ_v of $\hat{V}$ with eigenvalue $v > 0$, it also contains an eigenstate $\exp(-2iv\hat{p}_V/\hbar)\psi_v$ with eigenvalue $-v < 0$.

There are two possible solutions to this problem, which assume different viewpoints on the geometry of space and lead to inequivalent quantum representations. First, we can make sense of negative V if we define V as the oriented volume, that is define $V = |V|$ if the volume is $|V|$ and the orientation of space is right-handed, and $V = -|V|$ if the volume is $|V|$ and the orientation is left-handed. The orientation of space can be determined from the sign of the determinant of a matrix formed by the component of three basis vectors of the tangent space.

In Riemannian geometry, the basis vectors are usually expressed as a *triad*, e_I^i, where the index $I = 1, 2, 3$ labels the three independent vectors, and $i = 1, 2, 3$ labels the three components of each spatial vector. In order to normalize the basis, one usually imposes the relationship

$$q^{ij} = \sum_{I=1}^{3} e_I^i e_I^i \tag{4.45}$$

with the inverse spatial metric. Unlike the metric, the triad contains information about the orientation of space. The oriented volume can therefore be interpreted as an isotropic variable in a triad theory as opposed to a metric theory. In this form, it is implied by loop quantum cosmology (Bojowald 2002) because loop quantum gravity is based on a triad formulation of general relativity, as we will see in Section 5.2.6. (The appearance of a triad is also motivated by the fact that theories in which some of the matter content is described by fermions require a triad and cannot be formulated completely using just a metric.)

Second, it is possible to provide the correct degrees of freedom of an isotropic cosmological model by working with a non-canonical pair of basic variables, (V, D) such that $[\hat{V}, \hat{D}] \neq i\hbar$ in a quantization. Specifically, in order for the sign of $\hat{V}$-eigenvalues to be preserved, we should replace the translation operator obtained by exponentiating the canonical momentum $\hat{p}_V$ with an operator that, after exponentiating, provides a transformation that preserves the set of positive numbers. A simple transformation of this form is a dilation $\hat{D}$, such that $\exp(i\delta\hat{D}/\hbar)\psi(V) = \psi(\delta V)$. In terms of the canonical operators $\hat{V}$ and $\hat{p}_V$, a dilation operator is given by $\hat{D} = \hat{V}\hat{p}_V$, such that $[\hat{V}, \hat{D}] = i\hbar\hat{V}$ is indeed non-canonical. This operator generates dilations of V because

$$
\begin{aligned}
\hat{D}\psi(V) &= \frac{\hbar}{i} \lim_{\delta\to 0} \frac{d}{d\delta} \exp(i\delta\hat{D}/\hbar)\psi(V) \\
&= \frac{\hbar}{i} \lim_{\delta\to 0} \frac{d}{d\delta}\psi(\delta V) \\
&= V\frac{\hbar}{i}\frac{d\psi}{dV} = \hat{V}\hat{p}_V\psi(V).
\end{aligned}
\tag{4.46}
$$

The choice of $(\hat{V}, \hat{D})$ as basic operators leads to a representation on wave functions which is inequivalent to the standard representation of quantum mechanics based on a canonical pair. It is usually referred to as *affine quantization* (Klauder 2003, 2006) and has been applied to quantum cosmology (Bergeron et al. 2015a, 2015b, 2017, 2020).

The choice of basic commutators, or algebras, to be represented may have cosmological implications especially near zero volume, or close to the Big Bang singularity. If we use $(\hat{V}, \hat{D})$ in a version of affine quantum cosmology, the Hamiltonian constraint (4.31) in the isotropic dust model is replaced by

$$\hat{C}'_{H,1} = \hat{p}_{T_1} - \frac{6\pi G}{c^2}\frac{\hat{D}^2}{\hat{V}} \tag{4.47}$$

or its semiclassical version

$$\omega(\hat{C}'_{H,1}) = \omega(\hat{p}_{T_1})$$

$$- \frac{6\pi G}{c^2}\left(\frac{\omega(\hat{D})^2}{\omega(\hat{V})} + \frac{(\Delta_\omega D^2)}{\omega(\hat{V})} - \frac{\omega(\hat{D})}{\omega(\hat{V})^2}\Delta_\omega(VD) + \frac{\omega(\hat{D})^2}{\omega(\hat{V})^3}(\Delta_\omega V^2)\right). \quad (4.48)$$

(For these moment terms, we have assumed that the product of $\hat{D}^2$ and $\hat{V}^{-1}$ in Equation (4.47) is defined in completely symmetric ordering. If a different ordering is used, there are additional terms in which squared quantum fluctuations are replaced by explicit factors of $\hbar$. The following arguments for possible implications for large-fluctuation regimes are insensitive to such extra terms.)

Quantum corrections here provide two new repulsive terms,

$$\frac{(\Delta_\omega D^2)}{\omega(\hat{V})} \quad \text{and} \quad \frac{(\Delta_\omega V^2)\omega(\hat{D})^2}{\omega(\hat{V})^3}, \quad (4.49)$$

which are always positive in this representation. Moreover, the covariance $\Delta_\omega(VD)$ cannot be greater than $\Delta_\omega V \Delta_\omega D$ according to the uncertainty relation (3.25), which holds in the affine representation provided the right-hand side is replaced using the non-canonical commutator of $[\hat{V}, \hat{D}] = i\hbar\hat{V}$, such that

$$\left(\omega(\hat{V}^2) - \omega(\hat{V})^2\right)\left(\omega(\hat{D}^2) - \omega(\hat{D})^2\right)$$

$$- \left(\frac{1}{2}\omega(\hat{V}\hat{D} + \hat{D}\hat{V}) - \omega(\hat{V})\omega(\hat{D})\right)^2 \geqslant \frac{\hbar^2}{4}\omega(\hat{V})^2. \quad (4.50)$$

(The argument used here, however, does not depend on the specific right-hand side but only on its positivity.) Moreover, it is not possible for both $\Delta_\omega V$ and $\Delta_\omega D$ to vanish at the same time.

The repulsive terms are therefore dominant when quantum fluctuations are significant, preventing $\omega(\hat{V})$ in the denominators from approaching zero on any solution of the constraint, where $\omega(\hat{p}_{T_1})$ is a positive constant. As in this example, because an inequivalent representation implies a different behavior of quantum corrections, strong quantum regimes such as the Big Bang singularity may appear in a new light. Representation theory is therefore an important method in quantum cosmology; see Isham (1983) for a general application in quantization.

The new repulsive terms implied by using the dilation operator as a basic one are an example of a rather common possibility by which quantum cosmology may resolve the Big Bang singularity: If the range of $\hat{V}$-eigenvalues is bounded from below at $V = 0$, any non-zero volume fluctuation $\Delta_\omega V$ implies a positive volume expectation value $\omega(\hat{V})$, just because the wave function $\psi(V)$ is spread out and supported only on positive values. Zero $\omega(\hat{V})$ could be obtained from quantum evolution only if the state suddenly collapses to zero volume fluctuations right when it hits the boundary at zero volume. While such a behavior may be conceivable, it would not be generic in the space of all dynamical solutions, in contrast to the classical situation in which a singularity is reached for generic initial data (DeWitt 1967; Acacio de Barros et al. 1998; Falciano et al. 2007; Novello & Bergliaffa 2008). Recent results in this context can be found in Kiefer et al. (2019); Kiefer & Schmitz (2019).

4.3.2 Discrete Volume

A different choice of basic algebra makes it possible to keep the expected volume even further away from reaching zero. If we use an algebra such that its volume operator $\hat{V}$ has a discrete spectrum, a whole range of values around zero will be eliminated as possible outcomes of volume measurements.

We can make the volume spectrum discrete by choosing a periodic range for its momentum, such that only periodic functions of the momentum are allowed as wave functions in the momentum representation. For any fixed $0 \leqslant \epsilon < 1$, we may impose periodic boundary conditions

$$\psi(p_V + p_0) = e^{2\pi i \epsilon}\psi(p_V), \tag{4.51}$$

where p_0 is the period of the momentum range, usually assumed to be Planckian in this context. In quantum cosmology, the momentum (1.16) of V is proportional to the Hubble parameter times c^2/G, such that Gp_0^2/c^2 has the units of an energy density. As a specific example, one may therefore assume that Gp_0^2/c^2 equals the Planck density, $\rho_P = c^7/(\hbar G^2)$. Or, if p_0 corresponds to the value of p_V for a Planckian Hubble parameter close to $c/\ell_P = \sqrt{c^5/(G\hbar)}$, it equals

$$p_0 = \frac{c^2 H}{4\pi G} \sim \frac{1}{4\pi}\sqrt{\frac{c^9}{G^3\hbar}} = \frac{1}{4\pi}\frac{\hbar}{\ell_P^3}. \tag{4.52}$$

Every wave function that obeys the periodicity condition 4.51 is a superposition of wave functions

$$\psi_n(p_V) = \frac{1}{\sqrt{p_0}}\exp(-2\pi i(n - \epsilon)p_V/p_0) \tag{4.53}$$

with integers n. The spectrum of the volume operator, $\hat{V} = i\hbar\partial/\partial p_V$, is therefore discrete with eigenvalues $V_n = 2\pi(n - \epsilon)\hbar/p_0$. If p_0 is Planckian as in Equation (4.52), $V_n = 8\pi^2(n - \epsilon)\ell_P^3$ is quantized by multiples of the Planck volume.

In terms of algebras, we can implement the periodicity condition by replacing p_V, which would no longer be a global phase-space coordinate if its range is periodic, with the periodic function $S = \sin(2\pi p_V/p_0)$. The commutator of the corresponding operator with the volume operator equals $[\hat{V}, \hat{S}] = 2\pi i\hbar\hat{C}/p_0$ with $C = \cos(2\pi p_V/p_0)$. We therefore have linear commutator relationships involving three operators. They form a closed algebra with basic commutators

$$[\hat{V}, \hat{S}] = \frac{2\pi i\hbar}{p_0}\hat{C}, \quad [\hat{V}, \hat{C}] = -\frac{2\pi i\hbar}{p_0}\hat{S}, \quad [\hat{S}, \hat{C}] = 0 \tag{4.54}$$

which has a one-parameter family of inequivalent representations, given in terms of wave functions subject to the ϵ-dependent periodicity condition. These representations are inequivalent for different ϵ because the volume spectrum depends on ϵ.

In these representations, however, both positive and negative values are included in the spectrum of $\hat{V}$. While a significant range of small eigenvalues around zero may be eliminated by the discretization condition, the presence of positive and negative volume eigenvalues means that zero volume expectation values are allowed, indicating the possibility of a singularity. We can eliminate the negative eigenvalues if we combine the discretization condition with the replacement of translations in the volume by dilations, as in affine quantizations. Therefore, we should not only make the range of p_V periodic, but also replace periodic functions of p_V alone with such functions multiplied by the volume, mimicking our construction of the dilation operator $\hat{D}$ in Equation (4.46).

4.3.3 Loop Quantum Cosmology

In a discrete version of affine quantum cosmology, we work with three basic variables given by V together with

$$J_1 = V \sin(2\pi p_V / p_0) \quad \text{and} \quad J_2 = V \cos(2\pi p_V / p_0). \tag{4.55}$$

If we define operators as products of V with S and C, respectively, in the ordering in which $\hat{V}$ appears to the left, we still have linear relations

$$[\hat{V}, \hat{J}_1] = \frac{2\pi i \hbar}{p_0} \hat{J}_2, \quad [\hat{V}, \hat{J}_2] = -\frac{2\pi i \hbar}{p_0} \hat{J}_1, \quad [\hat{J}_1, \hat{J}_2] = -\frac{2\pi i \hbar}{p_0} \hat{V} \tag{4.56}$$

replacing Equation (4.54). They now belong to the Lie algebra $sl(2, \mathbb{R})$. This Lie algebra has different types of inequivalent representations (Bargmann 1947). Some of them, called the positive discrete series, are such that $\hat{V}$ has a discrete spectrum which is non-negative.

Algebras of this form have appeared in loop quantum cosmology (Bojowald 2015a, 2007b, 2007); and with a different interpretation of the basic operators in so-called *CVH*-models (Ben Achour & Livine 2017, 2019a, 2019b; Bodendorfer & Wuhrer 2019). (In *CVH*-models, the *V*olume, the *H*amiltonian, and a *C*omplexifier are used as basic variables.) They usually make it easier to obtain non-singular, bouncing solutions compared with non-discrete systems as seen in numerical examples of Ashtekar et al. (2006); but some, often implicit assumptions are still necessary. For instance, $sl(2, \mathbb{R})$ also has representations in which the volume spectrum contains both positive and negative values, called the continuous series. (Despite the name, the volume spectrum is still discrete in these representations.) Unless these representations can be ruled out, for which no convincing reason is known so far, one cannot be sure that generic solutions within this quantization scheme, based on the algebra (4.56), do bounce (Bojowald 2019a, 2019b). Moreover, the application of periodic functions means that it is no longer straightforward to quantize the Hamiltonian constraint (whose classical expression, like the Friedmann equation is quadratic and therefore non-periodic in p_V) by writing it in terms of basic operators.

Such models therefore work with modifications of the classical dynamics, replacing p_V^2 in the classical Hamiltonian constraint by a periodic function such as $p_0^2 \sin^2(2\pi p_V/p_0)/(4\pi^2)$ that can be expressed in terms of the basic variables. If p_0 is Planckian, the modification is tiny at low curvature, but it is very relevant in the Planck regime. In particular, because periodic functions are bounded, the modified Hamiltonian constraint,

$$\frac{p_0^2 \sin^2(2\pi p_V/p_0)}{4\pi^2} = \frac{c^2}{6\pi G}\rho_{\text{matter}}, \tag{4.57}$$

implies that the energy density is always bounded—provided additional quantum corrections from fluctuations and other moments remain small. (The maximum density allowed by the modified Friedmann equation is close to the Planck density if p_0 is Planckian.) Singularity avoidance based on bounce arguments may therefore be strengthened by placing strict upper bounds on the density. However, a major problem that cannot satisfactorily be addressed in minisuperspace models is whether the modifications can be consistent with general covariance. We will return to this question in Chapter 6.

The modified Friedmann Equation (4.57) can be expressed in more standard form by replacing p_V with a function proportional to the time derivative of the scale factor, derived from equations of motion. However, the modification not only changes the dynamics but also the usual relationship between p_V and $\dot{a}$ or H, given classically by Equation (1.16). Written as a Hamiltonian constraint, Equation (4.57) reads

$$C = -\frac{3G}{2\pi c^2}p_0^2 V \sin^2(2\pi p_V/p_0) + V\rho_{\text{matter}} = 0 \tag{4.58}$$

where C generates two Hamilton's equations:

$$\dot{V} = \frac{\partial C}{\partial p_V} = -\frac{6Gp_0}{c^2}V \sin(2\pi p_V/p_0)\cos(2\pi p_V/p_0) = -\frac{3Gp_0}{c^2}V \sin(4\pi p_V/p_0) \tag{4.59}$$

$$\dot{p}_V = -\frac{\partial C}{\partial V} = \frac{3Gp_0^2}{2\pi c^2} \sin^2(2\pi p_V/p_0) - \frac{\partial E_{\text{matter}}}{\partial V} = \frac{c^2}{6\pi G}\left(\rho_{\text{matter}} + P_{\text{matter}}\right) \tag{4.60}$$

with the matter pressure $P_{\text{matter}} = -\partial E_{\text{matter}}/\partial V$.

We can substitute Equation (4.59) in Equation (4.58) if we first write

$$\sin^2(2\pi p_V/p_0) = \frac{1}{2}(1 - \cos(4\pi p_V/p_0)) = \frac{1}{2}\left(1 - \sqrt{1 - \sin^2(4\pi p_V/p_0)}\right) \tag{4.61}$$

in the latter equation. Therefore, $C = 0$ is equivalent to

$$\frac{3Gp_0^2}{4\pi c^2}\left(1 - \sqrt{1 - \left(\frac{c^2 \dot{V}}{3Gp_0 V}\right)^2}\right) = \rho_{\text{matter}}, \tag{4.62}$$

or (Vandersloot 2005)

$$\left(\frac{\dot{a}}{a}\right)^2 = \frac{1}{9}\left(\frac{\dot{V}}{V}\right)^2 = \frac{G^2 p_0^2}{c^4}\left(1 - \left(1 - \frac{4\pi c^2 \rho_{\text{matter}}}{3Gp_0^2}\right)^2\right) = \frac{8\pi G}{3c^2}\left(\rho_{\text{matter}} - \frac{\rho_{\text{matter}}^2}{\rho_{\text{QG}}}\right) \tag{4.63}$$

with

$$\rho_{\text{QG}} = \frac{3Gp_0^2}{2\pi c^2}. \tag{4.64}$$

When $\rho_{\text{matter}} = \rho_{\text{QG}}$, Equation (4.63) implies that $\dot{a} = 0$. This extremum is a local minimum, as can be confirmed by computing the second derivative $\ddot{a}$ using a modified Raychaudhuri equation that follows from Equation (4.60).

The modified Friedmann Equation (4.63) therefore indicates that the singularity may be resolved, leading to a bounded energy density and a scale factor that never becomes zero. Although this outcome is encouraging, the derivation presented here shows that this result is merely a consequence of a discrete volume spectrum that prevents the wave function from being supported close to zero volume. Just looking at Equation (4.63), it may seem that there is a Planckian correction term at high curvature that is independent of quantum fluctuations. However, the new term is derived from the non-canonical algebra (4.56) and a correspondingly modified constraint, which incorporates the discreteness of $\hat{V}$ in quantum representations. Therefore, the mechanism by which Equation (4.63) may resolve the Big Bang singularity, although more intuitive in this form of a modified Friedmann equation, is nothing but the original DeWitt mechanism (DeWitt 1967) in disguise, strengthened by using a discrete volume spectrum.

Moreover, as shown in Figure 4.1; the presence of non-zero volume fluctuations might suggest a bouncing solution if one considers the absolute value $|V|$, even if the expectation value of the oriented volume V reaches zero. Lower bounds on the volume $|V|$ or related upper bounds on the energy density can therefore be misleading. A detailed analysis of quantum fluctuations related to all properties of minisuperspace states, including the sign of the volume, is therefore required. Such an analysis is still ongoing in loop quantum cosmology, but several basic features are presented in the next subsection.

4.3.4 Quantum Fluctuations

The observation that the high-density behavior of Equation (4.63) indirectly reflects the contribution of quantum fluctuations in a discrete volume space indicates that it may be sensitive to the inclusion of explicit quantum fluctuation terms in an effective Friedmann equation.

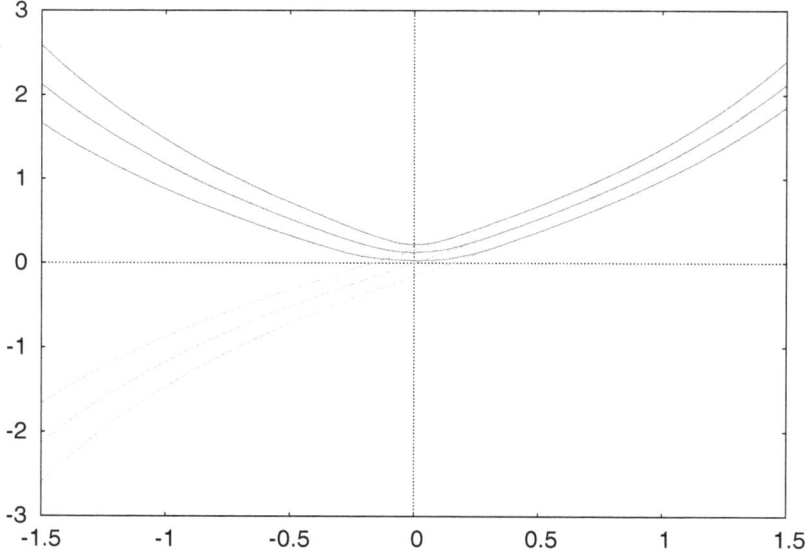

Figure 4.1. Illustration of a non-bouncing wave function for the volume (vertical) as a function of ϕ (horizontal). The variance is indicated by a range around the central line for the expectation value. The green curves show the expectation value and variance of the *oriented* volume V used often in loop quantum cosmology. In this state, the volume expectation value reaches zero. The red curves show the expectation value and variance of $|V|$ in the same state. For this positive variable, it seems that the state is bouncing just because a non-zero variance implies that $\omega(|\hat{V}|)$ is never zero even when $\omega(\hat{V}) = 0$.

Classical Solvable Model

Implications of quantum fluctuations can be illustrated in a model in which the modified Friedmann constraint (4.58) is evaluated for a free, massless scalar matter source with energy density

$$\rho_{\text{matter}} = \frac{c^2}{2} \frac{p_\phi^2}{V^2}. \tag{4.65}$$

Here, p_ϕ is the momentum of a homogeneous scalar field ϕ, and it is constant in time because there is no explicit ϕ-dependence in the resulting

$$C = -\frac{3G}{2\pi c^2} p_0^2 V \sin^2(2\pi p_V/p_0) + \frac{c^2}{2} \frac{p_\phi^2}{V}. \tag{4.66}$$

We can then solve the constraint equation $C = 0$ for p_ϕ,

$$p_\phi = \pm\sqrt{\frac{3Gp_0}{\pi c^4}} V \sin(2\pi p_V/p_0) = \pm\sqrt{\frac{3Gp_0}{\pi c^4}} J_1, \tag{4.67}$$

which is linear in the algebra generator J_1 seen in Equation (4.56). (More precisely, we may want to write $|J_1|$ instead of J_1 when we take a square root of J_1^2. This function is not linear, but deviations from linearity would be noticeable only around $J_1 = 0$.) The choice of $\pm$ indicates the two possible directions in which ϕ may change.

Therefore, the equations of motion with respect to ϕ, obtained by deparameterization, are linear:

$$\frac{dV}{d\phi} = \{V, p_\phi\} = \pm\sqrt{\frac{3Gp_0}{\pi c^4}} \{V, J_1\} = \pm 2\pi\sqrt{\frac{3G}{\pi c^4 p_0}} J_2 \tag{4.68}$$

$$\frac{dJ_2}{d\phi} = \{V, p_\phi\} = \pm\sqrt{\frac{3Gp_0}{\pi c^4}} \{J_2, J_1\} = \pm 2\pi\sqrt{\frac{3G}{\pi c^4 p_0}} V \tag{4.69}$$

$$\frac{dJ_1}{d\phi} = \{J_1, p_\phi\} = \pm\sqrt{\frac{3Gp_0}{\pi c^4}} \{J_1, J_1\} = 0, \tag{4.70}$$

using the Poisson brackets that correspond to the algebra relations (4.56). The first two equations are coupled to each other and are solved by

$$V(\phi) = A \exp\left(\sqrt{\frac{12\pi G}{c^4 p_0}} \phi\right) + B \exp\left(-\sqrt{\frac{12\pi G}{c^4 p_0}} \phi\right) \tag{4.71}$$

$$\pm J_2(\phi) = A \exp\left(\sqrt{\frac{12\pi G}{c^4 p_0}} \phi\right) - B \exp\left(-\sqrt{\frac{12\pi G}{c^4 p_0}} \phi\right) \tag{4.72}$$

with two integration constants A and B. The third equation tells us that J_1 is independent of ϕ. The value

$$J_1 = \sqrt{\frac{\pi c^4}{3G}} \frac{p_\phi}{p_0} \tag{4.73}$$

is determined by the constraint equation.

Depending on the signs of A and B, $V(\phi)$ may or may not reach the value zero. However, these choices are not arbitrary because the definitions of J_1 and J_2 imply that

$$J_1^2 + J_2^2 = V^2 \tag{4.74}$$

provided the momentum p_V is real. We therefore have a reality condition in addition to the constraint and equations of motion. Evaluating Equation (4.74) in Equations (4.73) and (4.72), we obtain the ϕ-independent condition

$$4AB = J_1^2 = \frac{\pi c^4}{3G} \frac{p_\phi^2}{p_0^2}. \tag{4.75}$$

Therefore, AB must be positive, such that A and B have the same sign. It is then impossible for Equation (4.71) to reach the value zero.

Quantum Solvable Model
Much of our classical analysis relied on the linear structure of Poisson brackets of V, J_1 and J_2, together with the linear p_ϕ (up to a possible absolute value). The same relationships are true for commutators instead of Poisson brackets. Therefore, we may view the solutions (4.71–4.73) as equations for evolving Heisenberg operators of the quantum model—or, if we put a state ω around V, J_1 and J_2, as equations for ϕ-dependent expectation values:

$$\omega(\hat{V})(\phi) = A \exp\left(\sqrt{\frac{12\pi G}{c^4 p_0}}\,\phi\right) + B \exp\left(-\sqrt{\frac{12\pi G}{c^4 p_0}}\,\phi\right) \tag{4.76}$$

$$\pm\omega(\hat{J}_2)(\phi) = A \exp\left(\sqrt{\frac{12\pi G}{c^4 p_0}}\,\phi\right) - B \exp\left(-\sqrt{\frac{12\pi G}{c^4 p_0}}\,\phi\right) \tag{4.77}$$

$$\omega(\hat{J}_1) = \sqrt{\frac{\pi c^4}{3G}\frac{p_\phi}{p_0}}. \tag{4.78}$$

There are no quantum corrections in these equations because they result from a linear model.

However, the reality condition (4.74) is quadratic. We should impose the related equation

$$\hat{J}_1^2 + \hat{J}_2^2 = \hat{V}^2 \tag{4.79}$$

for operators, implementing the condition that p_V now should quantize a real variable, and therefore be self-adjoint. Taking expectation values in a state ω, the reality condition implies

$$\omega(\hat{J}_1)^2 + (\Delta_\omega J_1)^2 + \omega(\hat{J}_2)^2 + (\Delta_\omega J_2)^2 = \omega(\hat{V})^2 + (\Delta_\omega V)^2 \tag{4.80}$$

with quantum corrections from fluctuations. Using the relationships between J_1 and p_ϕ and their squares implied by the constraint equation, we can write

$$4AB = \omega(\hat{V})^2 - \omega(\hat{J}_2)^2 = \frac{\pi c^4}{3G}\frac{\omega(\hat{p}_\phi)^2 + (\Delta_\omega p_\phi)^2}{p_0^2} + (\Delta_\omega J_2)^2 - (\Delta_\omega V)^2. \tag{4.81}$$

If volume fluctuations are sufficiently large, the right-hand side may be negative. It is then no longer guaranteed that the volume expectation value does not reach zero. Algebraically, such solutions may be possible in a representation of sl(2, $\mathbb{R}$) from the continuous series (Bojowald 2019a).

There is therefore an interesting and unexpected complexity in potential resolutions of the Big Bang singularity: Retracing our constructions of different algebras suitable for cosmological models, it seemed at first that a discrete spectrum of the volume might make it easier to avoid the singularity because a range around zero volume is then excluded for the support of a wave function. This argument should then be strengthened by combining a discrete volume spectrum with affine quantization, turning zero volume into an isolated boundary in a discrete space. However, detailed properties of the resulting algebra sl(2, $\mathbb{R}$) and its representations then show that the combination of discreteness with affine behavior leads to a much larger class of different inequivalent representations compared with a non-discrete affine quantization. Some of these representations are restricted to positive volume, but others are not and might reopen a passage to zero volume, or a Big Bang singularity.

As already mentioned, the modification used to achieve discreteness also makes it difficult to see whether the implied dynamics of an isotropic model can be part of a generally covariant full theory that includes equations for, at least, perturbative inhomogeneity. As we saw in Section 2.3, inhomogeneous models are subject to severe consistency conditions that implement general covariance. In Chapter 6 we will derive further surprising consequences of this modification on spacetime structure, and we will find several new obstacles to making the modified Friedmann equation part of a covariant theory.

4.4 Representations

The first statements about singularity avoidance based on the behavior of wave functions were made in DeWitt (1967). The avoidance can be strengthened by modifications suggested by quantum geometry, such as discrete volume spectra as we have seen, but there may be problems with general covariance as will be analyzed in more detail in Chapter 6. Alternatively, DeWitt (1967) has already suggested that we may work with the unmodified Friedmann equation but impose a boundary condition $\psi(0) = 0$ on wave functions $\psi(V)$, such that even the non-generic singular state which peaks sharply at $V = 0$ is ruled out. In a representation based on the dilation operator $\hat{D}$, this boundary condition is necessary for $\hat{D}$ to be self-adjoint. Technical questions about representations can therefore have important implications on the avoidance of singularities.

4.4.1 Differential and Difference equations

In a representation based on wave functions, the quantum Hamiltonian constraint $\hat{C}_{\mathrm{H}}$ is imposed by requiring that it annihilate any admissible wave function, $\hat{C}_{\mathrm{H}}\psi = 0$. The classical constraint equation on phase space is then replaced by a partial differential equation for wave functions, referred to as the *Wheeler–DeWitt equation*. Starting with Equation (4.31) and wave functions $\psi(V, T_1)$, for instance, the momenta are turned into partial derivatives while V acts by multiplication. The classical constraint $C_{\mathrm{H},1}(V, p_V; p_{T_1}) = 0$ is then quantized to the partial differential equation

$$\hat{C}_{\mathrm{H},1}\psi = \frac{\hbar}{i}\frac{\partial\psi}{\partial T_1} + \frac{6\pi G\hbar^2}{c^2}V\frac{\partial^2\psi}{\partial V^2} = 0, \qquad (4.82)$$

assuming a specific ordering of the non-commuting operators $\hat{V}$ and $\hat{p}_V^2$. In this deparameterized version, the Wheeler–DeWitt equation therefore appears in the form of a Schrödinger equation,

$$i\hbar\frac{\partial\psi}{\partial T_1} = \frac{6\pi G\hbar^2}{c^2} V\frac{\partial^2\psi}{\partial V^2}.$$ (4.83)

The inequivalent representation based on the dilation operator produces a similar equation, but with a different ordering. Since $\hat{D} = \hat{V}\hat{p}_V$, Equation (4.47) suggests a partial differential equation, such as

$$i\hbar\frac{\partial\psi}{\partial T_1} = \frac{6\pi G\hbar^2}{c^2} \frac{\partial}{\partial V}\left(V\frac{\partial\psi}{\partial V}\right)$$ (4.84)

or

$$i\hbar\frac{\partial\psi}{\partial T_1} = \frac{6\pi G\hbar^2}{c^2} \sqrt{V}\frac{\partial}{\partial V}\left(V\frac{\partial(\sqrt{V}\psi)}{\partial V}\right),$$ (4.85)

in which the combination $V\partial/\partial V$ appears and is not immediately brought to the form of a second-order derivative by V. (The two orderings (4.84) and (4.85) differ by a choice of ordering of $\hat{V}^{-1}$ and $\hat{D}^2$.) It would of course have been possible to choose such an ordering in the canonical quantization (4.83); but using the dilation operator as a basic one provides additional motivation that an ordering of the form (4.84) or (4.85) may be preferred.

If we use one of the algebras that imply discrete volume spectra, such as Equation (4.56), the differential nature of the Wheeler–DeWitt equation is turned into a difference equation. To be specific, we may use the same matter source provided by dust, but replace the term $V^{-1}D^2$ that gives rise to the differential operator in Equation (4.84) by $p_0^2 V^{-1}J_1^2/(4\pi^2)$ where $J_1 = V\sin(2\pi p_V/p_0)$ is defined in Equation (4.55). On a discrete volume eigenstate

$$\psi_n(p_V) = \exp(-2\pi i(n - \epsilon)p_V/p_0),$$ (4.86)

defined in Equation (4.53) with a parameter ϵ that labels inequivalent discrete representations, the operator $\hat{J}_1$ acts by

$$\hat{J}_1\psi_n = \frac{1}{2i}\hat{V}(\psi_{n-1} - \psi_{n+1}) = \frac{1}{2i}(V_{n-1}\psi_{n-1} - V_{n+1}\psi_{n+1})$$
$$= i\frac{\pi\hbar}{p_0}\left((n + 1 - \epsilon)\psi_{n+1} - (n - 1 - \epsilon)\psi_{n-1}\right),$$ (4.87)

while

$$\hat{V}^{-1}\hat{J}_1\psi_n = \frac{1}{2i}(\psi_{n-1} - \psi_{n+1}).$$ (4.88)

Combining these two equations, we obtain

$$\hat{V}^{-1}\hat{J}_1^2\psi_n = \frac{\pi\hbar}{2p_0}\big((n + 1 - \epsilon)(\psi_n - \psi_{n+2}) - (n - 1 - \epsilon)(\psi_{n-2} - \psi_n)\big). \qquad (4.89)$$

The modified Friedmann Equation (4.57) for dust matter is therefore quantized to the difference-differential equation

$$\begin{aligned}
i\hbar\frac{\partial\psi_n}{\partial T_1} &= \frac{3G\hbar p_0}{4c^2}\big((n + 1 - \epsilon)(\psi_n - \psi_{n+2}) - (n - 1 - \epsilon)(\psi_{n-2} - \psi_n)\big) \\
&= \frac{3}{4}c\ell_P^2 p_0\big((n + 1 - \epsilon)\psi_{n+2} - 2(n - \epsilon)\psi_n + (n - 1 - \epsilon)\psi_{n-2}\big).
\end{aligned} \qquad (4.90)$$

The coefficients in the second-order difference operator depend on the ordering chosen for the non-commuting operators $\hat{V}^{-1}$ and $\hat{J}_1$. For a Planckian value of p_0, such as $p_0 = (4\pi)^{-1}\hbar/\ell_P^3$ from Equation (4.52), the evolution equation is

$$it_P\frac{\partial\psi_n}{\partial T_1} = \frac{3}{16\pi}\big((n + 1 - \epsilon)\psi_{n+2} - 2(n - \epsilon)\psi_n + (n - 1 - \epsilon)\psi_{n-2}\big) \qquad (4.91)$$

with the Planck time $t_P = \ell_P/c$. Written with respect to the dimensionless time T_1/t_P, the coefficients of the evolution equation are therefore purely numerical.

4.4.2 Initial Conditions

Intuition about the beginning of the universe may suggest alternative definitions of initial conditions. The two most common examples are the tunneling proposal (Vilenkin 1984) which suggests that the wave function near zero volume should asymptotically be of the form of a plane wave moving toward positive volume. (Up to ordering choices, the Wheeler–DeWitt Equation (4.83) is of the form of a free-particle Schrödinger equation in the transformed variable $\sqrt{V}$.) Such a wave function cannot vanish at zero volume, but it contains only a mode that moves away from the singularity.

The second example, the no-boundary proposal (Hartle & Hawking 1983); uses geometrical intuition to suggest that the universe at early times might have been closed off like a sphere at its poles. There are no Lorentzian spacetime solutions that end smoothly at one time, closing off on themselves. However, if one replaces real time with imaginary time, as often done in evaluations of path integrals seen in Section 3.4.1, four-dimensional manifolds in the form of spheres exist and can be used to close off the universe at early times. The required ingredients are positive spatial curvature and a positive cosmological constant, Λ. If Λ is sufficiently large, the Euclidean universe at early times is closed off with a large radius. It does not reach high curvature, such that details of quantum gravity may be avoidable in this scenario of early-universe cosmology.

Let us first see why a transition to Euclidean space is necessary. With the specified ingredients of positive spatial curvature and a positive cosmological constant, as in Equation (1.34), the Hamiltonian constraint for Lorentzian signature is given by

$$C = -\frac{6\pi G}{c^2} V p_V^2 - \frac{3c^4 V^{1/3}}{8\pi G} + \frac{3c^4 V \Lambda}{8\pi G}. \qquad (4.92)$$

It implies the Friedmann equation

$$\left(\frac{1}{a}\frac{da}{d\tau}\right)^2 + \frac{c^2}{a^2} = c^2 \Lambda. \qquad (4.93)$$

At this stage it is convenient to transform from the scale factor a to a new variable $q = a^2$, as well as a time variable t that is related to proper time τ by

$$cd\tau = \frac{dt}{a(t)} = \frac{dt}{\sqrt{q(t)}}. \qquad (4.94)$$

With these choices, the Friedmann equation takes the form

$$\left(\frac{dq}{dt}\right)^2 = 4(\Lambda q - 1). \qquad (4.95)$$

Using the general relationship

$$cd\tau = N(t)dt \qquad (4.96)$$

between proper-time intervals and intervals of an arbitrary time t shows that the lapse function is here chosen as

$$N = \frac{1}{a} = \frac{1}{\sqrt{q}}. \qquad (4.97)$$

The classical picture of the no-boundary proposal requires an initial condition such that $q(0) = 0$, closing off space to a single point at the initial time. According to Equation (4.95), this condition does not imply a real solution for $q(t)$. In the quantum version, however, the same condition is imposed on paths integrated over in the Euclidean path integral. If the path integral is semiclassical, classical trajectories with the no-boundary condition $q(0) = 0$ will be the dominant contribution in a saddle-point approximation. The boundary condition inserted in Equation (4.95) leads to an imaginary dq/dt, which is indeed meaningful if we are using an imaginary time coordinate as defined in the Euclidean path integral.

The no-boundary condition is not compatible with the Friedmann Equation (4.95) if we try to preserve Lorentzian signature. Nevertheless, a Lorentzian version of the no-boundary path integral—based on an application of Picard–Lefschetz theory which, as discussed in Section 3.4.2, deforms the real integration contour in the complex plane instead of simply replacing it with the imaginary axis—is well-defined (Feldbrugge et al. 2017). In fact, since we have fixed time reparameterizations by introducing a specific time coordinate t, to be used for transition amplitudes in the path-integral formulation, we should no longer impose the Hamiltonian constraint or the Friedmann equation. We should only make sure that the second-order equation of motion, given by the Raychaudhuri equation

$$\frac{\mathrm{d}^2 q}{\mathrm{d}t^2} = 2\Lambda \qquad (4.98)$$

in the new variables, is satisfied.

This equation has the general solution $q(t) = \Lambda t^2 + c_1 t + c_2$, which allows the implementation of the no-boundary condition by a suitable choice of c_2, for given c_1. Any such solution can be brought to the form

$$q(t) = (q(t_1) - \Lambda t_1^2)\frac{t}{t_1} + \Lambda t^2 \qquad (4.99)$$

where the free parameter $q(t_1)$ is interpreted as a final condition at time t_1. Such a solution of the second-order equation, applied to the saddle-point approximation of the path integral, is called an instanton. Here, we are dealing with off-shell instantons because we do not impose the Hamiltonian constraint but only the second-order equation it generates.

4.4.3 Instability

The Lorentzian dynamics allows the implementation of the no-boundary initial condition for *off-shell instantons*, defined as solutions of the second-order equation of motion ignoring the first-order constraint (or the Friedmann equation). These classical solutions provide the dominant contribution to the corresponding Lorentzian path integral in a semiclassical or saddle-point approximation. The off-shell nature is achieved by making a specific choice for the lapse function N, which is then no longer free as a function of time. In a path integral, the constraint contribution NC to the action, which would implement the constraint $C = 0$ if

$$\int \mathcal{D}N \exp(iNC/\hbar) \propto \delta(C) \qquad (4.100)$$

is integrated over all independent $N(t)$, is then no longer fully imposed. Therefore, solving only the second-order equation of motion for $q(t)$ is consistent.

However, the choice (4.97) does not uniquely determine the full lapse function within this dynamical model, as shown by the presence of a free final time t_1 in the solution (4.99). The choice of t_1 can conveniently be rewritten as a multiplicative factor in the lapse function, $N(t) = M/\sqrt{q(t)}$, where M is a constant and does not depend on time. Since the value of M determines how quickly time progresses, or how quickly any given time such as t_1 will be reached, the choice of some $t_1 \neq 0$ can be mapped to fixing the value of some $M \neq 1$ with normalized $t_1 = 1$. Equation (4.99) then takes the form

$$q(t) = (q(1) - \Lambda M^2)t + \Lambda M^2 t^2, \qquad (4.101)$$

which can be obtained simply by rescaling t and t_1 in Equation (4.99) to Mt and Mt_1, respectively, and then setting $t_1 = 1$. (Upon rescaling, t loses its units while M, like q, has units of length squared.)

In Equation (4.100), the constraint remains unimposed even if the lapse function is not uniquely determined: The integration $\int \mathcal{D}N$ over all $N(t)$ at different times t is restricted to a single integration $\int \mathrm{d}M$ which is not sufficient to impose $C = 0$ at all times. Even with a free constant M that rescales the progress of time, we are therefore still dealing with off-shell instantons. The full path integral

$$\int \mathcal{D}q\mathcal{D}N \, \exp(iS/\hbar), \tag{4.102}$$

integrating over both metric components, q and N, of an isotropic geometry, is then reduced to a single integration: In a semiclassical approximation, path-integrating over $q(t)$ is replaced by evaluating the integrand at an off-shell instanton, giving a saddle point of the integrand. Path-integrating over $N(t)$ is reduced to a single standard integration over the time-independent M.

This remaining integration has an interesting consequence if one includes individual modes of inhomogeneous perturbations around the no-boundary instanton (Feldbrugge et al. 2017, 2018; Di Tucci & Lehners 2018). In the vacuum case, independent inhomogeneous degrees of freedom are given by tensor modes, which with our choice of time or lapse are subject to the wave equation

$$\frac{\partial^2 v}{\partial t^2} - \frac{\mathrm{d}^2 q}{\mathrm{d}t^2}\frac{v}{q} - \frac{M^2}{q^2}\Delta v = 0. \tag{4.103}$$

For convenience, the equation refers to $v = qh$ with a component h of the tensor mode that appears in Equation (1.77). In the present application, the Laplacian Δ is defined on a three-dimensional sphere because the no-boundary proposal requires background space with constant positive curvature.

On the sphere, Fourier modes, as eigenfunctions of the flat-space Laplacian, are replaced by spherical harmonics, labeled my multipole numbers. A given mode of multipole number ℓ on the background sphere implies $\Delta v_\ell = -\ell(\ell + 2)v_\ell$. Written in terms of modes, the wave equation is therefore reduced to an ordinary differential equation in time. For small t, close to the initial configuration of the no-boundary proposal, the equation can be approximated as

$$\frac{\mathrm{d}^2 v_\ell}{\mathrm{d}t^2} \approx -\frac{\ell(\ell + 2)M^2}{q(1)^2}\frac{v_\ell}{t^2}. \tag{4.104}$$

This equation is solved by superpositions of two independent solutions, $v_\pm = t^{(1\pm\gamma)/2}v(1)$ with

$$\gamma = \sqrt{1 - 4\frac{\ell(\ell + 2)M^2}{q(1)^2}}. \tag{4.105}$$

Transition amplitudes are determined by the M-integration that remains from the full path integral, after fixing the lapse function up to a scale parameter M and using a saddle-point approximation around off-shell instantons. For the saddle-point approximation, we should compute the action for modes, subject to Equation (4.104), evaluated in asymptotic solutions on a no-boundary background:

$$
\begin{aligned}
S &= \int_0^1 \left(\frac{1}{M} \left(\frac{dv}{dt} \right)^2 - \ell(\ell + 2)M\frac{v^2}{q^2} \right) dt \\
&= \frac{1}{M} \int_0^1 \left(\left(\frac{dv}{dt} \right)^2 + \frac{\gamma^2 - 1}{4t^2} v^2 \right) dt,
\end{aligned}
\tag{4.106}
$$

using the off-shell instanton with $q(t) \approx q(1)t$ in the second line. The dependence on M will be important, but it is not completely determined by the equation of motion (4.104). However, we can refer to the general form of gravitational actions, such that the potential term receives a factor of the rescaled lapse function M from the measure factor $\sqrt{-\det g}$, while the kinetic term receives a factor of $1/M$ from $\sqrt{-\det g}$ combined with the inverse metric component $g^{tt} \propto 1/M^2$ that multiplies second-order partial derivatives.

The two independent solutions for v give two different values of the action,

$$
S_{\pm} = \frac{1}{2M}(1 \pm \gamma)t^{\pm\gamma} \, |_{t=0}^1 \, v(1)^2,
\tag{4.107}
$$

respectively. Because S_- is not finite (using $\gamma > 0$), we should restrict all further considerations of the M-integral to the solution v_+.

Integrating $\exp(iS_+/\hbar)$ over M is non-trivial because γ, as a function of M defined in Equation (4.105), has a branch cut on the positive real axis. Picard–Lefschetz theory determines that the real contour should be deformed into the positive imaginary plane (Feldbrugge et al. 2018). For this deformation to be unobstructed, we should define the original integral such that the branch cut is circumvented in the positive imaginary half-plane of M. Then, however,

$$
S_+ \propto \frac{1}{M} = \frac{\operatorname{Re} M - i \operatorname{Im} M}{|M|^2}
\tag{4.108}
$$

has a negative imaginary part, which implies an unbounded $\exp(iS/\hbar)$. Therefore, the no-boundary proposal is unstable with respect to perturbations by inhomogeneous modes. (This conclusion has been contested by suggesting modifications of the original no-boundary proposal, such as different integration contours or final conditions in addition to the initial condition $q(0) = 0$ (Diaz Dorronsoro et al. 2017, 2018). These suggestions have been addressed in Feldbrugge et al. (2018).)

The universe, therefore, cannot be closed off in a smooth way because it is subject to unlimited inhomogeneous perturbations. This result is a detailed example of possibly pathological consequences of the negative kinetic energy assigned to gravity, as opposed to matter. Two different kinds of solutions have been proposed to counter the instability. The initial conditions can be adjusted such that perturbations become stable, or certain types of modified spacetime structures in models of quantum gravity can modify the mode equation such that the branch cut in M is moved away from the real axis. For instance, it is possible to overcome the conclusion of unstable perturbations if the initial condition is modified such that it includes a non-zero initial Hubble parameter (Vilenkin & Yamada 2018, 2019; Di Tucci & Lehners 2019). A possible quantum-gravity effect that could imply stability of the no-boundary proposal is given by physical signature change of the actual spacetime geometry (Bojowald & Brahma 2018); and not just a formal change to imaginary time in the path integral.

While a modified no-boundary proposal may then be meaningful, the necessity of motivating additional ingredients, usually obtained from quantum gravity, means that a detailed discussion of potential properties of such a theory cannot be avoided, even if one attempts to close off the universe at small curvature. In the next chapter we will therefore turn to a discussion of different approaches to quantum gravity. In addition, the treatment of the no-boundary path integral with a partial fixing of time reparameterization invariance, replacing a path integral over $N(t)$ with a single integration over M, raises the question whether this classical invariance is, in fact, realized also in a full path-integral treatment of gravity. The question of how general covariance can be realized in quantum gravity therefore reappears. While a complete answer to this question is, at present, unknown, several potential ingredients will be provided in the following two chapters.

References

Acacio de Barros, J., Pinto-Neto, N., & Sagiaro-Leal, M. A. 1998, PhLA, 241, 229

Anderson, E. 2012, in Classical and Quantum Gravity: Theory, Analysis and Applications, ed. V. R. Frignanni (New York: Nova)

Ashtekar, A., Pawlowski, T., & Singh, P. 2006, PhRvD, 73, 124038

Bargmann, V. 1947, AnMat, 48, 568

Ben Achour, J., & Livine, E. 2017, PhRvD, 96, 066025

Ben Achour, J., & Livine, E. 2019a, PhRvD, 99, 126013

Ben Achour, J., & Livine, E. 2019b, JCAP, 09, 012

Bergeron, H., Czuchry, E., Gazeau, J.-P., Malkiewicz, P., & Piechocki, W. 2015b, PhRvD, 92, 124018

Bergeron, H., Czuchry, E., Gazeau, J.-P., Malkiewicz, P., & Piechocki, W. 2015a, PhRvD, 92, 061302

Bergeron, H., Czuchry, E., Gazeau, J.-P., Malkiewicz, P., & Piechocki, W. 2017, PhRvD, 96, 043521

Bergeron, H., Czuchry, E., Gazeau, J.-P., Malkiewicz, P., & Piechocki, W. 2020, Univ, 6, 7

Bergmann, P. G. 1949, PhRv, 75, 680

Bergmann, P. G. 1961, RvMP, 33, 510

Bergmann, P. G., & Brunings, J. H. M. 1949, RvMP, 21, 480

Bodendorfer, N., & Wuhrer, D. 2019, arXiv: 1904.13269

Bojowald, M. 2002, CQGra, 19, 2717

Bojowald, M. 2007, PhRvD, 75, 123512

Bojowald, M. 2015a, RPPh, 78, 023901

Bojowald, M. 2007b, PhRvD, 75, 081301(R)

Bojowald, M. 2019a, arXiv: 1906.02231

Bojowald, M. 2019b, Mathematics, 7, 645

Bojowald, M., & Brahma, S. 2014, arXiv: 1411.3636

Bojowald, M., & Brahma, S. 2018, PhRvL, 121, 201301

Bojowald, M., & Halnon, T. 2018, PhRvD, 98, 066001

Bojowald, M., Höhn, P. A., & Tsobanjan, A. 2011a, CQGra, 28, 035006

Bojowald, M., Höhn, P. A., & Tsobanjan, A. 2011b, PhRvD, 83, 125023

Bojowald, M., & Tsobanjan, A. 2019, CMaPh, arXiv: 1906.04792

Brown, J. D., & Kuchař, K. V. 1995, PhRvD, 51, 5600

Coleman, S., & Weinberg, E. 1973, PhRvD, 7, 1888

DeWitt, B. S. 1967, PhRv, 160, 1113

Di Tucci, A., & Lehners, J.-L. 2018, PhRvD, 98, 103506

Di Tucci, A., & Lehners, J.-L. 2019, PhRvL, 122, 201302

Diaz Dorronsoro, J., Halliwell, J. J., Hartle, J. B., Hertog, T., & Janssen, O. 2017, PhRvD, 96, 043505

Diaz Dorronsoro, J., Halliwell, J. J., Hartle, J. B., Hertog, T., & Janssen, O. 2018, PhRvL, 121, 081302

Dirac, P. A. M. 1950, Canadian Journal of Mathematics, 2, 129

Dirac, P. A. M. 1958, RSPSA, 246, 326

Dirac, P. A. M. 1969, Lectures on Quantum Mechanics (New York: Yeshiva)

Falciano, F. T., Pinto-Neto, N., & Santini, E. S. 2007, PhRvD, 76, 083521

Feldbrugge, J., Lehners, J.-L., & Turok, N. 2017, PhRvD, 95, 103508

Feldbrugge, J., Lehners, J.-L., & Turok, N. 2017, PhRvL, 119, 171301

Feldbrugge, J., Lehners, J.-L., & Turok, N. 2018, arXiv: 1805.01609

Feldbrugge, J., Lehners, J.-L., & Turok, N. 2018, PhRvD, 97, 023509

Giacomini, F., Castro-Ruiz, A., & Brukner, C. 2019, NatCo, 10, 494

Gielen, S., & Menéndez-Pidal, L. 2020, arXiv: 2005.05357

Hartle, J. B., & Hawking, S. W. 1983, PhRvD, 28, 2960

Höhn, P. A., Kubalova, E., & Tsobanjan, A. 2012, PhRvD, 86, 065014

Husain, V., & Pawlowski, T. 2011b, PhRvL, 108, 141301

Husain, V., & Pawlowski, T. 2011a, CQGra, 28, 225014

Isham, C. J. 1983, in Relativity, Groups and Topology II, Lectures given at the 1983 Les Houches Summer School on Relativity, Groups and Topology

Isham, C. J. 1993, in Integrable Systems, Quantum Groups, and Quantum Field Theory (Dordrecht: Kluwer) 157

Kiefer, C., Kwidzinski, N., & Piontek, D. 2019, EPJC, 79, 686

Kiefer, C., & Schmitz, T. 2019, PhRvD, 99, 126010

Klauder, J. 2003, IJMPD, 12, 1769

Klauder, J. 2006, IJGMMP, 3, 81

Kuchař, K. V. 1992, in Proc. of the 4th Canadian Conf. on General Relativity and Relativistic Astrophysics, ed. G. Kunstatter, D. E. Vincent, & J. G. Williams (World Scientific: Singapore)

Malkiewicz, P. 2015, CQGra, 32, 135004

Malkiewicz, P., Peter, P., & Vitenti, S. D. P. 2020, PhRvD, 101, 046012

Novello, M., & Bergliaffa, S. E. P. 2008, PhR, 463, 127

Vandersloot, K. 2005, PhRvD, 71, 103506

Vanrietvelde, A., Hoehn, P. A., & Giacomini, F. 2018, arXiv: 1809.05093

Vanrietvelde, A., Hoehn, P. A., Giacomini, F., & Castro-Ruiz, E. 2020, Quantum, 4, 225

Vilenkin, A. 1984, PhRvD, 30, 509

Vilenkin, A., & Yamada, M. 2018, PhRvD, 98, 066003

Vilenkin, A., & Yamada, M. 2019, PhRvD, 99, 066010

Wendel, G., Martínez, L., & Bojowald, M. 2020, PhRvL, 124, 241301

Chapter 5

Quantum Gravity

Minisuperspace models can be modified in many ways, choosing different representations, initial or final conditions, or modified dynamics. An unlimited number of interesting effects can be produced in this way because the strong assumption of homogeneous space eliminates most (but not all) consistency conditions related to general covariance. Minisuperspace models are therefore not predictive, unless one can provide good reasons for a specific technical feature or modification to be suggested by some full theory of quantum gravity that does *not* rely on strong symmetries and is therefore much more tightly restricted.

In fact, minisuperspace models and full quantum gravity present two opposite extremes within the full range of cosmological discussions. Minisuperspace models are virtually unrestricted and can be used to resolve the Big Bang singularity in many different ways. Full quantum gravity is highly constrained by various consistency conditions, so much so that to date no version is known to be completely consistent and viable. In this chapter, we discuss and compare the main proposals, arranged in a perhaps unusual way that builds up non-standard features of spacetime structures. We will focus on properties relevant for cosmological models. The next and final Chapter 6 will then return to cosmological applications, but in contrast to Chapter 4 based on models that are inhomogeneous (while still partially symmetric in order to be tractable). These models, in which general covariance is a restrictive condition, will show crucial implications of consistency conditions (or their violations).

General covariance is not only an important consistency condition on potential theories of quantum gravity and their cosmological implications, it also determines what spacetime structures may be possible. Models in quantum cosmology occasionally assume, at least implicitly, that there is a Riemannian spacetime structure that includes Friedmann–Robertson–Walker line elements with a scale factor for which a modified or quantized Friedmann equation provides quantum-corrected dynamics. Possible new effects, such as a local minimum of the scale factor at non-zero value, are then interpreted in a standard geometrical picture, such as a bouncing

doi:10.1088/2514-3433/ab9c98ch5

spacetime that does not reach zero size. However, the assumption that Riemannian geometry can still be used to describe a quantized spacetime theory needs independent justification which cannot be achieved in the restricted setting of minisuperspace models. Such a justification, as well as an exploration of potential alternatives to Riemannian geometry, require information about consistent full versions of quantum gravity. It may then be possible that effective actions of quantum gravity are not solely described by higher-curvature terms, because higher curvature actions are based on the assumption of spacetime equipped with Riemannian geometry.

As long as the energy density or curvature remain sufficiently small, it may be sufficient to use an effective field theory that captures relevant quantum effects in a covariant setting without introducing the full complexity of quantum gravity. However, such an effective approach also encounters unique challenges in the field of quantum cosmology. For instance, the coefficients in an effective action depend on the scale used to average out microscopic degrees of freedom of a fundamental theory, or to define a mean-field approximation. As the scale is adjusted to the growing energy density during cosmic evolution (run backwards to understand how the Big Bang is approached) coefficients are renormalized because their values depend on how many momentum modes of each degree of freedom retain independent quantum degrees of freedom while others are being averaged. This time-dependent separation of scales is meaningful if evolution is adiabatic, for instance expressed as a condition that the averaged momenta (or a "momentum cutoff") must be much bigger than the rate of change of the background geometry described in the effective model. In cosmological models, the relevant rate of change is the Hubble parameter. If the Hubble parameter becomes Planckian deep in the high-curvature regime near the Big Bang, the adiabaticity condition is violated. It is then possible for some of the high-energy modes to participate in the dynamics, interacting with the background, and an averaged or mean-field description is no longer available.

This limitation of minisuperspace models, which are often believed to provide a mean-field description of full quantum gravity even though such a derivation can rarely be performed explicitly, is related to the importance of infrared renormalization already observed in Section 4.1.3: As a collapsing universe in backward evolution approaches high density, structure forms within co-moving regions. The range of inhomogeneity within which a homogeneous model may be used therefore has to be reduced as the universe collapses, even beyond the co-moving reduction of distance scales implied by a shrinking scale factor. Eventually, this infrared scale of a homogeneous model is pushed into the ultraviolet when the Planck regime is reached, and no modes remain that could reliably be described by a homogeneous model. Structure formation is also expected to lead to a multitude of different scales, complicating the application of a single-scale description as it is given by a minisuperspace model.

For these reasons, it is important to consider what different approaches to quantum gravity might be able to tell us about possible realizations of quantum spacetime and its evolution.

5.1 Classification and Comparison

The quantization of interactions in the standard model of particle physics is well understood and has been successfully tested by observations. This theory includes strong, self-interacting forces such as those in non-Abelian Yang–Mills theories, in particular in quantum chromodynamics. Also gravity is self-interacting, but by comparison with any other fundamental force it is very weak. One could therefore expect that it can be quantized in the established fashion based on perturbation theory. However, this is not possible because gravity differs from the standard-model interactions in two crucial ways.

First, gravity, as described by general relativity, is understood as a geometrical consequence of properties of spacetime curvature, rather than an independent force field on a pre-existing classical spacetime. When we quantize general relativity, we must therefore find a meaningful description of quantum spacetime. In an attempt to bring the potential quantum description of general relativity closer to the well-understood quantum field theories used in the standard model of particle physics, one may split the gravitational field, in general described by the complete spacetime metric $g_{\alpha\beta}$, into a fixed and simple background contribution, such as Minkowski spacetime $\eta_{\alpha\beta}$, and a dynamical field given by the difference $h_{\alpha\beta} = g_{\alpha\beta} - \eta_{\alpha\beta}$. Formally, it is possible to quantize only $h_{\alpha\beta}$ in a background-field quantization that keeps $\eta_{\alpha\beta}$ classical. However, there is nothing special about choosing $\eta_{\alpha\beta}$ as the background, except convenience implied by its high symmetry and mathematical simplicity.

A convincing theory of this form should therefore be able to demonstrate that its predictions do not depend on which geometry is chosen as the background. Such a demonstration is usually difficult after quantization. Depending on how this problem is addressed, the various approaches to quantum gravity are classified as *background-dependent* ones, in which a background spacetime is used and invariance with respect to changing it may or may not be shown; and *background-independent* ones, which intend to solve the invariance problem by trying to quantize directly the whole metric $g_{\alpha\beta}$ without introducing a background metric.

Background-independent approaches are free of the requirement to show independence of the choice of background, but all of them, sooner or later, run into serious problems because new methods have to be invented that allow one to eliminate the spacetime background required for all well-understood quantizations of field theories. Background-independent approaches to quantum gravity are usually more radical than background-dependent ones, because they may modify not only the dynamics of gravity but even the structure of spacetime. Also background-dependent approaches often lead to non-classical spacetime structures, but it is usually more difficult to uncover them because one must first be able to overcome the classical spacetime structure presupposed in the specified background. Examples of non-classical spacetime structures include varying and possibly fractional effective dimensions, non-commutative and non-associative features, or non-Riemannian structures that may lead to dynamical signature change. Approaches in the background-independent category, discussed below, are causal dynamical triangulations, fractional, non-commutative or non-associative geometry, canonical (and loop) quantum gravity, and causal-set theory. String theory has led to some of the same effects, but its formulation does not start out in a background-independent way.

Second, even if a background spacetime can be used, gravity is still crucially different from standard-model field theories because it is not renormalizable: A direct perturbative treatment would suggest that some of its quantum coupling constants are infinite. Three possible approaches to this problem have been proposed: The effective field-theory approach aims to work out low-energy implications that are reliable in spite of non-renormalizability. The lack of renormalizablity then implies that some coupling constants cannot be determined by the theory. Predictions are then not uniquely determined by properties of general relativity combined with quantum physics, but rather depend on several free parameters. The number of parameters increases without bound as higher and higher orders in Planck's constant are included as perturbative quantum corrections. The framework would therefore require an infinite number of measurements to test the theory, which cannot be falsifiable. Effective field theories in this context are viewed not as fundamental but rather as generic approximations of an unknown fundamental theory. While the fundamental theory remains unknown, low-energy properties accessible with effective theories may give useful information.

Another possible approach that might overcome the non-renormalizable nature of perturbative quantum gravity, called asymptotically safe quantum gravity, aims to test the possibility that general relativity may, after all, be renormalizable, but only if it is expanded perturbatively using coupling constants that differ from what the weak-field limit would suggest. Since so far we only know the weak-field behavior of gravity, suitable methods must be devised that allow one to find non-trivial values that allow a well-defined perturbative expansion. If such a point can be found, convergence properties of effective quantum gravity may be improved and the range of viable predictions could be extended, compared with a standard perturbative quantization of general relativity.

Finally, string theory addresses the non-renormalizability problem by embedding gravity in a larger, unified setting which modifies particle and spacetime pictures at the same time. By replacing pointlike test particles with test strings, crucial divergences in the usual perturbative quantization of general relativity can be removed. This replacement has several further, surprising consequences which have been derived from the implementation of various important consistency conditions, in particular by demanding general covariance: A stable theory subject to Lorentz symmetries can be achieved only if there are either more than four spacetime dimensions, or a large number of additional internal degrees of freedom, as well as supersymmetry. Since neither extra dimensions nor fundamental super-symmetry have yet been observed in experiments, string theory is often criticized for requiring these ingredients. However, compared with other approaches that seem to work without these features, string theory has only been the most successful approach in which it was possible to implement such a large number of consistency conditions. The contenders simply have not completed (and sometimes not even started) an investigation of similar general requirements that any approach to quantum gravity should eventually address. They may well have to face surprises once such an analysis is complete. A great success of string theory is showing that its consistency conditions can, in fact, be implemented, even if it comes at the "cost" (depending on one's viewpoint) of introducing exotic properties of spacetime.

5.2 Approaches

This section introduces the main approaches to quantum gravity largely independently of one another. Section 5.3 will then provide a brief comparison, as well as a proposed ranking based on their successes and open questions.

5.2.1 Perturbative Quantum Gravity as an Effective Theory

If one chooses a background metric, such as the Minkowski metric $\eta_{\alpha\beta}$, general relativity is formulated as a field theory for the perturbation field $h_{\alpha\beta} = g_{\alpha\beta} - \eta_{\alpha\beta}$ on the background. Perturbative methods of quantum gravity can then be applied, such that the classical $h_{\alpha\beta}$, subject to non-linear field equations, is replaced by a specific dynamics of self-interacting gravitons. The curvature term in the Einstein–Hilbert action (2.48) contains third and fourth-order terms as a polynomial in components of $h_{\alpha\beta}$. Feynman graphs of perturbative quantum gravity therefore contain three-valent and four-valent vertices of elementary graviton self-interactions.

Loop diagrams with these vertices provide quantum corrections to the classical self-interactions of the gravitational field in general relativity. If the corrections are covariant in Riemannian geometry, it must be possible to express them as implications of higher-curvature corrections added to the Einstein–Hilbert action. However, unambiguous vertices are obtained only if one fixes the gauge, which in gravity amounts to working with a specific set of coordinates. If a different choice is made, the Feynman rules change, and a complicated consistency condition is implied by general covariance.

Moreover, such expansions are performed around a background, usually Minkowski spacetime, and therefore background independence of quantum effects is another important question. It is related to covariance because a generic coordinate transformation changes the form of the background metric, and therefore the appearance of Feynman rules, even if the full invariant geometry determined by $g_{\alpha\beta}$ remains the same.

Quantum corrections of a covariant classical theory are, in general, not guaranteed to be covariant. In a path-integral quantization, for instance, the metric appears in the integrand through the classical action, which is covariant, and in the measure $\mathcal{D}g_{\alpha\beta}$. The latter is a measure on an infinite-dimensional space and therefore requires some care in its mathematical definition. Methods that allow one to define this measure do not always preserve symmetries of the classical theory. Covariance in a path-integral quantization therefore requires a dedicated analysis which, in full detail, has not been performed yet.

In fact, the path integral applied to an action with gauge symmetries is infinite because the integrand does not change in the direction of a symmetry transformation, such as a coordinate change. If one integrates over paths of the full metric tensor $g_{\alpha\beta}$, one therefore includes infinite integration ranges for constant integrands, which cannot be finite. Constant directions are usually removed from integrations by "fixing the gauge" that is, imposing conditions on the metric such that coordinate changes are no longer possible when the restricted form is to be preserved. Then, however, different gauge fixings, for which there are infinitely many choices, lead to different path integrals. Demonstrating covariance by showing that all these choices imply the same predictions is a complicated task.

If one assumes that quantum corrections preserve general covariance in perturbative quantum gravity, it is possible to express loop corrections with graviton self-interactions as higher-curvature actions. The resulting theory cannot be fundamental because there are infinitely many free coefficients in a generic higher-curvature effective action, most of which remain undetermined by loop corrections because they are based on diverging Feynman integrals. Renormalization methods allow one to replace diverging terms by undetermined coupling constants, but this procedure is not predictive if one has to postulate an infinite number of undetermined parameters of one's theory. (Perturbative quantum gravity is therefore not renormalizable.) Nevertheless, it is meaningful to use leading loop corrections in an analysis of possible implications of quantum gravity, as done in effective quantum gravity (Burgess 2004; Donoghue 1994a, 1994b).

5.2.2 Asymptotic Safety

The problem of perturbative quantum gravity not being renormalizable could be overcome if quantum gravity is *asymptotically safe* (Weinberg 1979). Self-interactions of gravitons then, by definition, modify the coupling constants of the theory such that a finite number of them is relevant at high energies or momenta. Such a theory can therefore be predictive and may be considered a fundamental theory of quantum gravity.

The coupling constants of quantum gravity in this setting include not only Newton's constant G and the cosmological constant Λ in Equation (2.48), but also all possible coefficients of higher-curvature terms, each of which we may schematically refer to as $A_{m,n}$ if it multiplies any mth order derivative of an nth order curvature term, such as $\nabla^m R^n$. (The covariant derivatives ∇ can be taken in any combination of spacetime coordinates, and R^n may be any scalar combination of n copies of the Riemann tensor (2.45). In addition, there may be non-local terms, such as $(\nabla^2)^{-1}R$, which are usually ignored in discussions of asymptotically safe quantum gravity.)

Most of the coupling constants have scaling dimensions, that is, they change if we expand all length (or time) units. For instance, Newton's constant is related to the Planck length by $G = c^3 \ell_P^2 / \hbar$ and therefore behaves like a length squared. If we have an inverse length scale k, usually interpreted as the limiting value of all wave numbers of modes considered in a derivation of quantum corrections, we can define the scaling-independent coupling constant

$$g_G = Gk^2. \tag{5.1}$$

Similarly, the cosmological constant (like a curvature term) scales like one over length squared, and therefore has the scaling-independent analog

$$g_\Lambda = \Lambda k^{-2}. \tag{5.2}$$

The remaining coupling constants of curvature terms have scaling-independent versions

$$g_{m,n} = A_{m,n} k^{2n+m-4}. \tag{5.3}$$

To see this, notice that these coupling constants contribute to the scaling-independent action via terms $A_{m,n} d^4 x \nabla^m R^n \sim A_{m,n} L^{4-m-2n}$. (In this action term, the usual scaling-dependent factor of $1/G$ is not included because it is considered the coupling constant only of the leading-order contribution, R.)

Among the scaling-independent constants $g_{m,n}$, there are infinitely many ones with a positive exponent of k, $2n + m - 4 > 0$. If the classical scaling (5.3) is used for a large range of k, starting at small wave number k or momentum $\hbar k$, infinitely many coupling constants increase as we move to larger momenta. The high-energy theory is then non-predictive because it requires us to determine an infinite number of free constants from observations.

However, as the energy or momentum scale is increased, coupling constants usually have to be adjusted by renormalization because interactions at different energies are sensitive to different degrees of freedom. Any perturbative expansion may then have to be reorganized if one progresses to higher energies, which can be rewritten as the original perturbative expansion of a theory but with modified, renormalized coupling constants. The coupling constants of gravity, in particular the $g_{m,n}$, therefore depend on the scale k in some form, which we can write generically as

$$g_{m,n}(k) = \sum_{i,j,l_{m,n}} f_{i,j,l_{m,n}}(k) G^i \Lambda^j A_{m,n}^{l_{m,n}} \tag{5.4}$$

with some coefficient functions $f_{i,j,l_{m,n}}(k)$ to be determined from self-interactions of the theory through loop diagrams. Unlike in the classical case, the scaling behaviors of different coupling constants are in general not independent of one another.

As k is increased, the coupling constants start deviating from their classical power-law behavior. A quantum field theory is defined to be asymptotically safe if its coupling constants have a fixed point at large k, such that $k dg_{m,n}(k)/dk \to 0$ for $k \to \infty$, with a *finite* number of negative scalings $k dg_{m,n}(k)/dk < 0$ for large finite k. The fixed-point values of the coupling constants then define a predictive fundamental theory because only finitely many of the infinite number of coupling constants $g_{m,n}$ increase as the momentum scale k is lowered toward a value that describes the typical scale of an observation.

General relativity could be an example of an asymptotically safe theory. The weak coupling at low energies, which at first might indicate a well-behaved perturbation expansion, would then be deceptive. Such an expansion leads to non-renormalizable behavior because infinitely many $g_{m,n}$ increase with the momentum scale, but an expansion around larger couplings near the fixed point at high energies may still be meaningful. In order to find these coupling constants, one should follow the k-dependent flow of the coupling constants, starting from small coupling, until they converge to a fixed point. One method usually used in this context, mainly by numerical evaluations, applies general flow equations for effective actions derived by Wetterich (1993).

An obvious problem is that there are infinitely many coupling constants to be analyzed before one finds the asymptotically safe fixed point—if the latter even exists for a specific theory like gravity. Most of these calculations have to be performed numerically, given the complexity of self-interactions. For each numerical test, or an analytical one if it is possible, one has to choose the possible higher-curvature corrections that may be generated at higher energies. Practical applications require that these be finitely many out of the infinitely many possible terms, such as all curvature terms up to a given maximal order n. Any such analysis therefore requires a truncation of possible higher-curvature terms. Nevertheless, some indications, obtained with different specific types of higher-curvature corrections, are promising (Niedermaier & Reuter 2006).

Methods from asymptotic safety have been used to determine properties of time-correlation functions of the volume (Knorr & Saueressig 2018), which can be compared with results from other cosmological models, in particular those derived from causal dynamical triangulations. Moreover, cosmological models based on asymptotic safety are sometimes formulated by using properties of the renormalization flow of Newton's constant, possibly changing the gravitational force at high density. It is, however, unclear how the momentum scale k should be related to a cosmological energy scale. Other problems include a lack of an understanding of the covariance question because applications of the asymptotic-safety scenario make use of a background expansion of the metric. Moreover, calculations are done in Euclidean signature, which makes it possible to associate large k with large energies while in Lorentzian signature a lightlike vector with small $\|k\|^2 = 0$ can easily correspond to a high-energy phenomenon. These and other questions have been discussed in Donoghue (2020); Bonanno et al. (2004); Kwapisz & Meissner (2005).

5.2.3 Causal Dynamical Triangulations

Like asymptotically safe gravity, causal dynamical triangulations (Ambjørn & Loll 1998; Ambjørn et al. 2012) are based largely on numerical calculations. In addition, there is a conceptual relationship because both approaches aim to find non-trivial fixed points of a non-perturbative theory of quantum gravity. However, a major new ingredient in causal dynamical triangulations is the implementation of an auxiliary discrete structure of spacetime, which makes it possible to perform detailed numerical calculations by replacing the classically continuous fields with a manageable finite but still large set of numbers. Periodic boundary conditions in space and time are used to model a finite spacetime region by a finite number of building blocks. Nevertheless, such a simulation may contain an entire model universe if the region is large enough.

The sought-after fixed point is required to have special properties (related to second-order phase transitions in statistical physics, as described in more detail below) such that a continuum limit can be taken where the elementary but auxiliary smallest discrete size approaches zero. If such a fixed point exists, the theory describes a continuum quantization of gravity and spacetime. Nevertheless, the underlying discreteness used in the setup may reveal physical implications about the structure of spacetime.

The discretization approach utilized in causal dynamical triangulations, called Regge calculus (Regge 1961), was originally developed for numerical simulations of classical spacetimes. In causal dynamical triangulations, it is combined with a quantum-statistics approach that allows one to draw conclusions about quantum spacetime. In particular, a large number of different discrete spacetimes is generated numerically by applying specific rules for adding or removing links in a triangulation, weighted by probabilities that are, as in a path integral, defined by the discretized action of general relativity.

These rules distinguish between spacelike and timelike links, characterizing these dynamical triangulations as causal ones. The resulting ensemble is then interpreted as a representative collection of "paths" in the set of all spacetimes, from which a path-integral can approximately be computed. The integrand, $\exp(iS/\hbar)$ or rather its Euclidean-time version $\exp(-S/\hbar)$ in order to obtain real-valued probabilities for the statistical interpretation, is adapted to discrete spacetimes according to the rules of Regge calculus. Like explicit calculations in asymptotically safe gravity, the approach therefore relies crucially on a transformation to Euclidean signature.

Deficit Angle and Regge Calculus
Regge calculus can be illustrated for two-dimensional surfaces, using the concept of the deficit angle as a measure of curvature. A smooth curved surface can be approximated by a collection of Euclidean triangles if its continuous curvature is pushed into conical points. Any "bulge" of a surface of positive curvature is then represented by a small cone attached to the rest of the surface. Moreover, each such cone can be approximated by gluing together a collection of triangles as shown in Figures 5.1 and 5.2. In the triangulated approximation of the surface, each triangle is assumed to be flat in Euclidean space because curvature is represented only by the deficit angle at the conical tip. One can therefore use Euclidean triangle rules to compute angles or distances.

Curvature at the tip of a cone is expressed through the *deficit angle* as shown in the figures. If one follows a closed curve around the tip and translates the initial tangent vector (or any other vector at the initial point) such that it always remains parallel to itself—again referring to flat Euclidean geometry in the embedding space —it comes back rotated by a certain angle after completing the loop. This angle defines the deficit angle at the tip of the cone.

If all triangles meeting at a conical point contribute an angle of ϕ, the deficit angle corresponding to the missing triangle in Figure 5.1 is given by $\epsilon = 2\pi - \phi$. A two-dimensional sphere, for instance, can be triangulated as a single tetrahedron. At each corner of the tetrahedron, three equilateral triangles meet each of which contributes

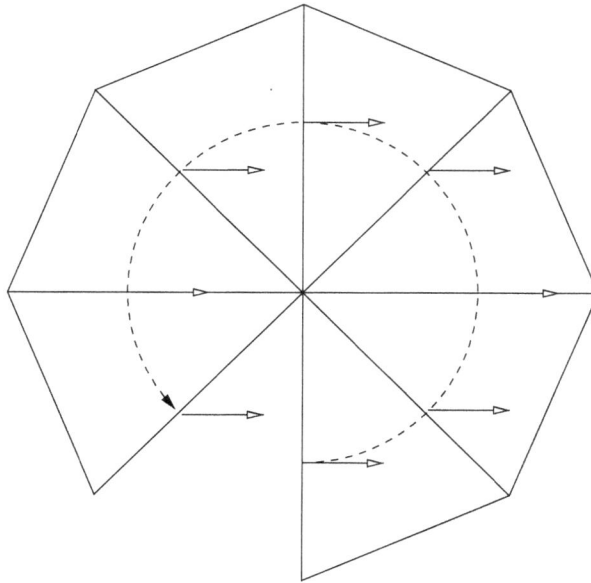

Figure 5.1. Deficit angle as a measure of curvature in Regge calculus. The seven triangles shown here represent an approximation of a closed surface in which the curvature is concentrated in a conical point at the center. To construct the surface, one cuts out the collection of triangles and glues together (or, in mathematical language, identifies) the two sides of the missing triangle that meet at the center. The angle of the missing triangle at the center is the deficit angle of the linear approximation. As shown here and in the next figure, it is equal to the angle between the initial and final directions of a vector carried along a closed curve around the conical center: Moving through a triangle in the flat representation used in this figure, the vector is parallel translated by keeping its direction fixed. However, the missing triangle implies that the loop is not complete in the flat plane. After cutting out the triangles and identifying the required sides, the missing segment means that the vector is not rotated by a full 360° at the final point of the loop. This deficit angle is directly related to the curvature of the conical surface obtained by identifying the missing triangle sides; see Figure 5.2.

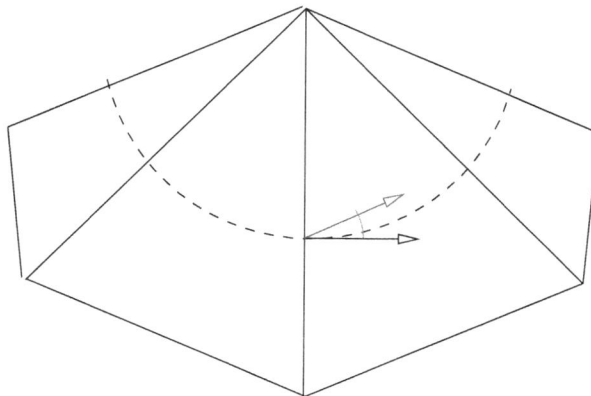

Figure 5.2. Conical surface obtained by identifying two triangle sides shown in Figure 5.1. The deficit angle is equal to the rotation angle (red) of a vector carried along a closed loop around the tip of the cone (final vector in blue).

an angle of $\frac{1}{3}\pi$. Since there are three such triangles meeting at the corner, the deficit angle is $\epsilon = 2\pi - \pi = \pi$.

On two-dimensional surfaces, the deficit angle of a triangulation is proportional to the surface integral of the Ricci scalar. Qualitatively, any such relationship has to involve an area integral of curvature because an angle does not have units while curvature has units of one over length squared. In order to determine the coefficient, we continue our example of the sphere and now compute its Ricci scalar. The line element $ds^2 = r^2(d\vartheta^2 + \sin^2\vartheta \, d\varphi^2)$ of a sphere with radius r implies a Christoffel symbol (2.26) with non-zero coefficients

$$\Gamma^\vartheta_{\varphi\varphi} = -\sin\vartheta\cos\vartheta, \quad \Gamma^\varphi_{\vartheta\varphi} = \Gamma^\varphi_{\varphi\vartheta} = \frac{\cos\vartheta}{\sin\vartheta}. \tag{5.5}$$

The derivation is very similar to the related example of spatially flat Friedmann–Robertson–Walker models, Equation (2.50), except that we now have ϑ instead of proper time as the coordinate that appears in metric components, and there is only one further coordinate, φ, instead of x, y and z. Also the derivation of the Ricci tensor is very similar to Equations (2.52) and (2.53), leading to

$$R_{\vartheta\vartheta} = -\frac{\partial\Gamma^\varphi_{\varphi\vartheta}}{\partial\vartheta} - \left(\Gamma^\varphi_{\varphi\vartheta}\right)^2 = 1 \tag{5.6}$$

$$R_{\varphi\varphi} = \frac{\partial\Gamma^\vartheta_{\varphi\varphi}}{\partial\vartheta} - \Gamma^\vartheta_{\varphi\varphi}\Gamma^\varphi_{\vartheta\varphi} = \sin^2\vartheta \tag{5.7}$$

as the non-zero components.

The Ricci scalar is therefore

$$R = R_{\vartheta\vartheta} + \frac{1}{r^2\sin^2\vartheta}R_{\varphi\varphi} = \frac{2}{r^2}. \tag{5.8}$$

Integrating this constant curvature over a quarter sphere, we obtain

$$\int R dA = 2\pi = 2\epsilon, \tag{5.9}$$

using our result for the deficit angle $\epsilon = \pi$ in the last step. This calculation determines the relationship between the surface integral of curvature and the deficit angle also for more complicated surfaces. If the surface does not have constant curvature, one would integrate only over a subregion corresponding to one third of the triangle area meeting at the corner where the deficit angle is taken. The remaining two thirds are attributed to other corners of the triangles where there may be further deficit angles.

The deficit angle can be positive or negative. Our example showed a positive deficit angle implied by a missing triangle around a conical point. If the deficit angle is negative, there is a surplus triangle which, when dangling sides are identified, requires the surface to buckle and form a saddle. Such a surface has negative curvature, such that the relationship $\int R dA = 2\epsilon$ remains meaningful. Moreover,

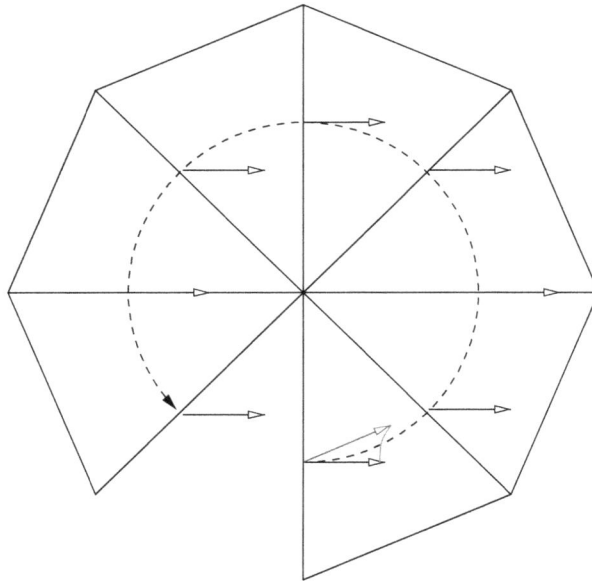

Figure 5.3. Hyperbolic deficit angle in Minkowski spacetime, measured as arc length along a unit hyperbola (red). The blue arrow is the rotated (or rather, boosted) initial arrow after sides of the missing triangle are identified. In this case, the deficit angle is negative because the piece of the red hyperbola between the two arrows is timelike, even though there is a missing triangle which would imply a positive deficit angle in Euclidean signature.

the deficit angle can be defined for surfaces in Lorentzian signature, in which case it refers to a hyperbolic angle: Such an angle is defined by the arc length measured on a unit hyperbola in two-dimensional Minkowski spacetime, replacing arc length on a unit circle in the Euclidean case. Since the Minkowski line element may be negative, a missing triangle can imply a positive or a negative deficit angle; see Figures 5.3 and 5.4.

Regge calculus is crucially based on the deficit angle, which can be defined in any dimension, including four for spacetime. Instead of triangulations of a two-dimensional surface one then uses the more general concept of a simplicial decomposition, where an n-*simplex* is an n-dimensional version of a triangle, such as a tetrahedron for $n = 3$. Combinatorially, an n-simplex can be drawn in a two-dimensional projection by choosing $n + 1$ points on the plane and connecting any possible pair of these points by a straight line. For $n = 2$ and $n = 3$, respectively, one obtains standard triangles and tetrahedra. For $n = 4$, an example is shown in Figure 5.5. In general, an n-simplex, according to its definition, contains $n + 1$ corners and $\binom{n + 1}{2} = \frac{1}{2}n(n + 1)$ one-dimensional edges (or one-simplices).

Simplicial decompositions in which n-simplices are glued together on their $(n - 1)$-simplicial sides can be constructed in any dimension. Their deficit angles should still be related to surface integrals of curvature because curvature has units of one over length squared in any dimension. In an n-dimensional triangulation,

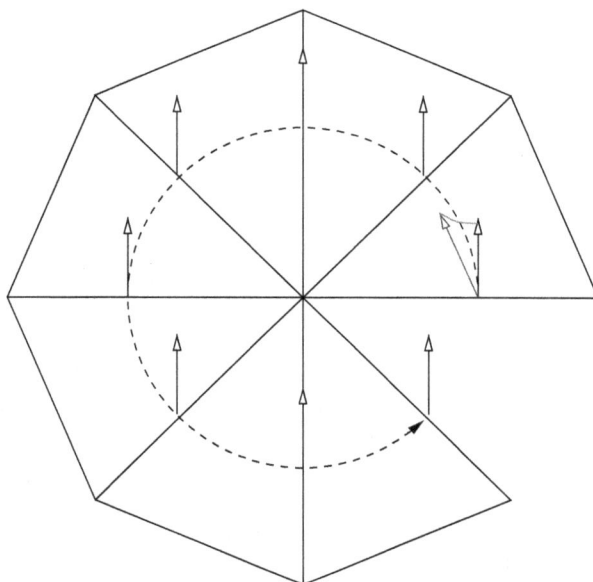

Figure 5.4. Hyperbolic deficit angle in Minkowski spacetime, as in Figure 5.3 but with a missing triangle in a spacelike direction. Here, the deficit angle is positive because the red piece of a hyperbola is spacelike and defines a positive hyperbolic angle through arc length.

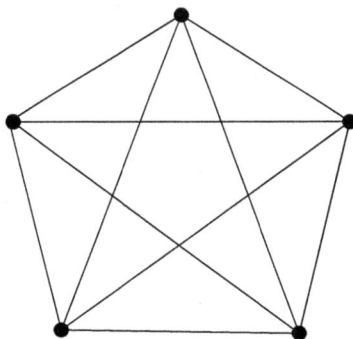

Figure 5.5. Example of a four-simplex projected on the plane.

curvature is therefore located on the $(n - 2)$-simplices. A two-dimensional hypersurface transversal to an $(n - 2)$-simplex can then be used to relate the deficit angle to an area integral of curvature.

In order to reduce almost all information contained in simplicial decompositions to the deficit angles, Regge calculus assumes that all one-simplices contained in a decomposition, such as the sides of triangles or the edges of tetrahedra, have the same length, a, in an embedding in $n + 1$-dimensional flat space. The remaining parameter a can then be changed in order to improve the approximation for smaller a, which requires a larger number of simplices and therefore more calculations of

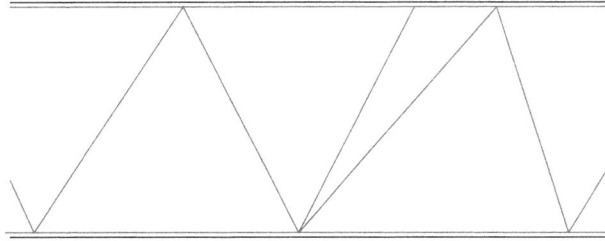

Figure 5.6. Example of a two-dimensional causal dynamical triangulation with two horizontal spatial slices (black) connected by triangles (blue).

deficit angles. In Lorentzian signature, there are spacelike and timelike one-simplices, which cannot have the same length because their line elements have opposite signs. One may then assume that timelike one-simplices have imaginary Minkowski length ia, while spacelike ones still have length a. More generally, causal dynamical triangulations introduce a free parameter α such that timelike one-simplices have length αa. Initially, α is imaginary, but it is changed to a real number when the Lorentzian path integral is replaced by a Euclidean version.

Triangulating Spacetime
Causal dynamical triangulations apply Regge calculus to manifolds with a causal structure, modeled on spacetime with Lorentzian signature. To do so, one must know which sides of simplices correspond to timelike and which ones to spacelike directions. By construction, causal dynamical triangulations therefore implement a causal structure which is not present in general triangulations. Examples are illustrated in Figure 5.6 for two-dimensional spacetimes, in Figure 5.7 for three-dimensional spacetimes, and in Figure 5.8 for four-dimensional spacetimes. A timelike side or edge then has length αa with some parameter α that determines the difference between spacelike and timelike length. In particular, α is imaginary in Lorentzian signature.

In two dimensions, there is only one type of triangle because any causal triangle must have two corners on one spatial slice and the remaining one on the next or preceding slice. In three dimensions, however, there are two different types because the four corners of a tetrahedron can be split into three corners on one slice and the remaining one on another slice (forming a (3, 1)-tetrahedron), or into two corners on each one of two successive spatial slices (forming a (2, 2)-tetrahedron). In four dimensions there are also two types, given by (4, 1)-simplices and (3, 2)-simplices.

Another difference between two dimensions on one hand and three or four on the other can be seen in Regge discretizations of the gravitational action

$$S = \frac{c^4}{16\pi G} \int \mathrm{d}t \mathrm{d}^{n-1}x (R - 2\Lambda) \tag{5.10}$$

in n spacetime dimensions. In two dimensions, the area integral $\mathrm{d}t\mathrm{d}xR$ of the Ricci scalar does not depend on the geometry (or the metric) but only on the topology of the manifold: According to the Gauss–Bonnet theorem, it is equal to 2π times the

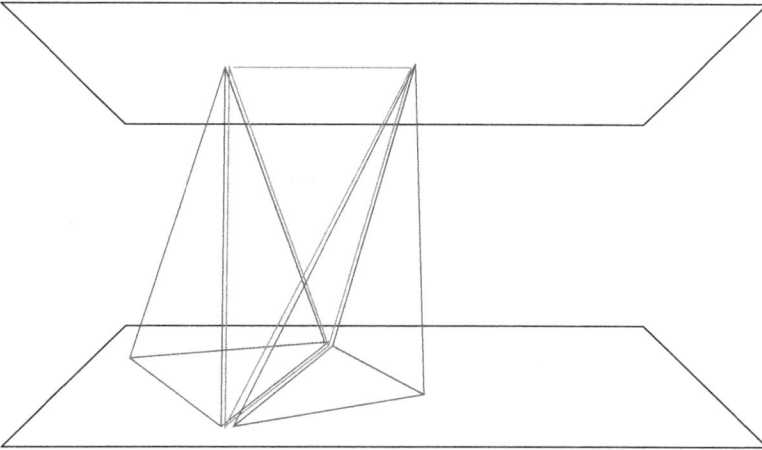

Figure 5.7. Example of a three-dimensional causal dynamical triangulation. In order to show the tetrahedra more clearly, a viewpoint from between two spatial slices has been assumed. The two different types of tetrahedra are shown: two (3, 1)-tetrahedra (blue) with three corners on one spatial slice and one on the other, one and (2, 2)-tetrahedron (red) with two corners on each spatial slice.

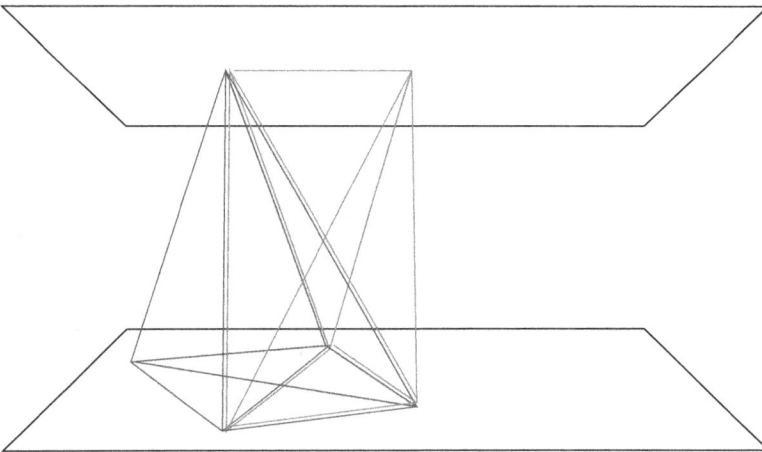

Figure 5.8. Example of a four-dimensional causal dynamical triangulation, shown in a three-dimensional projection. As in Figure 5.7; a viewpoint from between two spatial slices has been assumed. The two different types of tetrahedra in four dimensions are shown: a (4, 1)-simplex (blue) with three corners on one spatial slice and one on the other, glued along a three-dimensional tetrahedral side to a (3, 2)-simplex (red) with three corners on one spatial slice and two on the other.

Euler characteristic χ of the two-dimensional manifold, which latter is defined as $\chi = 2(1 - g)$ with the genus g of the manifold (equal to the number of handles it contains). If the topology of spacetime is fixed, the action of a two-dimensional causal dynamical triangulation is then simply given by a constant plus the cosmological-constant term. The latter depends only on the volume of the triangulated spacetime, and is therefore related to the number N of triangles in a given triangulation: $S = A + B\Lambda N$. The constant A is determined by the Euler characteristic, while the constant B depends on the Minkowski area of a triangle with one spacelike side length a and two timelike side lengths αa. We will not need the precise values.

In three dimensions, any edge or one-simplex σ_1 of a tetrahedron σ_3 contains curvature defined through the deficit angle of triangles given by cross-sections of all tetrahedra that contain this edge. According to Equation (5.9), this deficit angle is equal to one half the area integral of the Ricci scalar over the same cross-section. In the three-dimensional action, however, we do not have an area integral of curvature but the spacetime integral $\int \mathrm{d}t\mathrm{d}^2 x R$, which in Regge calculus is approximated by

$$\int \mathrm{d}t\mathrm{d}^2 x R = 2 \sum_{\sigma_1} \epsilon_{\sigma_1} V_{\sigma_1}, \tag{5.11}$$

summing over all one-simplices in a given triangulation, each of which has a deficit angle ϵ_{σ_1} and volume (length) V_{σ_1}. Similarly, in four dimensions the action is related to a sum over deficit angles times the areas of two-simplices.

The cosmological-constant term can be included in the same sum as in Equation (5.11) if we decide how to compute a three-dimensional spacetime volume assigned to an edge of a tetrahedron in a triangulation, for instance by the average

$$\bar{V}_{\sigma_1} = \frac{1}{6} \sum_{\sigma_3 \ni \sigma_1} V_{\sigma_3} \tag{5.12}$$

defined by contributions of one sixth of the volume V_{σ_3} of any tetrahedron that shares the given one-simplex. (It is only one sixth of the volume because each of these parent tetrahedra has five additional one-simplices that require equal attention.) The three-dimensional gravitational action of a given triangulation $\mathcal{T}$ is therefore Regge discretized by

$$S[\mathcal{T}] = \frac{c^4}{8\pi G} \sum_{\sigma_1} (\epsilon_{\sigma_1} V_{\sigma_1} - \Lambda \bar{V}_{\sigma_1}) = \frac{\hbar c}{8\pi \ell_{\mathrm{P}}^2} \sum_{\sigma_1} (\epsilon_{\sigma_1} V_{\sigma_1} - \Lambda \bar{V}_{\sigma_1}). \tag{5.13}$$

Since each of the two types of tetrahedra, the version (3, 1) and the version (2, 2) in three dimensions, has the same shape—determined by three spacelike edges of length a and three timelike edges of length αa in the first case, and two spacelike edges of length a and four timelike edges of length αa in the second case—the Minkowski line element can be used to compute the volumes and deficit angles of any configuration that appears in a given triangulation. The result depends on the numbers of all possible types of simplices:

- The number N_0 of 0-simplices, or vertices.
- The number N_1^{s} of spacelike one-simplices, or edges contained in a spatial slice.
- The number N_1^{t} of timelike one-simplices, or edges stretched between two spatial slices.

- The number N_2^s of spacelike two-simplices, or sides contained in a spatial slice.
- The number N_2^t of timelike two-simplices, or sides stretched between two spatial slices.
- The number $N_3^{(3,\,1)}$ of tetrahedra of type (3, 1) (blue in Figure 5.7).
- The number $N_3^{(2,\,2)}$ of tetrahedra of type (2, 2) (red in Figure 5.7).

These seven parameters are not independent. For instance, the N_2^s triangles in each spatial slice, taken independently, have $3N_2^s$ sides. Since each side is shared by two triangles, we have

$$2N_1^s = 3N_2^s. \tag{5.14}$$

Similarly, the $N_3 = N_3^{(3,\,1)} + N_2^{(2,\,2)}$ tetrahedra, taken together, have $4N_3$ side triangles. Again, each triangle is shared by two tetrahedra, such that $2N_2 = 4N_3$, where $N_2 = N_2^s + N_2^t$. Thus,

$$N_2^s + N_2^t = 2\left(N_3^{(3,\,1)} + N_3^{(2,\,2)}\right). \tag{5.15}$$

Another independent relationship is obtained because each tetrahedron of type (3, 1) has one spacelike triangle as a side, which is shared with another tetrahedron:

$$2N_2^s = N_3^{(3,\,1)}. \tag{5.16}$$

A final relationship

$$N_0 - N_1 + N_2 - N_3 = \chi \tag{5.17}$$

relates the numbers of simplices in any triangulation to the Euler characteristic χ of the manifold. Since this relation is topological, it does not distinguish between spacelike and timelike simplices. The Euler characteristic of the three-dimensional spacetime manifold is equal to the Euler characteristic of the two-dimensional spatial slices Σ because the one-dimensional time direction T is assumed to be independent: $M = T \times \Sigma$ is the topology of the three-manifold used in causal dynamical triangulations.

These four relationships leave three numbers undetermined. We can therefore eliminate all of them but, say, N_0, $N_3^{(3,\,1)}$ and $N_3^{(2,\,2)}$. As in two dimensions, the volumes and deficit angles of all simplices can be computed in Minkowski spacetime because the one-dimensional side lengths, a for spacelike ones and αa for timelike ones, are assumed to be fixed. The gravitational action then takes the form

$$S[\mathcal{T}] = \frac{\hbar c}{16\pi\ell_P^2}\left(AN_0 + BN_3^{(3,\,1)} + CN_3^{(2,\,2)}\right) - \frac{2\Lambda\hbar c}{16\pi\ell_P^2}\left(DN_3^{(3,\,1)} + EN_3^{(2,\,2)}\right) \tag{5.18}$$

with coefficients A, B, C, D, and E that depend on a and α. (The cosmological term is proportional to the three-dimensional spacetime volume in which $D \neq E$ in general because of different contributions from spacelike and timelike directions.) Two of these five parameters, such as C and E, can be eliminated by combining them with the constants G and Λ. Similar relationships hold in four dimensions (Ambjørn et al. 2012).

Euclidean Time and Statistical Physics

We have now reached the stage at which causal dynamical triangulations commonly transform from Lorentzian to Euclidean signature. At first sight, this step seems unmotivated because one has taken much care to distinguish between spacelike and timelike directions and to compute different volume contributions for them. However, since the aim is to use triangulations as representations of the metric or geometry in a path integral, Euclidean time is preferred (though not necessary) in order to improve the convergence properties of a Euclidean-time integrand $\exp(-S/\hbar)$, as opposed to a Lorentzian-time $\exp(iS/\hbar)$. While this step may introduce other unwanted features because we would rather like to understand Lorentzian quantum gravity (about which we will say more below), the Euclidean formalism can more easily be implemented computationally.

Moreover, in spite of the introduction of Euclidean signature, there is a difference between causal dynamical triangulations of a Lorentzian manifold with a preferred spatial slicing, and (non-causal) dynamical triangulations obtained had we done all constructions in Euclidean signature: Even after transforming to Euclidean time, causal dynamical triangulations are distinct from unrestricted triangulations both in their combinatorics and in their elementary geometrical relations between different simplices. A combinatorial example is the possibility of "baby universes" splitting off from an original connected spatial slice if a preferred slicing with fixed spatial topology is not required. Moreover, a three-dimensional triangulation in which we do not distinguish between spacelike and timelike simplices is characterized by four numbers (N_0, N_1, N_2 and N_3) subject to two relations, Equation (5.17) and a three-dimensional version of Equation (5.14). It would have only two free numbers, and therefore a smaller number of coefficients compared with Equation (5.18). Causal dynamical triangulations and unrestricted, Euclidean dynamical triangulations (Ambjørn & Jurkiewicz 1992) therefore define distinct path integrals, even though both are implemented with a Euclidean path integral. Euclidean dynamical triangulations have been found to lack a suitable continuum limit (Agishtein & Migdal 1992; Catterall et al. 1994; Bialas et al. 1996; de Bakker 1996).

The transformation to Euclidean time can be implemented by turning the imaginary α for timelike one-simplices into a real number. If $\alpha = 1$, formerly timelike and spacelike one-simplices now contribute in the same way, such that tetrahedra of types (3, 1) and (2, 2) contribute the same volume. In terms of coefficients in Equation (5.18), this symmetry implies $B = C$ and $D = E$. The number of coefficients then equals the number in Euclidean dynamical triangulations, but combinatorially one is still dealing with a smaller set of triangulations and therefore a different path integral because baby universes are ruled out. In addition to G and Λ, which in Equation (5.18) may be identified with C and E, respectively, only one free coupling constant, A, remains. However, the deviation Δ of α from the symmetric value $\alpha = 1$ can be reintroduced, and indeed provides an important parameter in causal dynamical triangulations that can be tuned in attempts to find phase transitions using the resulting partition function. The action can then schematically be written as

$$S[\mathcal{T}] = \frac{\hbar c}{16\pi\ell_P^2}\left(A(\Delta)N_0 + B(\Delta)N_3^{(3,\,1)} + N_3^{(2,\,2)}\right) - \frac{2\Lambda\hbar c}{16\pi\ell_P^2}\left(D(\Delta)N_3^{(3,\,1)} + N_3^{(2,\,2)}\right) \quad (5.19)$$

with functions $A(\Delta)$, $B(\Delta)$, and $D(\Delta)$ that can be determined from Lorentzian Regge calculus. (There are no coefficients of $N_3^{(2,\,2)}$ and $N_3^{(2,\,2)}$ because two coupling constants can be chosen to be given by $1/G$ and Λ, respectively.)

In Euclidean time, the approach is close to statistical physics, in which a periodic identification of the time direction (which is useful for numerical purposes because it implies a finite universe) can be interpreted as the inverse temperature. The Euclidean action S_E is defined such that $iS = -S_E$. Since Euclidean time is introduced as $t_E = it$, see Equation (3.173) in Section 3.4.1, we have

$$\begin{aligned}
S_E &= -iS = -i\int_{-\infty}^{\infty}\left(\frac{1}{2}m\left(\frac{dx}{dt}\right)^2 - V(x(t))\right)dt \\
&= -\int_{-\infty}^{\infty}\left(\frac{1}{2}m\left(\frac{dx}{dt}\right)^2 - V(x(t))\right)dt_E \quad (5.20) \\
&= \int_{-\infty}^{\infty}\left(\frac{1}{2}m\left(\frac{dx}{dt_E}\right)^2 + V(x(t))\right)dt_E = \int_{-\infty}^{\infty} H_E dt_E
\end{aligned}$$

in an example of classical mechanics, with the Euclidean Hamiltonian H_E. Using time independence in an equilibrium configuration, $\exp(-S_E/\hbar) = \exp(-H_E\Delta t_E/\hbar) = \exp(-\beta H_E)$ provides the statistical weights of a distribution at inverse temperature $\beta = 1/(k_B T)$. Thus, the periodicity in Euclidean time implies a temperature $T = \hbar/(k_B\Delta t_E)$.

The probabilities $\exp(-S_E(\mathcal{T})/\hbar)$ are used to set up a Monte-Carlo evaluation of the partition function: Using a small number of elementary changes of a triangulation, called *Pachner moves* (Pachner 1978), an initial triangulation can be successively manipulated to produce a large class of (hopefully) generic triangulations. Each Pachner move changes the triangulation and therefore the Euclidean action, which can be used to assign a probability for the move being performed (or rejected). The result is a statistical ensemble of causal triangulations $\mathcal{T}$ in which each $\mathcal{T}$ is represented with probability $\exp(-S_E(\mathcal{T})/\hbar)$. Summing over an observable over this ensemble, such as the spatial volume at a given time slice in each triangulation, is then equivalent to evaluating the path integral with the observable inserted in the integrand, giving its expectation value.

Causal dynamical triangulations therefore explore quantum cosmology of universes in an ensemble described by a thermal state, which is not pure but rather a mixture of many quantum configurations. In thermal states, small-scale details, such as the positions of individual molecules, or here the discrete building blocks in Regge calculus, usually do not matter. It is therefore possible for physical properties in such a state to be largely insensitive to the underlying discretization.

This independence on microscopic details can be strengthened if the temperature or other thermodynamical parameters are tuned such that the system is at a second-order phase transition. At a first-order phase transition, such as evaporating water, properties of a substance change locally and are described by microscopic processes such as weakening bonds. By contrast, a second-order phase transition, such as the spontaneous magnetization of a ferromagnetic material when the temperature drops below the Curie temperature, creates long-range order or long-range correlations between the microscopic constituents.

Phase Transitions

Numerical evaluations of causal dynamical triangulations are able to explore the phase structure of spacetime in a specific microscopic discretization (Ambjørn et al. 2004, 2005). "Thermodynamical" parameters that can be tuned in this theory include the cosmological constant. A second-order phase transition makes sure that there is a long-range continuum limit of the theory, which then approaches continuum spacetime as we perceive it with long-range coherence rather than "evaporating" or disintegrating into its microscopic constituents. The presence and order of a phase transition depend on detailed physical properties and therefore constitute an important test of the dynamics. There are indications that second-order phase transitions indeed exist in causal dynamical triangulations (Ambjørn et al. 2011).

The phases of causal dynamical triangulations are defined by different types of spacetime geometries. For a positive cosmological constant in Euclidean signature, a macroscopic universe is expected to expand for a long time and then recollapse, forming the shape of a four-dimensional sphere of positive curvature. Such configurations have indeed been found for certain ranges of the parameters A and Δ in Equation (5.19), forming a phase diagram.

Qualitative properties of different phases and the transitions between them can be understood based on properties of the Euclidean deficit angle, in particular its sign. A macroscopic causal triangulation close to a four-dimensional sphere, illustrated in Figure 5.9; is obtained if most deficit angles are positive. Therefore, compared with a flat triangulation, there are missing four-simplices around each two-simplex. A different picture is obtained if a large number of deficit angles are negative, implying locally negative curvature and therefore many saddle points. As shown in Figure 5.10, such a triangulation does not seem to form a single, macroscopic universe because it recollapses soon after space has expanded just a little bit. The large number of negative deficit angles implies that there are surplus four-simplices compared with a flat triangulation. One can therefore distinguish between these two configurations by the ratio of the number of four-simplices over the number of lower-dimensional simplices, such as two-simplices or 0-simplices. This ratio can be used as an order parameter that takes recognizably different values in the two phases —phase A in which non-classical configurations dominate and phase C in which

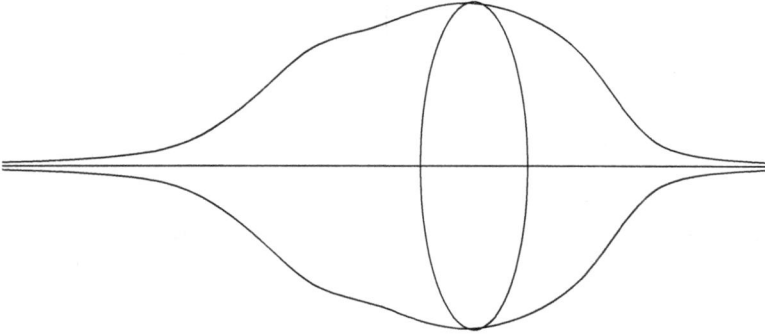

Figure 5.9. Three-dimensional sketch of a triangulation that determines a macroscopic universe approximating a four-dimensional sphere. Positive curvature implies positive deficit angles, and therefore a reduced number of four-simplices compared with a flat triangulation. Such triangulations appear in a phase, traditionally called "phase C," in which classical general relativity may be realized as a limit.

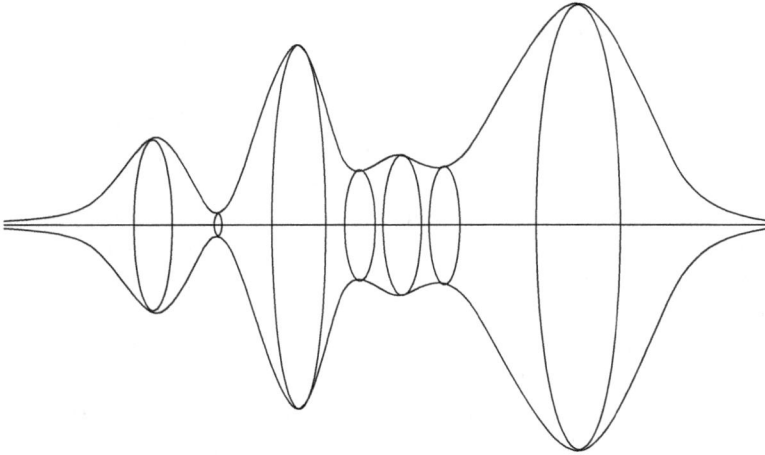

Figure 5.10. Three-dimensional sketch of a triangulation in which any expansion is soon followed by collapse, such that no macroscopic universe can form. At local minima of the spatial volume the curvature and therefore deficit angles are negative, implying a surplus of four-simplices. This phase, traditionally called "phase A," does not represent the classical limit.

classical-type configurations dominate—characterized by the presence or absence of macroscopic triangulations.

According to the action (5.19), dynamical contributions from lower-dimensional simplices, such as N_0, are controlled by the coupling constant A. For large A, the number of lower-dimensional simplices is suppressed by a factor of $\exp(-cAN_0/(16\pi\ell_P^2))$ in the probability that results from $\exp(-S_E(\mathcal{T})/\hbar)$. It is therefore expected that phase A is located at larger A, while phase C appears in a range of smaller A. This expectation has indeed been confirmed by numerical simulations (Ambjørn et al. 2004, 2005).

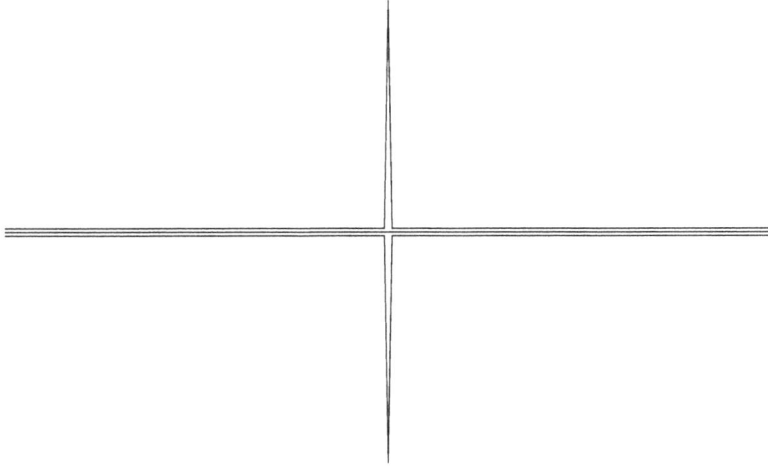

Figure 5.11. Three-dimensional sketch of a triangulation that expands and collapses in a few time steps. This phase, traditionally called "phase B," does not provide enough time for a recognizable classical-type configuration to result.

The third phase, phase B, is found for all A, but at small or even negative values of the asymmetry parameter Δ. This range corresponds to values of the length factor α of timelike one-simplices close to one or greater than one. A single time step in the triangulation then steps over a long spacetime distance, making it impossible to see most of the expansion and collapse of a four-dimensional sphere, as indicated in Figure 5.11. Since the coefficient of N_0 in Equation (5.19), given by $A(\Delta)$, also depends on Δ, the same order parameter, given by the number of 0-simplices divided by the number of four-simplices, can be used to distinguish a transition from phase A or C to phase B. (For computational reasons, the transition from phase A to phase B has been explored in less detail than the transitions from phase C to phase A or B.) More recently, it was found that the phase-space region initially assigned to a single phase C in fact consists of two different phases, both of which allow macroscopic geometries but with different degrees of homogeneity (Ambjørn et al. 2014, 2015; Coumbe et al. 2016; Ambjørn et al. 2017).

The existence of different phases has therefore been demonstrated in causal dynamical triangulations. In addition, it is of considerable interest to determine their order. A second-order phase transition is characterized by long-range order or correlations; the second-order nature can therefore be ascertained through correlation functions. In spacetime, order is then maintained not only in spacelike directions but also in time. One can therefore use time-correlation functions, defined in Section 3.3.6, to test whether long-range order can be established for certain values of the independent parameters in Equation (5.19) that span the phase diagram, given by the asymmetry parameter Δ that determines how much α differs from one, and by the coefficient A.

In order to compute a time-correlation function in a field theory with independent spatial directions, one should first spatially average the field, schematically written as

$$x(t) = \sum_{t=\text{const}} \phi(x^i) \tag{5.21}$$

for some field ϕ of a discretized fundamental theory. The spatial coordinates x^i are defined on the spatial slice that belongs to the fixed time value t. A common choice for $x(t)$ in four-dimensional causal dynamical triangulations is the number $N_3(t)$ of spacelike three-simplices on the spatial slice determined by t, which can be viewed as a spatial average of the local density $\phi(t)$ of three-simplices. Since each spatial three-simplex has the same volume, $N_3(t)$ is proportional to the spatial volume at time t of a triangulation. Quantum statistics comes in by averaging this value over a large number of triangulations, yielding the ensemble average $\langle N_3(t)\rangle$.

As a first result, it has been found that, in one of the phases C, the ensemble average $\langle N_3(t)\rangle$ is very close to a time-dependent solution of the Euclidean Friedmann equation for isotropic models (Ambjørn et al. 2008, 2011):

$$\langle N_3(t)\rangle \sim \cos^3(t/\zeta) \tag{5.22}$$

with some characteristic time $\zeta \propto N_4^{1/4}$ that is related to the number of four-simplices in the triangulation (or the four-volume of the universe). In contrast to minisuperspace truncations, quantum cosmological dynamics is here obtained from a fundamental theory without suppressing degrees of freedom. The quantum nature of these solutions is shown by non-zero fluctuations $(\Delta N_3(t))^2 = \langle N_3(t)^2\rangle - \langle N_3(t)\rangle^2$.

The volume variable N_3 can also be used to compute time-correlation functions,

$$C_{N_3}(\Delta t) = \langle N_3(t) N_3(t + \Delta t)\rangle - \langle N_3(t)\rangle \langle N_3(t + \Delta t)\rangle. \tag{5.23}$$

No t-dependence is included in the argument of C_{N_3} because the correlation function is expected to be time-independent in a time-translation invariant theory. According to our previous result (3.164) for the harmonic oscillator, C_{N_3} as a function of Δt is then expected to be of the form

$$C_{N_3}(\Delta t) \propto \exp(-|\Delta t|/\xi) \tag{5.24}$$

with some constant ξ, called the *correlation time* of N_3. (Comparing Equation (5.23) with the definition of Equation (3.164), notice that no product of expectation values is subtracted in the latter expression. This is so only because we previously computed the time-correlation function in energy eigenstates of the harmonic oscillator, which have vanishing position expectation values.)

For a free field theory, the harmonic result (3.164) should be exactly realized, while interacting theories may have different time-correlation functions. Nevertheless, the characteristic behavior (5.24) can be found to a good approximation for small intervals Δt, during which interactions do not have much time to act. The equation therefore provides a general definition of the correlation time.

The correlation time is related to fundamental parameters of the theory, such as the frequency $\nu \propto 1/\xi$ for the harmonic oscillator. Similarly, a free field theory, which we discretize here as a model for causal dynamical triangulations, has a correlation time related to the mass of the field. A discretized Hamiltonian could be written as

$$H(\phi) = \frac{1}{2} \sum_{\sigma_0} \left(\frac{(\phi(\sigma_0 + \delta t) - \phi(\sigma_0))^2}{c^2 \delta t^2} + \frac{c^2 m^2}{\hbar^2} \phi(\sigma_0)^2 \right) \delta V \tag{5.25}$$

if a scalar field ϕ of mass m is put on the 0-simplices of a causal dynamical triangulation. Adding δt to a 0-simplex σ_0 indicates that we jump to the next spatial slice in order to compute a discretized time derivative for the kinetic energy. Compared with the Hamiltonian of a harmonic oscillator, the product $mc^2/\hbar \sim \nu$ plays the role of the frequency. Since the latter is proportional to the inverse correlation time according to Equation (3.164), we obtain $\xi \sim \hbar/(mc^2)$. In the discrete setting, the unitless correlation length

$$\xi_\delta(\delta t) = \frac{\xi(\delta t)}{\delta t} \sim \frac{\hbar}{\delta t m(\delta t) c^2} \tag{5.26}$$

is more interesting because it determines the correlation between distant discrete building blocks. As δt is varied to explore possible continuum limits, usually both ξ and m depend on δt if the latter is used as a renormalization scale

Here, causal dynamical triangulations make an important connection with the proposal of asymptotic safety: If the theory has a non-trivial ultraviolet fixed point, the renormalized mass $m(\delta t)$, like the coupling constants, approaches a constant value in the ultraviolet limit $\delta t \to 0$. According to Equation (5.26), this is possible only if the correlation length diverges for $\delta t \to 0$. The existence of an ultraviolet fixed point therefore requires long-range correlations, or a second-order phase transition. Causal dynamical triangulations provide independent methods to test this proposal.

In causal dynamical triangulations, δt can be adjusted by changing the length a of spacelike one-simplices, which then adjusts the length αa of timelike simplices. Since the aim is to find evidence for the possibility of macroscopic, non-collapsed states of spacetime in a continuum limit, it is convenient to look for a fixed point of the spacetime volume V_4 in addition to fixed-point masses or coupling constants, a process that defines *finite-size scaling*. The spacetime volume of a causal triangulation is related to the basic length a through the number N_4 of four-simplices, $V_4 \propto N_4 a^4$. Since a is proportional to δt, Equation (5.26) implies $V_4 \propto N_4/(m^4 \xi_\delta^4)$. If both V_4 and m are at a fixed point, the correlation length is related to the number of four-simplices by $\xi_\delta \propto N_4^{1/4}$. A diverging correlation length is then equivalent to an infinite number of four-simplices while the four-dimensional volume remains finite, clearly demonstrating a continuum limit. (The limit of an infinite number of four-simplices plays the role of the "thermodynamic limit" of a diverging number of constituents, which is required for phase transitions to exist in a formal mathematical sense.)

Cosmological Implications and Open Questions
The thermal state selected by causal dynamical triangulations is unlike states usually assumed in cosmological minisuperspace models, which in most cases are pure. Moreover, the choice of initial conditions for the dynamical evolution of a wave

function is replaced by the procedure of setting up a thermal ensemble. These features require a different viewpoint on cosmological applications compared with the more traditional approach to quantum cosmology based on the Wheeler–DeWitt equation or path-integral versions such as the no-boundary proposal that are designed to provide initial conditions.

Nevertheless, there is an interesting relationship between implications of causal dynamical triangulations compared with the no-boundary proposal, as well as other models of quantum cosmology. Unlike what has been observed for the no-boundary proposal (Feldbrugge et al. 2017, 2018), the existence of a continuum limit in causal dynamical triangulations suggests that inhomogeneities are controlled in this approach because the most likely configurations in the ensemble of its thermal states are close to a continuum distribution of discrete building blocks, at least in the two phases C. This result constitutes an important test that has been passed successfully by causal dynamical triangulations.

Another successful test is given by the derivation of Friedmann-like solutions from time-dependent expected three-volumes in causal dynamical triangulations, as already mentioned in Equation (5.22). According to this result, it is possible to study sufficiently simple cosmological models in causal dynamical triangulations by averaging over a large number of microscopic states in an ensemble without eliminating any degrees of freedom from the fundamental equations. Detailed tests of the dynamics and of minisuperspace methods can therefore be performed, akin to computer experiments that analyze "predictions" made within a simplified mini-superspace setting (Baytaş et al. 2020).

However, several questions remain to be addressed. Starting with the second test passed by causal dynamical triangulations, although it is reassuring to know that minisuperspace results can sometimes be viewed as good approximations of fundamental calculations with large numbers of degrees of freedom, the almost exact realization of Friedmann-like solutions is puzzling from the perspective of the long-standing averaging problem of cosmology (Buchert 2011): First, given a general spacetime (or a general causal triangulation in an ensemble) it is difficult to determine whether it is approximately homogeneous in space because doing so would require one to construct spatial slices in which this condition is fulfilled. Second, because Einstein's equation is highly non-linear, even if a given spacetime can be approximated by a spatially homogeneous one with some spatial slicing, the time-dependent spatial metric defined by this slicing is not expected to obey the equations of an exactly homogeneous model: The given spacetime metric $g_{\alpha\beta}$ in this construction is a solution of Einstein's full equation. Its Einstein tensor can be averaged spatially using the same slicing, but because of non-linearity, this average is not equal to the Einstein tensor $G_{\alpha\beta} = R_{\alpha\beta} - \frac{1}{2}Rg_{\alpha\beta}$ of a spatially averaged metric: $\langle G_{\alpha\beta}[g] \rangle \neq G_{\alpha\beta}[\langle g \rangle]$.

In causal dynamical triangulations, the expectation value (5.22) is very close to an exact solution of the Friedmann equation. There are also large fluctuations in the ensembles of rather small four-dimensional spacetimes in which this result has been obtained so far. But it is somewhat puzzling that the time-dependence of the expectation value does not deviate more from the exact Friedmann result, given both the presence of fluctuations and the highly non-trivial nature of spatially averaging a non-linear spacetime theory. Perhaps, the dynamics implemented in causal dynamical triangulations somehow suppresses non-linearity or inhomogeneity, or a suppression of inhomogeneity may be an artifact of the relatively small linear size for which full four-dimensional simulations are currently possible.

The analytical form of Friedmann-like expectation values (5.22) is not very distinctive because it is determined by the three lowest-order terms,

$$S_a \propto a^3 \left(A_1 \left(\frac{\dot{a}}{a} \right)^2 + A_2 \frac{k}{a^2} + A_3 \right) \tag{5.27}$$

that can be expected from a leading-order derivative expansion of any a-dependent action compatible with the two transformations that remain of general covariance in an isotropic setting: rescaling the averaging volume and reparameterizing time, as discussed in detail in Section 2.2.1. These transformations preserve a preferred spatial slicing as assumed in causal dynamical triangulations. Any spatial average of a simulation with spherical spatial topology is therefore expected to be compatible with an action principle of the form (5.27); which then implies solutions of the form (5.22).

Results are more distinctive in recent studies of causal dynamical triangulations with toroidal (or flat) spatial topology. The curvature term in Equation (5.27) then does not contribute because $k = 0$, and it is possible to detect small a-dependent contributions that would be hidden by the curvature term in spherical simulations (Ambjørn et al. 2016, 2017). These new terms have been found to obey unexpected power-law behaviors, which cannot easily be explained by the usual quantum effects in minisuperspace models such as semiclassical corrections from back-reaction of quantum fluctuations or factor-ordering choices in Hamilton operators (Baytaş et al. 2020). This difficulty might indicate that the dynamics realized by causal dynamical triangulations differs from general relativity not only by standard quantum effects but also by new phenomena, perhaps related to the preferred spatial slicing.

This observation brings us back to the averaging problem. The main mathematical difficulty of the averaging problem—finding a spatial slicing in which one can assume homogeneity or successfully average a given spacetime—is avoided in causal dynamical triangulations because the approach has a built-in preferred slicing. Only this slicing (or its time translations, such as averaging between two spatial slices on which 0-simplices are positioned) has been used in calculations of time-dependent expectation values or correlation functions. A generic spacetime, however, be it classical or quantum, may be spatially homogeneous even if it appears very inhomogeneous in some slicing defined by a random choice of time coordinate. Causal dynamical triangulations do not make use of coordinates, but this does not imply that they are covariant because it is still possible to probe different slicings; see Figure 5.12. The result that the preferred slicing always turns out to be close to a homogeneous one suggests that slicing independence needs to be studied carefully in this approach—unless inhomogeneity is suppressed in the same way on all slicings

Figure 5.12. Different slicings in a causal triangulation (blue and red) in addition to the preferred slicing (black). The timelike gaps in the blue and red slicings do not necessarily violate the spacelike condition, provided the pieces where they stay on the preferred slicing are long enough—in particular in the continuum limit in which the timelike distance between slices approaches zero.

even for semiclassical properties, which also would be in disagreement with general relativity. One might argue that causal dynamical triangulations are covariant because they work with large ensembles of many configurations, and therefore implicitly average over different choices of spacetime slicings. However, it has not been explored in detail whether the realization of different slicings is complete. A systematic approach that weakens the reliance on a fixed slicing has been constructed in Jordan & Loll (2013a, 2013b), but it requires a more involved computational implementation.

Another question that remains to be addressed is the current use of Euclidean time without transforming back to Lorentzian time. Even though causal dynamical triangulations implement some spacetime properties by constructing suitable sub-classes of general Euclidean triangulations, Euclidean time is still used in several crucial places. A formulation of causal dynamical triangulations in a Lorentzian path integral (Feldbrugge et al. 2018; as initially applied to the no-boundary proposal, based on Picard–Lefshetz theory as discussed in Section 3.4.2) could test whether there is a physical difference between these two time choices. Since instabilities in the no-boundary proposal were found only after a formulation in a Lorentzian path integral was available, the apparent success of finding stable macroscopic geometries in causal dynamical triangulations might, for all we know, be a result of hiding potential instabilities by using a Euclidean formulation.

The role of the deficit angle and its sign in explaining the different phases, as shown in Figures 5.9–5.10; also suggests crucial differences in the Lorentzian case. In Euclidean signature, the sign of the deficit angle is strictly related to the surplus or lack of four-simplices. In Lorentzian signature, by contrast, the sign of the deficit angle also depends on whether a four-simplex is missing in a spacelike or timelike direction. It is therefore unclear whether the number of lower-dimensional simplices relative to the number of four-simplices can still be used as an order parameter to distinguish between different phases. Even a surplus of four-simplices would be suppressed if it appears in a spacelike direction because the action is then positive. A surplus of four-simplices is then statistically preferred in timelike directions, which may contribute to accelerated expansion, on average, as expected for a Lorentzian universe with positive cosmological constant. The cosine in Equation (5.22) could perhaps be replaced correctly by its hyperbolic version. However, it has not been explored yet whether there are other, undesired consequences of the different behavior of signs of the deficit angle in Lorentzian signature.

Other puzzling features reside in the technical details of discretized action principles used in causal dynamical triangulations. As we saw in the example of three-dimensional causal triangulations, Equation (5.18) which has a similar qualitative form in four dimensions, there are three independent fundamental parameters, given by the numbers of certain simplices in a triangulation, which reduce to just two in the symmetric case of $\alpha = 1$. The remaining parameters then specify the four-volume and the number of 0-simplices. Qualitatively, this number is equal to the number of independent covariant integrations of the metric and up to second-order derivatives, $\int \sqrt{-\det g}\, \mathrm{d}t \mathrm{d}^3 x$ and $\int \sqrt{-\det g}\, R \mathrm{d}t \mathrm{d}^3 x$.

However, it is possible to choose $\alpha \neq 1$, and such a value is in fact preferred in searches for macroscopic triangulations and second-order phase transitions (phases C). For $\alpha \neq 1$, there are three independent fundamental parameters in the action (5.18) or its four-dimensional analog. The spacetime asymmetry suggested by $\alpha \neq 1$ indicates that a non-covariant term may be present in the action, such as an integral of the form $\int \sqrt{-\det g}\, \dot{g}^2 \mathrm{d}t \mathrm{d}^3 x$ in which time derivatives are not covariantly paired with space derivatives. While such a result, if it can be confirmed, may reveal an interesting relationship with Hořava–Lifshitz gravity (Hořava 2009), it would also highlight a breakdown of covariance.

Finally, the continuum integrals possibly corresponding to the discrete terms in Equation (5.18) depend non-quadratically on the metric $g_{\alpha\beta}$, usually taken as the fundamental field in continuum theories of gravity. These degrees of freedom are local, and they self-interact non-linearly. Local degrees of freedom are therefore replaced by non-local parameters such as N_0 in a causal triangulation, and any possible self-interactions are hidden because the action (5.18) is linear in the number parameters. Unless one can find non-linear relationships between these parameters (or between the parameters that appear in the action and another number, such as N_3 used in the volume expectation value $\langle N_3(t) \rangle$) one could conclude that non-linearity is indeed suppressed in causal dynamical triangulations, as discussed previously in the context of the averaging problem.

In spite of these open questions, causal dynamical triangulations are one of the most successful approaches to quantum gravity. This framework has been able to address questions that can (and should) be posed in any approach to quantum gravity, rather than just solving problems of technical nature that are implied by the formulation of the given approach. In fact, several open problems mentioned here are suggested by general questions about quantum gravity. It is a success that they can meaningfully be discussed thanks to the specific results found in causal dynamical triangulations.

5.2.4 Non-commutative, Non-associative, and Fractional Geometries

Any two functions, f_1 and f_2, on a spacetime manifold, usually assumed to take real or complex values, can be multiplied with each other in a pointwise fashion: the product $f_1 f_2$ is another function which takes values $(f_1 f_2)(x) = f_1(x) f_2(x)$ at any point x on the manifold. This operation is commutative, as a direct consequence of the commutative product on the real or complex numbers.

If we consider the set of all functions together with this product as well as a similar definition of addition, it forms an abstract algebra in which multiplication is defined as an operation $(f_1, f_2) \mapsto f_1 f_2$ for any pair of algebra elements. Although it may seem somewhat inconvenient, this algebra can in principle be defined without a direct reference to points x, just by specifying the result $f_1 f_2$ for any pair (f_1, f_2). It turns out that the manifold on which these algebra elements are defined can be reconstructed from the algebra: There is a unique manifold, called the Gel'fand spectrum of the algebra, such that any algebra element is uniquely identified with a function on this manifold and all algebra products agree with function products.

If we already have a manifold on which our algebra elements are defined as functions, a point x defines a one-dimensional complex representation of the algebra, or a *character*: Evaluation of functions at x is the same as a mapping $x: f \mapsto f(x) \in \mathbb{C}$. (Interpreting x as a mapping from functions to the complex numbers can be written as the trivial-looking but nevertheless deep Gel'fand homomorphism, $x(f) = f(x)$.) Pointwise definitions of the usual operations on function spaces ensure that the evaluation mapping preserves the additive and multiplicative structures, and is therefore a representation (or an algebra homo-morphism). A mathematical theorem states that the set of points on a manifold is equivalent to the set of characters of the algebra of functions on the manifold. If we have a commutative algebra, the set of its characters can therefore always be viewed as the manifold on which algebra elements are defined as functions.

Characters can be defined on any algebra. However, if the algebra is not commutative ($ab \neq ba$ in general) or not even associative ($a(bc) \neq (ab)c$ in general), algebra elements cannot possibly be functions on a manifold with the standard pointwise product. The Gel'fand spectrum, which is still defined in these cases through characters of the algebra, is then interpreted as a non-standard manifold on which, through the algebra, a non-commutative or non-associative version of geometry and physics are defined (Connes 1996, 1994; Doplicher et al. 1995). Since we should preserve the classical limit, relevant algebras are usually defined in terms of *star-products*, providing an $\hbar$-dependent multiplication $f_1 \star_\hbar f_2$ such that $\lim_{\hbar \to 0} f_1 \star_\hbar f_2 = f_1 f_2$ is the usual commutative and associative pointwise product on a manifold. However, for non-zero $\hbar$ the star-commutator

$$[f_1, f_2]_\star = f_1 \star_\hbar f_2 - f_2 \star_\hbar f_1 \tag{5.28}$$

or star-associator

$$[f_1, f_2, f_3]_\star = f_1 \star_\hbar (f_2 \star_\hbar f_3) - (f_1 \star_\hbar f_2) \star_\hbar f_3 \tag{5.29}$$

may be non-zero.

Much of the usual tensor calculus or differential forms used in Riemannian geometry can be reformulated algebraically. For instance, a vector field v on a standard manifold is a mapping from functions to functions, $v: f_1 \mapsto v(f_2)$ such that the Leibniz rule

$$v(f_1 f_2) = f_1 v(f_2) + v(f_1) f_2 \tag{5.30}$$

holds. The mapping is usually defined using partial derivatives,

$$v(f) = \sum_{\alpha=0}^{3} v^{\alpha} \frac{\partial f}{\partial x^{\alpha}}, \tag{5.31}$$

but independently it may be viewed as an abstract algebraic operation subject to the condition (5.30). In the latter form, it is defined without reference to coordinates and can also be used in non-commutative and non-associative geometry, such as

$$v \star_{\hbar} (f_1 \star_{\hbar} f_2) = f_1 \star_{\hbar} v \star_{\hbar} (f_2) + v \star_{\hbar} (f_1) \star_{\hbar} f_2. \tag{5.32}$$

(The symbol $\star_{\hbar}$ after the v may seem redundant, but it indicates that the application of a vector field in a non-commutative setting may depend on whether it is applied from the left or from the right: In component form, there may be a difference between $\sum_{\alpha=0}^{3} v^{\alpha} \star_{\hbar} \partial f / \partial x^{\alpha}$ and $\sum_{\alpha=0}^{3} (\partial f / \partial x^{\alpha}) \star_{\hbar} v^{\alpha}$.) Higher-order tensor operations are more involved in the latter cases because one has to keep track of the order or bracketing of all factors.

With non-commutative or non-associative tensor calculus it is possible to define the curvature in non-commutative or non-associative geometry. Instead of tedious component expressions as in Section 2.2.1 with many factors of components and partial derivatives, a compact notation can be introduced based on linear maps, which are well-suited for an algebraic formulation. For instance, the covariant derivative ∇ is algebraically defined as a map $\nabla : (v, f) \mapsto \nabla_v f$ from a pair of a vector field v, also defined algebraically as above, and a function f to a new function, $\nabla_v f$, which is linear in v and obeys a Leibniz rule (5.30) in f:

$$\nabla_{gv} f = g \nabla_v f \quad \text{and} \quad \nabla_v (gf) = g \nabla_v f + (\nabla_v g) f \tag{5.33}$$

for any other function g. By putting $\star_{\hbar}$ between all symbols and keeping the ordering as written, the definition can be generalized to non-commutative and non-associative algebras. (The latter also require additional parentheses in an equation such as (5.33); which in any specific proposal has to be checked for consistency.)

Given a covariant derivative in algebraic form, the Riemann curvature tensor is defined as a mapping from three vectors, u, v, and w, to a fourth vector $R(u, v, w)$, such that

$$R(u, v, w) = \nabla_u \nabla_v w - \nabla_v \nabla_u w - \nabla_{[u,v]} w \tag{5.34}$$

where the vector-field commutator $[u, v]$ is defined in components as

$$[u, v]^{\alpha} = \sum_{\beta=0}^{3} \left(u^{\beta} \frac{\partial v^{\alpha}}{\partial x^{\beta}} - v^{\beta} \frac{\partial u^{\alpha}}{\partial x^{\beta}} \right). \tag{5.35}$$

(This commutator is in general non-zero even on a commutative manifold.) Expressed in a basis, the components of the vector $R(u, v, w)$ defined in Equation (5.34) equal the components (2.45) of the Riemann tensor if basis vectors are inserted for u, v, and w.

Like the covariant derivative, the definition (5.34) can, with some care, be generalized to non-commutative and non-associative manifolds. Given the Riemann tensor, versions of the Ricci tensor and the Ricci scalar then follow. The version that corresponds to the Ricci scalar can finally be used to define the actions of non-commutative (Aschieri et al. 2005, 2006; Alvarez-Gaume et al. 2006; Steinacker 2007, 2010; Dimitrijević Ćirić et al. 2020) or non-associative gravitational theories (Blumenhagen & Plauschinn 2011; Blumenhagen et al. 2011, 2014; Blumenhagen & Fuchs 2016). However, extensions from commutative multiplication to star products can be made in different ways, implying ambiguities in the resulting physics. It is also a difficult and still ongoing process to implement all the required consistency conditions, making sure that compatible results are made for predictions of the same quantity derived with different orderings or bracketings of operations. Similar constructions based on generalized tensor calculus can be made to account for a possible fractional, non-integer powers in the expressions of calculus and differential geometry (Calcagni 2010, 2013).

In order to simplify calculations, many studies use a test-particle picture to probe non-commutative or non-associative theories. The star product is then restricted to a limited set of functions that are thought to represent coordinates on a non-commutative or non-associative manifold. As a consequence, not only a position and a momentum component have a non-zero commutator, but even different components of the position (or momentum) do so. In the non-commutative case, calculations in such theories can be performed in much the same way as in standard quantum mechanics, but non-associative derivations remain challenging.

While non-zero commutators can well be described by a standard quantum-mechanics description, setting up a non-associative version of quantum mechanics encounters several difficulties. First, the action of operators on wave functions is by definition associative, and therefore wave functions and the superposition principle they fulfill cannot be used. An algebraic formulation as developed in Chapter 3 may still be applicable, but it requires modifications.

For instance, the commutator in a non-associative theory does not act like a classical derivative because the previous derivation (3.11) of a Leibniz rule for $[\hat{A}, \hat{B}\hat{C}]$ is now modified to (Mylonas et al. 2014)

$$
\begin{aligned}
[\hat{A}, \hat{B}\hat{C}] &= \hat{A}(\hat{B}\hat{C}) - (\hat{B}\hat{C})\hat{A} \\
&= \hat{A}(\hat{B}\hat{C}) - ((\hat{B}\hat{A})\hat{C} - (\hat{B}\hat{A})\hat{C}) - (\hat{B}\hat{C})\hat{A} \\
&= ((\hat{A}\hat{B})\hat{C} + [\hat{A}, \hat{B}, \hat{C}]) - (\hat{B}\hat{A})\hat{C} \\
&\quad + (\hat{B}(\hat{A}\hat{C}) - [\hat{B}, \hat{A}, \hat{C}]) - (\hat{B}(\hat{C}\hat{A}) - [\hat{B}, \hat{C}, \hat{A}]) \\
&= [\hat{A}, \hat{B}]\hat{C} + \hat{B}[\hat{A}, \hat{C}] + [\hat{A}, \hat{B}, \hat{C}] - [\hat{B}, \hat{A}, \hat{C}] \\
&\quad + [\hat{B}, \hat{C}, \hat{A}],
\end{aligned}
\tag{5.36}
$$

using an associator as defined in Equation (5.29). Therefore, there is no direct correspondence between commutators with a Hamilton operator and Hamilton's equations in classical mechanics. There are specific, physically motivated examples

(Mylonas et al. 2014; Blumenhagen & Fuchs 2016; Bojowald et al. 2017) in which there is a difference even between multiple products of the same algebra element, such as

$$[\hat{A}, \hat{A}^2] = \hat{A}(\hat{A}^2) - (\hat{A}^2)\hat{A} = [\hat{A}, \hat{A}, \hat{A}] \neq 0, \tag{5.37}$$

in which case the algebra is said not to be power-associative. A definition of powers of operators or of their exponential function as a power series, and therefore time evolution, is then ambiguous.

A physical, though somewhat exotic, example in which non-associativity is required of a quantum theory is given by a charged particle moving in a background density of magnetic monopoles. While fundamental magnetic monopoles have not been observed and there are strong upper limits on their density (Vant-Hull 1968; Palmer & Taylor 1968; Bojowald et al. 2018), there are condensed-matter systems that can serve as analog models (Gingras 2009). An isolated magnetic monopole can be described without non-associativity, provided its magnetic charge g is quantized in relation to the fundamental electric charge e of a proton, such that $eg = N\hbar c$ for some half-integer N, as derived by Dirac (1931). Because the electric fine structure constant, $\alpha = e^2/(\hbar c) = 1/137$, is small, the fundamental magnetic charge allowed by Dirac's quantization condition is large. Accordingly, a fundamental Dirac monopole in elementary particles would significantly alter atomic spectra (Eliezer & Roy 1962) and is therefore ruled out. However, smaller fundamental magnetic charges may still be possible, provided non-associative quantum mechanics is allowed (Bojowald et al. 2018).

The relationship between magnetic monopole densities and non-associativity can already be seen at the classical level. The classical analog is a violation of the Jacobi identity of Poisson brackets, which when quantized leads to non-zero associators (Günaydin & Zumino 1986; Jackiw 1985, 2004; Günaydin et al. 1978). An analysis of Poisson brackets requires a canonical description of a charged particle in a background magnetic field. Starting with the Lagrangian

$$L = \frac{1}{2}m|\vec{v}|^2 + q\vec{v} \cdot \vec{A} \tag{5.38}$$

in which an electric charge q couples to a background vector potential $\vec{A}$, we first compute the canonical momentum

$$\vec{p} = m\vec{v} + q\vec{A}. \tag{5.39}$$

Crucially, the canonical momentum does not equal the kinematical momentum $m\vec{v}$. Also crucially, and somewhat surprisingly, the Hamiltonian

$$H = \vec{p} \cdot \frac{d\vec{v}}{dt} - L = \frac{1}{2}m|\vec{v}|^2 \tag{5.40}$$

contains just the kinetic energy and no coupling term. However, when expressed in terms of the canonical momentum, the Hamiltonian

$$H = \frac{1}{2m} | \vec{p} - q\vec{A} |^2 \tag{5.41}$$

does depend on the vector potential.

Hamilton's equations of motion confirm that the system indeed describes a charged particle subject to the Lorentz force: We have the standard result

$$\frac{dx^i}{dt} = \frac{\partial H}{\partial p^i} = \frac{1}{m}(p^i - qA^i) = v^i, \tag{5.42}$$

while

$$\frac{dp^i}{dt} = -\frac{\partial H}{\partial x^i} = \frac{q}{m}(\vec{p} - q\vec{A}) \cdot \frac{\partial \vec{A}}{\partial x^i} = q\vec{v} \cdot \frac{\partial \vec{A}}{\partial x^i}. \tag{5.43}$$

The last equation describes the correct Lorentz force because

$$m\frac{d^2x^i}{dt^2} = \frac{dp^i}{dt} - q\frac{dA^i}{dt} = q\left(\vec{v} \cdot \frac{\partial \vec{A}}{\partial x^i} - \vec{v} \cdot \vec{\nabla} A^i\right) = q(\vec{v} \times \vec{B})^i \tag{5.44}$$

using the magnetic field $\vec{B} = \vec{\nabla} \times \vec{A}$ determined by the vector potential.

While the canonical momentum p^i has, by definition, canonical Poisson brackets with the position components x^i, the kinematical momentum does not: We compute

$$\{mv^i, mv^j\} = \{p^i - qA^i, p^j - qA^j\} = q\left(\frac{\partial A^j}{\partial x^i} - \frac{\partial A^i}{\partial x^j}\right) = q\sum_{\ell=1}^{3} \epsilon_{ij\ell}B^\ell \tag{5.45}$$

with the completely antisymmetric tensor ϵ_{ijk}. A second Poisson bracket then implies that

$$\{\{mv^i, mv^j\}, mv^k\} = q\sum_{\ell=1}^{3} \epsilon_{ij\ell}\{B^\ell, p_k - qA_k\} = q\sum_{\ell=1}^{3} \epsilon_{ij\ell}\frac{\partial B^\ell}{\partial x^k}. \tag{5.46}$$

The antisymmetrization of the left-hand side is defined as the Jacobiator,

$$J = \frac{1}{6}\sum_{i=1}^{3}\sum_{j=1}^{3}\sum_{k=1}^{3} \epsilon_{ijk}\{\{mv^i, mv^j\}, mv^k\} = \frac{1}{3}q \, \text{div} \, \vec{B}. \tag{5.47}$$

The Jacobiator vanishes if there are no magnetic monopoles, such that div $\vec{B} = 0$. The bracket $\{\cdot, \cdot\}$ is then indeed a Poisson bracket, and $\vec{x}$ and $\vec{p}$ can be represented as operators as in standard quantum mechanics. But if there are magnetic monopoles, div $\vec{B} \neq 0$ means that the vector potential and therefore the canonical momentum are not available, because any $\vec{A}$ such that $\vec{B} = \vec{\nabla} \times \vec{A}$ would imply that div $\vec{B} = 0$. A quantum theory would therefore have to be built from operators for $\vec{x}$ and the kinematical momentum $m\vec{v}$.

However, as a second consequence of a non-zero div $\vec{B}$, the Jacobiator (5.47) is not zero, such that the bracket $\{\cdot, \cdot\}$ cannot be a strict Poisson bracket. (It is a twisted

Poisson bracket; Park 2000; Klimcik & Strobl 2002; Severa & Weinstein 2001). A quantum version of the non-zero Jacobiator is a non-zero associator because

$$\sum_{i=1}^{3}\sum_{j=1}^{3}\sum_{k=1}^{3} \epsilon_{ijk}[[\hat{O}^i, \hat{O}^j], \hat{O}^k] = 2 \sum_{i=1}^{3}\sum_{j=1}^{3}\sum_{k=1}^{3} \epsilon_{ijk}[\hat{O}^i, \hat{O}^j, \hat{O}^k] \tag{5.48}$$

for any operators $\hat{O}^i$ such as the kinematical momenta, where we have replaced Poisson brackets with commutators on the left-hand side. (This identity follows directly if all commutators are expanded according to their definition.) The associator therefore has a completely antisymmetric contribution proportional to div $\vec{B}$, and must be non-zero in the presence of magnetic monopoles.

Even if non-associative quantum theories can be constructed, or if one contents oneself with non-commutativity, point-particle versions of non-commutativity raise questions. One should be careful with interpretations of such models because they may just rewrite standard quantum mechanics in a less-familiar form: According to Darboux's theorem (Arnold 1997), for any non-degenerate basic commutator (or its classical version, the Poisson bracket) $[z_i, z_j] = P_{ij}(z_k)$, where $z_1 = x$ and $z_2 = p$ or an analogous higher-dimensional version, can be mapped to canonical variables x' and p' with standard brackets, $[x', p'] = i\hbar$, while in the case of more than one dimension all the components of position commute with one another. Non-commutative position brackets, therefore, sometimes only require a re-interpretation of basic variables in order to map them to the standard commutative case. Their physical implications are then unclear.

A surprising mathematical result had initially led to the first branch of non-commutative geometry (Connes 1996). The Seely–DeWitt theorem (Seeley 1967) shows that the classical gravitational action is intimately related to properties of a Dirac operator defined on the spacetime manifold. A Dirac operator, as originally introduced by Dirac himself to be used in his equation of motion for fermions, is a first-order differential operator which squares to the Laplacian. In general, it is a matrix-valued operator, such that it can act on non-commutative, matrix-valued functions. If one sums up all its eigenvalues up to a certain upper bound, such as the Planck energy, one obtains a function of the upper bound that also depends on the geometry of the manifold on which the operator is defined. According to the Seely–DeWitt theorem, an expansion of this function by the inverse upper bound has curvature invariants as coefficients, the first one of which equals the Einstein–Hilbert action. Generalized versions of the Dirac operator can therefore be used to define new actions of non-commutative gravity. Cosmological implications have, for instance, been described by Sakellariadou (2012).

5.2.5 String Theory

The basic premise of string theory (Polchinski 1998; Lüst & Theisen 1989) is that elementary excitations of objects moving in spacetime are not described by point particles but by one-dimensional strings. In addition to the average position and momentum of a string, which approximate the motion of a point particle, a string

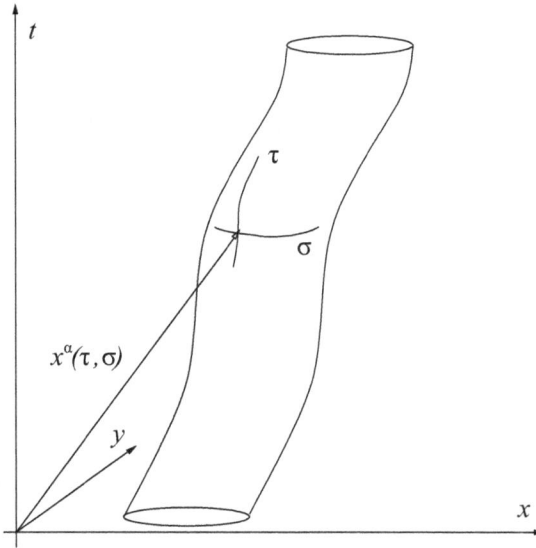

Figure 5.13. A closed string moving through spacetime $x^\alpha = (t, x, y, \ldots)$ sweeps out a cylindrical worldsheet, replacing worldlines of point particles. The worldsheet is parameterized by two coordinates, τ and σ, such that every event on it is described by a mapping $x^\alpha(\tau, \sigma)$.

can vibrate internally or wind itself around structures in spacetime or around topologically non-trivial dimensions. A single string can therefore describe a multitude of different phenomena in a unified way. Moreover, it turns out that the quantum description of an interacting string field is less divergent than the quantum field theory of point particle. For this reason, string theory may be able to describe quantum gravity in a consistent perturbative way without having to appeal to asymptotic safety.

As shown in Figure 5.13; a string sweeps out a two-dimensional worldsheet in spacetime, replacing the one-dimensional worldlines of point particles. Interactions can be visualized as worldsheets branching off, still forming smooth surfaces in contrast to the abrupt intersections of interacting timelike worldlines. We will describe here only the case of closed strings, while open strings behave similarly but require more detailed discussions of boundary conditions. Moreover, gravitational excitations in string theory are contained in the closed sector, while the open sector provides additional matter or radiation fields that may be used for unification of the fundamental interactions.

Worldsheet Actions

A two-dimensional worldsheet, embedded in spacetime with metric $g_{\alpha\beta}$, acquires an induces geometry on its surface. The surface metric h_{ab} can be derived using the mapping $x^\alpha(\tau, \sigma)$ that describes the embedding in terms of surface coordinates, $(\tau, \sigma) = y^a$, $a = 0$, 1: The chain rule implies

$$ds^2\big|_{x^\alpha = x^\alpha(y^a)} = \sum_{\alpha=0}^{3} \sum_{\beta=0}^{3} g_{\alpha\beta} dx^\alpha(y^a) dx^\beta(y^b)$$

$$= \sum_{\alpha=0}^{3} \sum_{\beta=0}^{3} g_{\alpha\beta} \left(\sum_{a=0}^{1} \frac{\partial x^\alpha}{\partial y^a} dy^a \right) \left(\sum_{b=0}^{1} \frac{\partial x^\beta}{\partial y^b} dy^b \right) \tag{5.49}$$

$$= \sum_{a=0}^{1} \sum_{b=0}^{1} h_{ab} dy^a dy^b$$

with

$$h_{ab} = \sum_{\alpha=0}^{3} \sum_{\beta=0}^{3} g_{\alpha\beta} \frac{\partial x^\alpha}{\partial y^a} \frac{\partial x^\beta}{\partial y^b}. \tag{5.50}$$

If the dynamics of a string is geometrical, as required for a string-description of spacetime, it should be determined by the induced metric h_{ab}. The *Polyakov action* expresses this dynamics with respect to a separate reference metric, m_{ab}, that has been chosen on the worldsheet and is unrelated to the spacetime metric $g_{\alpha\beta}$:

$$S[x^\alpha] = -\frac{1}{2\alpha'} \int_{-\infty}^{\infty} \int_0^1 \sqrt{-\det m} \sum_{a=0}^{1} \sum_{b=0}^{1} \sum_{\alpha=0}^{3} \sum_{\beta=0}^{3} m^{ab} \frac{\partial x^\alpha}{\partial y^a} \frac{\partial x^\beta}{\partial y^b} g_{\alpha\beta} d\tau d\sigma. \tag{5.51}$$

The *string tension*, $1/\alpha'$, traditionally replaces the inverse of Newton's constant as the multiplier of the action. As indicated by the integration ranges, we allow all real numbers for τ but restrict the bounded set of σ for a closed string to range from zero to one. The latter choice is arbitrary, but it can always be mapped to this form by suitably rescaling σ as well as the spatial metric $q = m_{11}$ defined by m_{ab} for constant τ, maintaining a fixed worldsheet circumference, $C = \int_0^1 \sqrt{q} d\sigma$.

Any action has units of $\hbar$, while the units of $\sqrt{-\det m}\, d\tau d\sigma$ and $m^{ab}(\partial/\partial y^a)(\partial/\partial y^b)$ cancel out for any two-dimensional metric, and the remaining $g_{\alpha\beta}(\partial x^\alpha)(\partial x^\beta)$ has units of length squared. The string tension therefore has units of $[\alpha'] = L^2/[\hbar]$, and it defines a length parameter $\ell_s = \sqrt{\hbar\alpha'}$ called the *string length*. Conceptually, this parameter is analogous to the Planck length $\ell_P = \sqrt{\hbar G/c^3}$, defined with Newton's constant that appears in the action of general relativity just like α' appears in the Polyakov action. A characteristic mass parameter of strings with tension $1/\alpha'$ is given by $m_s = \hbar/(c\ell_s) = \sqrt{\hbar/(c^2\alpha')}$.

The Polyakov action depends on the reference metric m_{ab} in addition to the worldsheet vector x^α. However, because there are no derivatives of m_{ab} in the action, its Euler–Lagrange equations obtained by varying m_{ab} can be solved algebraically, such that m_{ab} can be eliminated from the action. In order to take a derivative of the action by m_{ab}, it is useful to write the common expression for the inverse m^{ab} of a 2×2-matrix m_{cd} as

$$m^{ba} \det m = \sum_{c=0}^{1} \sum_{d=0}^{1} \epsilon^{ac} \epsilon^{bd} m_{cd} \tag{5.52}$$

with the completely antisymmetric tensor ϵ^{ab} in two dimensions, defined such that $\epsilon^{01} = 1$ and $\epsilon^{10} = -\epsilon^{01} = -1$, while $\epsilon^{00} = 0 = \epsilon^{11}$. Using these components, Equation (5.52) indeed implies the usual components of

$$\begin{pmatrix} m^{00} & m^{01} \\ m^{10} & m^{11} \end{pmatrix} = \begin{pmatrix} m_{00} & m_{01} \\ m_{10} & m_{11} \end{pmatrix}^{-1} = \frac{1}{\det m} \begin{pmatrix} m_{11} & -m_{01} \\ -m_{10} & m_{00} \end{pmatrix}. \tag{5.53}$$

The determinant of the inverse metric can therefore be written as

$$\det m^{-1} = \frac{1}{2} \sum_{a=0}^{1} \sum_{b=0}^{1} \sum_{c=0}^{1} \sum_{d=0}^{1} \epsilon_{ac} \epsilon_{bd} m^{ab} m^{cd}, \tag{5.54}$$

where ϵ_{ab} with subscript has *the same* components as ϵ^{ab}, in any coordinate system. This definition of components of ϵ_{ab} and ϵ^{ab} is compatible with the tensor transformation law (1.26) if ϵ_{ab} has density weight -1: The general tensor-transformation law for a 2-tensor with subscripts and density weight -1,

$$\epsilon_{a'b'} = \det(\partial y^{c'}/\partial y^{c}) \sum_{a=0}^{1} \sum_{b=0}^{1} \frac{\partial y^{a}}{\partial y^{a'}} \frac{\partial y^{b}}{\partial y^{b'}} \epsilon_{ab}, \tag{5.55}$$

implies

$$\epsilon_{a'b'} = \delta_{a}^{a} \delta_{b}^{b} \epsilon_{ab} \tag{5.56}$$

thanks to the definition of the determinant and antisymmetry of ϵ_{ab}. The additional determinant factor in Equation (5.55), compared with Equation (1.26), takes into account the density weight. (By definition, using the three-dimensional setting as an example, $\sqrt{\det q}$ transforms with density weight one, while $\mathrm{d}^3 x$ transforms with weight -1 such that $\sqrt{\det q}\, \mathrm{d}^3 x$ is invariant. The determinant factor in Equation (5.55) is therefore the same as in Equation (1.22).)

The basic derivatives

$$\frac{\partial m^{ab}}{\partial m^{cd}} = \frac{1}{2} \left(\delta_{c}^{a} \delta_{d}^{b} + \delta_{c}^{b} \delta_{d}^{a} \right) \tag{5.57}$$

for independent components of a symmetric matrix, $m^{ab} = m^{ba}$, then imply the derivative

$$\frac{\partial \sqrt{-\det m}}{\partial m^{ab}} = \frac{\partial}{\partial m^{ab}} \frac{1}{\sqrt{-\det m^{-1}}} = \frac{1}{2} (-\det m)^{3/2} \frac{\partial \det m^{-1}}{\partial m^{ab}}$$

$$= \frac{1}{2} (-\det m)^{3/2} \sum_{c=0}^{1} \sum_{d=0}^{1} \epsilon_{ac} \epsilon_{bd} m^{cd} = -\frac{1}{2} m_{ba} \sqrt{-\det m}, \tag{5.58}$$

using the subscript version of Equation (5.52).

These results imply that the functional derivative of $S[x^\alpha]$ by m^{ab} is equal to

$$\frac{\delta S[x^\alpha]}{\delta m^{ab}} = -\frac{\sqrt{-\det m}}{2\alpha'} \sum_{\alpha=0}^{3} \sum_{\beta=0}^{3} \left(-\frac{1}{2} m_{ba} \sum_{c=0}^{1} \sum_{d=0}^{1} m^{cd} \frac{\partial x^\alpha}{\partial y^c} \frac{\partial x^\beta}{\partial y^d} + \frac{\partial x^\alpha}{\partial y^a} \frac{\partial x^\beta}{\partial y^b} \right) g_{\alpha\beta}. \quad (5.59)$$

The right-hand side is zero according to Euler–Lagrange equations for the action (5.51), such that

$$m_{ba} \sum_{c=0}^{1} \sum_{d=0}^{1} \sum_{\alpha=0}^{3} \sum_{\beta=0}^{3} m^{cd} \frac{\partial x^\alpha}{\partial y^c} \frac{\partial x^\beta}{\partial y^d} g_{\alpha\beta} = 2 \sum_{\alpha=0}^{3} \sum_{\beta=0}^{3} \frac{\partial x^\alpha}{\partial y^a} \frac{\partial x^\beta}{\partial y^b} g_{\alpha\beta} \quad (5.60)$$

or

$$m_{ba} X = 2 \sum_{\alpha=0}^{3} \sum_{\beta=0}^{3} \frac{\partial x^\alpha}{\partial y^a} \frac{\partial x^\beta}{\partial y^b} g_{\alpha\beta} \quad (5.61)$$

where

$$X = \sum_{c=0}^{1} \sum_{d=0}^{1} \sum_{\alpha=0}^{3} \sum_{\beta=0}^{3} m^{cd} \frac{\partial x^\alpha}{\partial y^c} \frac{\partial x^\beta}{\partial y^d} g_{\alpha\beta}. \quad (5.62)$$

This result, in turn, implies that the integrand of Equation (5.51) takes the form

$$\sqrt{-\det m} \sum_{a=0}^{1} \sum_{b=0}^{1} \sum_{\alpha=0}^{3} \sum_{\beta=0}^{3} m^{ab} \frac{\partial x^\alpha}{\partial y^a} \frac{\partial x^\beta}{\partial y^b} g_{\alpha\beta}$$

$$= X\sqrt{-\det m} = \sqrt{-\frac{1}{2} \sum_{c=0}^{1} \sum_{d=0}^{1} \epsilon^{ab} \epsilon^{cd} X m_{ac} X m_{bd}} \quad (5.63)$$

$$= \sqrt{-2 \sum_{a=0}^{1} \sum_{b=0}^{1} \sum_{c=0}^{1} \sum_{d=0}^{1} \sum_{\alpha=0}^{3} \sum_{\beta=0}^{3} \sum_{\gamma=0}^{3} \sum_{\delta=0}^{3} \epsilon^{ac} \epsilon^{bd} \frac{\partial x^\alpha}{\partial y^a} \frac{\partial x^\beta}{\partial y^b} \frac{\partial x^\gamma}{\partial y^c} \frac{\partial x^\delta}{\partial y^d} g_{\alpha\beta} g_{\gamma\delta}},$$

using the subscript version of Equation (5.54). The right-hand side no longer depends on the reference metric m_{ab}.

Moreover, we can factorize the expression in the square root:

$$2 \sum_{a=0}^{1} \sum_{b=0}^{1} \sum_{c=0}^{1} \sum_{d=0}^{1} \sum_{\alpha=0}^{3} \sum_{\beta=0}^{3} \sum_{\gamma=0}^{3} \sum_{\delta=0}^{3} \epsilon^{ac} \epsilon^{bd} \frac{\partial x^{\alpha}}{\partial y^{a}} \frac{\partial x^{\beta}}{\partial y^{b}} \frac{\partial x^{\gamma}}{\partial y^{c}} \frac{\partial x^{\delta}}{\partial y^{d}} g_{\alpha\beta} g_{\gamma\delta}$$

$$= 2 \sum_{\alpha=0}^{3} \sum_{\beta=0}^{3} \sum_{\gamma=0}^{3} \sum_{\delta=0}^{3} \left(\sum_{a=0}^{1} \sum_{c=0}^{1} \epsilon^{ac} \frac{\partial x^{\alpha}}{\partial y^{a}} \frac{\partial x^{\gamma}}{\partial y^{c}} \right) \left(\sum_{b=0}^{1} \sum_{d=0}^{1} \epsilon^{bd} \frac{\partial x^{\beta}}{\partial y^{b}} \frac{\partial x^{\delta}}{\partial y^{d}} \right) g_{\alpha\beta} g_{\gamma\delta}$$

$$= 2 \sum_{\alpha=0}^{3} \sum_{\beta=0}^{3} \sum_{\gamma=0}^{3} \sum_{\delta=0}^{3} \left(\frac{\partial x^{\alpha}}{\partial \tau} \frac{\partial x^{\beta}}{\partial \sigma} - \frac{\partial x^{\alpha}}{\partial \sigma} \frac{\partial x^{\beta}}{\partial \tau} \right) \left(\frac{\partial x^{\gamma}}{\partial \tau} \frac{\partial x^{\delta}}{\partial \sigma} - \frac{\partial x^{\gamma}}{\partial \sigma} \frac{\partial x^{\delta}}{\partial \tau} \right) g_{\alpha\beta} g_{\gamma\delta}$$

$$= 4 \left(\left(\sum_{\alpha=0}^{3} \sum_{\beta=0}^{3} \frac{\partial x^{\alpha}}{\partial \tau} \frac{\partial x^{\beta}}{\partial \tau} g_{\alpha\beta} \right) \left(\sum_{\alpha=0}^{3} \sum_{\beta=0}^{3} \frac{\partial x^{\alpha}}{\partial \sigma} \frac{\partial x^{\beta}}{\partial \sigma} g_{\alpha\beta} \right) - \left(\sum_{\alpha=0}^{3} \sum_{\beta=0}^{3} \frac{\partial x^{\alpha}}{\partial \tau} \frac{\partial x^{\beta}}{\partial \sigma} g_{\alpha\beta} \right)^{2} \right).$$

Inserting the last two results in the Polyakov action (5.51) transforms it to the equivalent *Nambu–Goto action*

$$S[x^{\alpha}] = \frac{1}{\alpha'} \int_{-\infty}^{\infty} \int_{0}^{1} \sqrt{-\det h[x^{\alpha}]} \, d\tau d\sigma \qquad (5.64)$$

with the determinant of the induced metric h_{ab},

$$\det h[x^{\alpha}] = \left(\sum_{\alpha=0}^{3} \sum_{\beta=0}^{3} \frac{\partial x^{\alpha}}{\partial \tau} \frac{\partial x^{\beta}}{\partial \tau} g_{\alpha\beta} \right) \left(\sum_{\alpha=0}^{3} \sum_{\beta=0}^{3} \frac{\partial x^{\alpha}}{\partial \sigma} \frac{\partial x^{\beta}}{\partial \sigma} g_{\alpha\beta} \right)$$
$$- \left(\sum_{\alpha=0}^{3} \sum_{\beta=0}^{3} \frac{\partial x^{\alpha}}{\partial \tau} \frac{\partial x^{\beta}}{\partial \sigma} g_{\alpha\beta} \right)^{2}. \qquad (5.65)$$

While the Nambu–Goto action is independent of a reference metric, it is non-polynomial in x^{α} and therefore more complicated than the Polyakov action. It has, however, the advantage that it can be interpreted more easily: Its integrand is equal to the square root of the determinant of the induced metric h_{ab}, Equation (5.50), which determines the area element of a worldsheet embedded in the background spacetime with metric $g_{\alpha\beta}$. The Nambu–Goto action, and therefore the equivalent Ployakov action, are simply given by the area of the string worldsheet in the embedding spacetime, multiplied by the string tension.

Canonical Formulation
We return to the Polyakov action (5.51) and start a canonical analysis in preparation for quantization. To do so, we first write the reference metric m_{ab} on the worldsheet split into its time and space components, which is conveniently done in **ADM** form (Arnowitt et al. 1962) expressing its time-related components by a lapse function N and a one-dimensional shift vector M through the line element

$$ds^2 = -N^2 d\tau^2 + q(d\sigma + M d\tau)^2$$
$$= (-N^2 + qM^2)d\tau^2 + 2qM d\tau d\sigma + q d\sigma^2. \tag{5.66}$$

All functions—N, M and the one-dimensional spatial metric q—may be functions of τ and σ. Expressed as a matrix, the reference metric is given by

$$(m_{ab}) = \begin{pmatrix} -N^2 + qM^2 & qM \\ qM & q \end{pmatrix} \tag{5.67}$$

with determinant

$$\det(m_{ab}) = -N^2 q \tag{5.68}$$

and inverse

$$(m^{ab}) = \begin{pmatrix} -1/N^2 & M/N^2 \\ M/N^2 & 1/q - M^2/N^2 \end{pmatrix}. \tag{5.69}$$

The Polyakov action then takes the form

$$S[x^\alpha] = \frac{1}{2\alpha'} \int_{-\infty}^{\infty} \int_0^1 N\sqrt{q}$$
$$\times \sum_{\alpha=0}^{3} \sum_{\beta=0}^{3} \left(\frac{1}{N^2} \frac{\partial x^\alpha}{\partial \tau} \frac{\partial x^\beta}{\partial \tau} - 2\frac{M}{N^2} \frac{\partial x^\alpha}{\partial \tau} \frac{\partial x^\beta}{\partial \sigma} \right. \tag{5.70}$$
$$\left. - \left(\frac{1}{q} - \frac{M^2}{N^2} \right) \frac{\partial x^\alpha}{\partial \sigma} \frac{\partial x^\beta}{\partial \sigma} \right) g_{\alpha\beta} d\tau d\sigma,$$

and we compute the momenta of the components x^α as

$$P_\alpha = \frac{\delta S}{\delta(\partial x^\alpha/\partial \tau)} = \frac{\sqrt{q}}{\alpha' N} \sum_{\beta=0}^{3} g_{\alpha\beta} \left(\frac{\partial x^\beta}{\partial \tau} - M\frac{\partial x^\beta}{\partial \sigma} \right). \tag{5.71}$$

With these ingredients, a Legendre transformation takes us to the Hamiltonian,

$$\int_{-\infty}^{\infty} H_S d\tau = \int_{-\infty}^{\infty} \int_0^1 \sum_{\alpha=0}^{3} P_\alpha \frac{\partial x^\alpha}{\partial \tau} d\tau d\sigma - S[x^\alpha]$$
$$= \int_{-\infty}^{\infty} \int_0^1 \left(\frac{N}{2\sqrt{q}} \sum_{\alpha=0}^{3} \sum_{\beta=0}^{3} \left(\alpha' P_\alpha P_\beta g^{\alpha\beta} + \frac{1}{\alpha'} \frac{\partial x^\alpha}{\partial \sigma} \frac{\partial x^\beta}{\partial \sigma} g_{\alpha\beta} \right) \right. \tag{5.72}$$
$$\left. + M \sum_{\alpha=0}^{3} P_\alpha \frac{\partial x^\alpha}{\partial \sigma} \right) d\tau d\sigma.$$

Since the reference metric, and therefore the components N, M, and q, are arbitrary, physical implications should be independent of their choice. Therefore, the (functional) derivatives of H_S by N, M, and q should be zero. It is easy to take these derivatives because Equation (5.72) is linear in $N/\sqrt{q}$ and M. Moreover, the absence of any independent appearance of N and q other than in the form $N/\sqrt{q}$ implies that the three vanishing derivatives imply only two conditions:

$$H = \frac{1}{2} \sum_{\alpha=0}^{3} \sum_{\beta=0}^{3} \left(\alpha' p_\alpha p_\beta g^{\alpha\beta} + \frac{1}{\alpha'} \frac{\partial x^\alpha}{\partial\sigma} \frac{\partial x^\beta}{\partial\sigma} g_{\alpha\beta} \right) = 0 \tag{5.73}$$

and

$$D = \sum_{\alpha=0}^{3} p_\alpha \frac{\partial x^\alpha}{\partial\sigma} = 0 \tag{5.74}$$

when evaluated at any (τ, σ). In integral form, reinstating the multipliers $N/\sqrt{q}$ and M, these conditions are equivalent to requiring that

$$H[N/\sqrt{q}] = \frac{1}{2} \int_0^1 \frac{N}{\sqrt{q}} \sum_{\alpha=0}^{3} \sum_{\beta=0}^{3} \left(\alpha' p_\alpha p_\beta g^{\alpha\beta} + \frac{1}{\alpha'} \frac{\partial x^\alpha}{\partial\sigma} \frac{\partial x^\beta}{\partial\sigma} g_{\alpha\beta} \right) d\sigma = 0 \tag{5.75}$$

and

$$D[M] = \int_0^1 M \sum_{\alpha=0}^{3} p_\alpha \frac{\partial x^\alpha}{\partial\sigma} d\sigma = 0 \tag{5.76}$$

vanish for all functions $N/\sqrt{q}$ and M.

Hypersurface Deformations on a Worldsheet
We have used the same symbols H and D that in Section 2.3 denoted the normal and spatial generators of hypersurface deformations because the same brackets (2.154), (2.156), and (2.158) are fulfilled here. Computing Poisson brackets and integrating by parts shows that

$$\{H[N_1/\sqrt{q}], H[N_2/\sqrt{q}]\} = D[q^{-1}(N_1\partial N_2/\partial\sigma - N_2\partial N_1/\partial\sigma)] \tag{5.77}$$

$$\{H[N/\sqrt{q}], D[M]\} = H[(N/\sqrt{q})\partial M/\partial\sigma - M\partial(N/\sqrt{q})/\partial\sigma] \tag{5.78}$$

$$\{D[M_1], D[M_2]\} = D[M_1\partial M_2/\partial\sigma - M_2\partial M_1/\partial\sigma]. \tag{5.79}$$

Only Equation (5.78) differs slightly from the previous Equation (2.158) because the multiplier in H always appears in the form $N/\sqrt{q}$ which has a density weight of -1. (When deriving these Poisson brackets, one should keep in mind that the spacetime metric $g_{\alpha\beta}$ and inverse metric $g^{\alpha\beta}$ depend on x^γ, and therefore have non-zero Poisson brackets with p_δ.)

The two-dimensional nature of hypersurface deformations on a worldsheet greatly simplifies their behavior compared with the general form seen in Section 2.3. In particular, the inverse of the corresponding one-dimensional spatial metric is simply a factor of $1/q$ in Equation (5.77), rather than a tensor as in Equation (2.155). It is therefore possible to absorb the metric dependence in Equation (5.77) by using $N_1/\sqrt{q}$ and $N_2/\sqrt{q}$ instead of N_1 and N_2, respectively, as they indeed appear in Equation (5.75). (Spatial derivatives of q, which are initially obtained after replacing N_1 and N_2 with $N_1/\sqrt{q}$ and $N_2/\sqrt{q}$ in the right-hand side of Equation (5.77), cancel out in the antisymmetric combination.) It is therefore possible to express the worldsheet hypersurface-deformation brackets (5.77)–(5.79) of string theory in terms of a Lie algebra with constant (q-independent) coefficients, while higher-dimensional hypersurface deformations are not of this form. (They rather form a Lie algebroid.) Representations of Lie algebras then determine possible properties of quantizations.

In fact, in spite of the different forms of H[N] in Equation (5.75) and D[M] in Equation (5.76), the brackets involving H[N] can be written in a way closely related to the simpler D[M]. The bracket (5.79) belongs to the Lie algebra of one-dimensional diffeomorphisms because these are the spatial deformations on a two-dimensional worldsheet. This algebra is usually referred to as the *Virasoro algebra*. The bracket is unrelated to the dynamics because it only describes the mathematical behavior of deformations of space.

Nevertheless, it is possible to express also the dynamical generator H[$N/\sqrt{q}$] of time deformations in the same form if we apply a linear transformation of the generators from H[$N/\sqrt{q}$] and D[M] to

$$V^{\pm}[\lambda] = H[\lambda] \pm D[\lambda]. \tag{5.80}$$

It follows directly from Equations (5.77)–(5.79) that these new generators have brackets

$$\{V^{\pm}[\lambda_1],\ V^{\pm}[\lambda_2]\} = \pm 2V^{\pm}[\lambda_1\partial\lambda_2/\partial\sigma - \lambda_2\partial\lambda_1/\partial\sigma] \tag{5.81}$$

while

$$\{V^{\pm}[\lambda_1],\ V^{\mp}[\lambda_2]\} = 0. \tag{5.82}$$

The algebra of two-dimensional hypersurface deformations is therefore just a double copy of spatial deformations, described by the Virasoro algebra.

The generators of these two copies are usually expressed in "conformal gauge," setting $N = 1$, $M = 0$ and $q = 1$. In this gauge, the Virasoro generators have the simple form

$$V^{\pm}[\lambda] = \frac{1}{2\alpha'} \int_0^1 \lambda(\sigma) \sum_{\alpha=0}^{3} \sum_{\beta=0}^{3} g_{\alpha\beta}\left(\frac{\partial x^{\alpha}}{\partial\tau} \pm \frac{\partial x^{\alpha}}{\partial\sigma}\right)\left(\frac{\partial x^{\beta}}{\partial\tau} \pm \frac{\partial x^{\beta}}{\partial\sigma}\right)d\sigma \tag{5.83}$$

which integrates the square of the spacetime length of the vector $\partial\vec{x}/\partial\tau \pm \partial\vec{x}/\partial\sigma$. For our purposes, it will not be necessary to assume this gauge. We will instead continue to work with unrestricted hypersurface deformations on a worldsheet in order to highlight the relationships of string quantizations with the canonical treatment shown in Section 2.3.

Quantized Strings

For a given spacetime metric $g_{\alpha\beta}$, the Polyakov action (5.51) determines the dynamics of four functions on the worldsheet, given by $x^\alpha(\tau, \sigma)$, $\alpha = 0, 1, 2, 3$. The presence of an arbitrary worldsheet metric implies two constraints, $H[N/\sqrt{q}] = 0$ and $D[M] = 0$ or their linear combinations $V^\pm(\lambda) = 0$, which can be used to solve for two of the four functions (and their momenta). However, because the constraints depend on x^α only through derivatives of these functions by τ (via p_α) or σ, constant contributions to all four x^α are unconstrained. A string can therefore be strung at any event in spacetime.

Because constant contributions to x^α are unconstrained, while the constraints

$$H[1] = \frac{1}{2} \int_0^1 \sum_{\alpha=0}^3 \sum_{\beta=0}^3 \left(\alpha' p_\alpha p_\beta g^{\alpha\beta} + \frac{1}{\alpha'} \frac{\partial x^\alpha}{\partial \sigma} \frac{\partial x^\beta}{\partial \sigma} g_{\alpha\beta} \right) \frac{d\sigma}{\sqrt{q}} = 0 \qquad (5.84)$$

and

$$D[1] = \int_0^1 \sum_{\alpha=0}^3 p_\alpha \frac{\partial x^\alpha}{\partial \sigma} d\sigma = 0 \qquad (5.85)$$

evaluated with constant multipliers $N/\sqrt{q}$ and M, respectively, do not identically vanish, two integrated worldsheet constraints remain even if we eliminate two of the four functions x^α. These constraints should be imposed explicitly on the two functions that remain of the x^α.

In order to solve the remaining constraints, we parameterize the canonical functions $x^\alpha(\sigma)$ and $p_\alpha(\sigma)$ on an initial worldsheet slice of constant τ in terms of their Fourier modes on a circle given by the range of σ. Using a generic spatial worldsheet metric q, Fourier decomposition is defined in terms of the functions

$$e_n(\sigma) = \frac{1}{\sqrt{C}} \exp\left(2\pi i n C^{-1} \int_0^\sigma \sqrt{q}\, d\tilde\sigma \right) \qquad (5.86)$$

with integer n and the worldsheet circumference $C = \int_0^1 \sqrt{q}\, d\sigma$, such that

$$\int_0^1 e_{n_1}(\sigma)^* e_{n_2}(\sigma) \sqrt{q}\, d\sigma = \frac{1}{2\pi} \int_0^{2\pi} \exp(-i(n_1 - n_2)\phi)\, d\phi = \delta_{n_1, n_2} \qquad (5.87)$$

after substituting $\phi(\sigma) = 2\pi C^{-1} \int_0^\sigma \sqrt{q}\, d\tilde\sigma$.

Expressed in this orthonormal basis of functions on the circle, we write

$$x^\alpha(\sigma) = \sum_{n=-\infty}^\infty x_n^\alpha e_n(\sigma) \qquad (5.88)$$

and the momentum functions

$$p_\alpha(\sigma) = \sqrt{q} \sum_{n=-\infty}^\infty p_{n\alpha} e_{-n}(\sigma) \qquad (5.89)$$

with spatial density weight one. Signs and coefficients in these expressions are chosen such that $(x_n^\alpha, p_{n\alpha})$ are canonically conjugate for any fixed n and α:

$$\{x_n^\alpha, p_{m\beta}\} = \left\{\int_0^1 e_n(\sigma_1)^* x^\alpha(\sigma_1)\sqrt{q(\sigma_1)}\,d\sigma_1, \int_0^1 e_{-m}(\sigma_2)^* p_\beta(\sigma_2)d\sigma_2\right\}$$

$$= \int_0^1 \int_0^1 e_n(\sigma_1)^* e_m(\sigma_2)\delta(\sigma_1 - \sigma_2)\sqrt{q(\sigma_1)}\,d\sigma_1 d\sigma_2 \delta_\beta^\alpha \tag{5.90}$$

$$= \delta_{n,m}\delta_\beta^\alpha.$$

Because $x^\alpha(\sigma)$ and $p_\alpha(\sigma)$ take real values, we have

$$x_{-n}^\alpha = (x_n^\alpha)^* \quad \text{and} \quad p_{-n\alpha} = p_{n\alpha}^*. \tag{5.91}$$

Only the unrestricted constant modes, $p_{0\alpha}$ and x_0^α which are real, and the positive modes with $n > 0$ are therefore independent (and in general complex). The latter are subject to the remaining constraints, which we are able to implement now.

The Virasoro constraints $V^\pm[\lambda] = 0$ imply that the two spacetime vectors $\mathbf{v}_\pm$ with components

$$v_\pm^\alpha = \sqrt{\alpha'}\sum_{\beta=0}^3 g^{\alpha\beta}p_\beta \pm \frac{1}{\sqrt{\alpha'}}\frac{\partial x^\alpha}{\partial\sigma} \tag{5.92}$$

are lightlike, $\sum_{\alpha=0}^3\sum_{\beta=0}^3 v_\pm^\alpha v_\pm^\beta g_{\alpha\beta} = 0$. Particular solutions of these conditions are given by $\mathbf{v}_\pm = 0$, or

$$\sum_{\beta=0}^3 g^{\alpha\beta}p_\beta = \mp\frac{1}{\alpha'}\frac{\partial x^\alpha}{\partial\sigma} \tag{5.93}$$

for so-called left-moving and right-moving string excitations. Using our mode decompositions of $p_\alpha(\sigma)$ and $x^\alpha(\sigma)$, we obtain

$$\sum_{\beta=0}^3 g^{\alpha\beta}\sum_{n=-\infty}^\infty \sqrt{q}\,p_{n\beta}e_{-n}(\sigma) = \mp\frac{2\pi i}{C\alpha'}\sum_{n=-\infty}^\infty n x_n^\alpha\sqrt{q}\,e_n(\sigma) \tag{5.94}$$

from Equation (5.93). Comparing coefficients, we derive the equations

$$\sum_{\beta=0}^3 g^{\alpha\beta}p_{n\beta} = \pm\frac{2\pi in}{C\alpha'}x_{-n}^\alpha \tag{5.95}$$

which allow us to solve for the imaginary components

$$\sum_{\beta=0}^3 g^{\alpha\beta}\,\text{Im}\,p_{n\beta} = \pm\frac{2\pi n}{C\alpha'}\,\text{Re}\,x_{-n}^\alpha = \pm\frac{2\pi n}{C\alpha'}\,\text{Re}\,x_n^\alpha \tag{5.96}$$

and

$$\text{Im } x_n^\alpha = \pm \frac{C\alpha'}{2\pi n} \sum_{\beta=0}^{3} g^{\alpha\beta} \text{ Re } p_{-n\beta} = \pm \frac{C\alpha'}{2\pi n} \sum_{\beta=0}^{3} g^{\alpha\beta} \text{ Re } p_{n\beta}. \tag{5.97}$$

The constraint (5.84) expressed in terms of modes x_n^α and $p_{n\alpha}$ is given by

$$\begin{aligned}
H[1] &= \frac{1}{2}\alpha' \sum_{\alpha=0}^{3} \sum_{\beta=0}^{3} \sum_{n=-\infty}^{\infty} \left(p_{n\alpha}p_{-n\beta} g^{\alpha\beta} + \frac{4\pi^2 n^2}{C^2\alpha'^2} x_n^\alpha x_{-n}^\beta g_{\alpha\beta} \right) \\
&= \frac{1}{2}\alpha' \sum_{\alpha=0}^{3} \sum_{\beta=0}^{3} \left(p_{0\alpha}p_{0\beta} g^{\alpha\beta} + 2 \sum_{n=1}^{\infty} \left(p_{n\alpha}p_{n\beta}^* g^{\alpha\beta} + \frac{4\pi^2 n^2}{C^2\alpha'^2} x_n^\alpha (x_n^\beta)^* g_{\alpha\beta} \right) \right).
\end{aligned} \tag{5.98}$$

In the second step, we have used the reality conditions (5.91) and separated the zero mode of p_α, while the zero mode of x^α is eliminated by a factor of n^2.

Mass Spectrum
Since the constraint $H[1] = 0$ has to vanish, we can use Equation (5.98) to compute the allowed string mass values m, which are related to the average momentum

$$P_\alpha = \int_0^1 p_\alpha(\sigma)\mathrm{d}\sigma = \sqrt{C} p_{0\alpha} \tag{5.99}$$

carried by the string, using

$$\int_0^1 e_n(\sigma)\sqrt{q}\,\mathrm{d}\sigma = C^{-1/2} \int_0^1 \sqrt{q}\,\mathrm{d}\sigma\delta_{n,0} = \sqrt{C}\delta_{n,0}. \tag{5.100}$$

The mass is therefore relativistically determined by the zero mode of the momentum, $p_{0\alpha}$, via

$$-m^2 c^2 = \|P\|^2 = C \sum_{\alpha=0}^{3} \sum_{\beta=0}^{3} p_{0\alpha}p_{0\beta} g^{\alpha\beta}. \tag{5.101}$$

Using the constraint (5.98); we obtain

$$m^2 c^2 = 2C \sum_{\alpha=0}^{3} \sum_{\beta=0}^{3} \sum_{n=1}^{\infty} \left(p_{n\alpha}p_{n\beta}^* g^{\alpha\beta} + \frac{4\pi^2 n^2}{C^2\alpha'^2} x_n^\alpha (x_n^\beta)^* g_{\alpha\beta} \right). \tag{5.102}$$

If the string is moving in Minkowski spacetime, with constant $g_{\alpha\beta} = \eta_{\alpha\beta}$, the right-hand side is a series of harmonic-oscillator Hamiltonians, which can easily be quantized. (In curved spacetimes, the metric components $g_{\alpha\beta}$ are functions of x^α, implying complicated interacting Hamiltonians.) Because of the indefinite sign of norms in Lorentzian signature, one of the mode series contributes a negative Hamiltonian, which is in danger of violating positivity required of the left-hand side of Equation (5.102), $m^2 c^2$. However, in Equation (5.102) we have so far written all modes even though some of the coefficients can be eliminated by solving the constraints H[$N/\sqrt{q}$] and D[M] with non-constant $N/\sqrt{q}$ and M. If we use the two constraints to solve for the time components x_n^0 and one of the spacelike ones, say x_n^1, the corresponding momenta, p_{n0} and p_{n1}, will be eliminated by the corresponding gauge transformations. (This solution procedure is usually performed by choosing a specific gauge, called the light-cone gauge.) The time component of the zero-mode p_{00} remains unrestricted because we are eliminating components based only on constraints with non-constant N and M, while the constant $N = 1$ gave us the mass condition (5.102).

Eliminating constrained modes, we obtain the mass condition

$$m^2c^2 = 2C \sum_{\alpha=2}^{3} \sum_{n=1}^{\infty} \left(|p_{na}|^2 + \frac{4\pi^2n^2}{C^2\alpha'^2}|x_n^\alpha|^2 \right) \tag{5.103}$$

which is guaranteed to be non-negative. In this expression, we have used the Euclidean metric for the remaining spatial components, $\alpha = 2$ and $\alpha = 3$ still assuming that the string is moving in Minkowski spacetime. The complex-valued p_{na} and x_n^α amount to two degrees of freedom for fixed n and α, which can be shown explicitly by performing a canonical transformation to their real and imaginary parts (divided by $\sqrt{2}$ to ensure the canonical nature of the transformation),

$$x_n^{\alpha,\,1} = \frac{1}{\sqrt{2}}(x_n^\alpha + (x_n^\alpha)^*), \quad x_n^{\alpha,\,2} = \frac{1}{\sqrt{2}\,i}(x_n^\alpha - (x_n^\alpha)^*) \tag{5.104}$$

canonically conjugate to

$$p_{na,1} = \frac{1}{\sqrt{2}}\left(p_{na} + p_{na}^*\right), \quad p_{na,2} = \frac{1}{\sqrt{2}\,i}\left(p_{na} - p_{na}^*\right). \tag{5.105}$$

Therefore,

$$2\left(|p_{na}|^2 + \frac{4\pi^2n^2}{C^2\alpha'^2}|x_n^\alpha|^2\right) = p_{na,\,1}^2 + \frac{4\pi^2n^2}{C^2\alpha'^2}\left(x_n^{\alpha,\,1}\right)^2 + p_{na,\,2}^2 + \frac{4\pi^2n^2}{C^2\alpha'^2}\left(x_n^{\alpha,\,2}\right)^2. \tag{5.106}$$

The two oscillators for fixed n and α are identified as a single one if we use the conditions (5.96) and (5.97) imposed by the restriction to left-moving and right-moving modes:

$$2\left(|p_{na}|^2 + \frac{4\pi^2n^2}{C^2\alpha'^2}|x_n^\alpha|^2\right)\Bigg|_{\text{left/right-moving}} = 2\left(p_{na,\,1}^2 + \frac{4\pi^2n^2}{C^2\alpha'^2}(x_n^{\alpha,\,1})^2\right). \tag{5.107}$$

When quantized, each harmonic oscillator with Hamiltonian H $= \frac{1}{2}m_{\text{nonrel}}^{-1}p^2 + \frac{1}{2}m_{\text{nonrel}}\nu^2x^2$ has energy eigenvalues $E_j = (j + \frac{1}{2})\hbar\nu$, where ν is the frequency and m_{nonrel} the non-relativistic mass in the quantum-mechanics analogy. Here, according to Equation (5.107), the non-relativistic mass is $m_{\text{nonrel}} = 1$, while the actual relativistic mass m of the string is determined by the harmonic energies, Equation (5.103). The frequency according to Equation (5.107) is mode-dependent, $\nu = 2\pi n/(C\alpha')$. The mass spectrum given by a quantization of Equation (5.103) is therefore

$$m_j^2 c^2 = \frac{4\pi\hbar}{\alpha'} \sum_{\alpha=2}^{3} \sum_{n=1}^{\infty} \left(j_{n\alpha} + \frac{1}{2} \right) n \tag{5.108}$$

with infinitely many quantum numbers, $j_{n\alpha}$, which are non-negative integers. Notice that the string circumference C has canceled out. The quantum result is therefore independent of the reference metric on the worldsheet.

Using the string mass $m_s = \sqrt{\hbar/(c^2\alpha')}$, we can write

$$m_j^2 = 4\pi m_s^2 \sum_{\alpha=2}^{3} \sum_{n=1}^{\infty} \left(j_{n\alpha} + \frac{1}{2} \right) n. \tag{5.109}$$

The mass spectrum is therefore quantized, with a characteristic scale given by the string mass. However, Equation (5.109) has a problem because the contribution $\sum_{n=1}^{\infty} n$ is infinite, summing the zero-point energies (3.84) of infinitely many harmonic oscillators.

Such infinities are always produced in an oscillator treatment of quantum field theories because the classical picture of a field as infinitely many oscillators does not tolerate a direct transfer to a quantum theory. One possibility that can be applied here is to "regularize" the infinite sum by interpreting it as the value of an analytic continuation of the Riemann ζ-function, defined for positive integers z as $\zeta(z) = \sum_{n=1}^{\infty} n^{-z}$, to the non-positive $z = -1$ through a continuous curve of complex values z. This procedure leads to a finite value, $\zeta(-1) = -1/12$. We therefore obtain the regularized mass spectrum

$$m_j^2 = 4\pi m_s^2 \sum_{\alpha=2}^{3} \left(\sum_{n=1}^{\infty} j_{n\alpha} n - \frac{1}{24} \right). \tag{5.110}$$

This result is finite for finitely-excited string states (that is, states in which only finitely many $j_{n\alpha}$ are non-zero). It also shows that low excitations (small $j_{n\alpha}$) are relevant because any excitation that produces a non-zero $\sum_{\alpha=2}^{3} \left(\sum_{n=1}^{\infty} j_{n\alpha} n - 1/24 \right)$ results in a mass on the order of m_s. If the string length is microscopic, which in quantum-gravitational physics means that it is within a few orders of magnitude of the Planck length, such excitations describe Planck-mass particles that are not relevant for current particle phenomenology. We should therefore analyze low-lying excitations.

An analysis of low-lying excitations reveals two further problems: (i) The ground state, obtained for all $j_{n\alpha} = 0$, has a negative m_j^2. (It is called a tachyon.) The quantization is therefore unstable because a relativistic field with negative mass squared has a potential unbounded from below, such as $\frac{1}{2}c^2 m^2 \phi^2/\hbar^2 < 0$ for a scalar field ϕ. (ii) There are no massless states and therefore no gravitons that could quantize gravitational waves, which travel at the speed of light. This problem indicates that the quantization violates covariance because mass terms for the graviton, described classically by perturbations of the metric around a background spacetime, are prohibited by general covariance: As shown by a general analysis, the contributions to a metric action consistent with general covariance do not include a term quadratic in the metric; see Chapter 2.

The second problem can be solved by introducing extra dimensions. If the string moves in d-dimensional spacetime instead of our usual four dimensions, the sum over α in Equation (5.110) runs from $\alpha = 2$ to $\alpha = d$. The subtraction of 1/24 is then replaced by a subtraction of $(d - 2)/24$. If $d = 26$, the singly excited states with just one non-zero $j_{1\alpha} = 1$ are then massless and can be used to construct graviton states. Recall that, according to Equation (5.107), a single quantum number $j_{n\alpha}$ represents two oscillators for left-moving and right-moving modes. States with the two quantum numbers $j_{1\alpha 1} = j_{1\alpha 2} = 1$ can therefore be used to construct excitations of a two-index tensor. It is interesting to note that all numerical factors, including $\zeta(-1) = -1/12$, work out just in the right way for an integer dimension to result.

The tachyonic ground state remains in the spectrum even with extra dimensions, but it turns out that it can be eliminated by introducing supersymmetry. The string coordinates $x^{\alpha}(\sigma)$ are then combined with suitable fermion fields on the worldsheet, which can be transformed into one another while preserving the equations of motion. Combining bosonic and fermionic versions of oscillators leads to a completely non-negative mass spectrum. (The introduction of supersymmetry also reduces the required number of spacetime dimensions from $d = 26$ to $d = 10$.) In this way, string theory is led to extra dimensions and supersymmetry by implementing important consistency conditions, related to covariance and stability, that have not been addressed in such detail in other approaches to quantum gravity.

Compact Extra Dimensions
The need to include extra dimensions and supersymmetry in string quantizations greatly multiplies the number of independent degrees of freedom, not only on the worldsheet but also in effective spacetime theories for the background metric $g_{\alpha\beta}$ of the embedding spacetime. Because we perceive only four spacetime dimensions, the six surplus dimensions cannot lead to observable extensions or motion in their directions. One possibility is that they are compact like circles and of small circumference, such that we cannot resolve different points on them. They may nevertheless have implications on potentially observable spectra because a string moving in a tiny compact dimension would have momentum that contributes to its energy, even if we cannot directly detect its motion. Moreover, components of the spacetime metric in directions of these extra dimensions may in general depend on the four spacetime coordinates we know. From our perspective, they therefore represent independent fields in addition to our four-dimensional spacetime metric.

The existence of a compact spatial extra dimension, here referred to simply as x, has two implications, one of which is general and one is characteristic of strings. The first is that any wave function $\psi(x)$ that describes a dynamical quantum object, be it a particle or a string, must be such that it implies a periodic probability density,

$|\psi(x)|^2$, in this direction. (In general, the wave function may also depend on the other dimensions, but we will explicitly write only the relevant, compact one.)

If a compact extra dimension has circumference $C = 2\pi R$, as measured with a flat background metric $g_{\alpha\beta}$, any wave function consistent with the periodicity condition $|\psi(x + 2\pi R)|^2 = |\psi(x)|^2$ must be a superposition of plane waves

$$\psi_N(x) = \frac{1}{\sqrt{2\pi R}} e^{i(N+\epsilon)x/R} \tag{5.111}$$

with integer n and some fixed $0 \leqslant \epsilon < 1$, as in Equation (4.53) where a momentum representation was used.

These wave functions define momentum eigenstates:

$$\hat{P}\psi_N(x) = \frac{\hbar}{i} \frac{d\psi_N(x)}{dx} = (N + \epsilon)\frac{\hbar}{R}\psi_N(x) \tag{5.112}$$

with eigenvalues

$$P_N = (N + \epsilon)\frac{\hbar}{R}. \tag{5.113}$$

Time-reversal symmetry requires that for any eigenvalue P_N in the momentum spectrum, $-P_N$ is also an eigenvalue. Therefore, the two values $\epsilon = 0$ and $\epsilon = 1/2$ are possible. Only the first value, $\epsilon = 0$, includes eigenstates with zero momentum and is therefore preferred. (However, even for $\epsilon = 1/2$, states with zero momentum expectation value but non-zero momentum fluctuations exist, such as $(\psi_0 + \psi_{-1})/\sqrt{2}$.)

Non-zero momentum in an invisible compact dimension contributes to the mass spectrum of quantized strings because it implies additional energy in the relativistic equation $E^2 = m^2c^4 + |\vec{P}|^2 c^2$. If the corresponding motion is invisible, we would attribute its energy to an additional rest mass. Using $\epsilon = 0$, the mass spectrum (5.109) is therefore modified to

$$\begin{aligned}
m_{j,N}^2 &= \frac{N^2\hbar^2}{c^2R^2} + 4\pi m_s^2 \sum_{\alpha=2}^{d} \sum_{n=1}^{\infty} \left(j_{n\alpha} + \frac{1}{2}\right)n \\
&= m_s^2\left(N^2\frac{\ell_s^2}{R^2} + 4\pi \sum_{\alpha=2}^{d} \sum_{n=1}^{\infty} \left(j_{n\alpha} + \frac{1}{2}\right)n\right).
\end{aligned} \tag{5.114}$$

The new quantum number N describes the momentum of a string moving in a compact dimension. In addition, a string can wind around a compact dimension, which implies another energy contribution because the string has tension $1/\alpha'$. If a string winds around an extra dimension with coordinate $x = x^\alpha$ for some fixed α, the corresponding spacetime image of the worldsheet, $x(\tau, \sigma)$ at fixed τ, has an extra term compared with the previous mode decomposition (5.88): We now have

$$x(\sigma) = 2\pi R M \sigma + \sum_{n=-\infty}^{\infty} x_n e_n(\sigma) \tag{5.115}$$

with the winding number M, an integer. The function $x(\sigma)$ for a coordinate on a compact dimension is no longer required to be periodic in σ even for a closed string because this dimension itself is periodic. The coordinate values x and $x + 2\pi R M$, for any integer M, therefore describe the same point in spacetime, and the string is indeed closed.

The σ-derivative of $x(\sigma)$ then has a new constant contribution $2\pi R M$, which contributes a term

$$\frac{1}{c^2 \alpha'^2}\left(\frac{\partial x}{\partial \sigma}\right)^2 = \left(\frac{2\pi R M}{c\alpha'}\right)^2 = \left(\frac{2\pi R M m_s}{\ell_s}\right)^2 \tag{5.116}$$

to the mass spectrum derived from the mass condition (5.101) together with the constraint (5.84). (In a flat compact direction, we may assume a constant $\sqrt{q} = C$, such that these factors cancel out in a combination of Equation (5.101) with Equation (5.84). If $\sqrt{q}$ is not constant, we should replace $2\pi\sigma$ in Equation (5.115) with the angle $\phi(\sigma) = 2\pi C^{-1} \int_0^\sigma \sqrt{q}\,d\tilde{\sigma}$, which has also been used to show orthonormality of the mode functions in Equation (5.87).)

Combining Equation (5.114) with Equation (5.116), the full mass spectrum in the presence of a compact extra dimension is therefore

$$m_{j,\,N,\,M}^2 = m_s^2\left(N^2 \frac{\ell_s^2}{R^2} + 4\pi^2 M^2 \frac{R^2}{\ell_s^2} + 4\pi \sum_{\alpha=2}^{d} \sum_{n=1}^{\infty}\left(j_{n\alpha} + \frac{1}{2}\right)n\right). \tag{5.117}$$

Its invariance under the transformation $N \leftrightarrow M$ combined with $R \mapsto \ell_s^2/(2\pi R)$ is called *T-duality* (Sathiapalan 1987; Candelas et al. 1985). Based on the mass spectrum (and, as it turns out, all other string observables), it is impossible to distinguish between motion in an extra dimension of circumference $2\pi R$ and winding around an extra dimension with circumference ℓ_s^2/R.

This symmetry may be interpreted as implying a minimal non-zero length: If we start with an extra dimension of large detectable circumference R, we can see motion (momentum modes with quantum numbers N) but winding modes require too much energy to be excited. As the circumference is decreased by some process, it will be more and more difficult to perceive motion because we cannot resolve different points on the compact dimension, but it will become easier to excite and observe energy from winding modes. In fact, trying to increase the position resolution requires more energy so as to have a sufficiently small Compton wave length. This energy will, for sufficiently small R, start exciting winding modes. These additional modes will eventually dominate the mass spectrum, effectively applying a transformation of *T*-duality. The transition between perceiving momentum modes and perceiving winding modes happens when *T*-duality acts as an identity on the radius

R, that is $R = \ell_s^2/(2\pi R)$ or $R = \ell_s/\sqrt{2\pi}$. The string length, up to a factor of $1/\sqrt{2\pi}$, therefore determines the minimal resolvable distance.

In a more dynamical picture, T-duality has been used to suggest a cosmological scenario of pre-Big Bang cosmology (Veneziano 1991, 1999; Gasperini & Veneziano 1993, 2003). The T-dual phase is then viewed as a collapsing universe before the Big Bang.

Non-metric Fields

Eliminating the reference metric from the Polyakov action (5.51) shows, via the Nambu–Goto action (5.64); that the string action is proportional to the area swept out by the string in a background spacetime. In Euclidean space, it is possible to determine the area of a surface in two different ways: Describing the surface by a mapping $x^i(y^a)$ from surface coordinates y^a, $a = 1, 2$, to space coordinates x^i, $i = 1, 2, 3$, we have the area element

$$\mathrm{d}A = \frac{1}{2} \sum_{i=1}^{3} \sum_{j=1}^{3} \sum_{a=1}^{2} \sum_{b=1}^{2} \delta_{ij} \frac{\partial x^i}{\partial y^a} \frac{\partial x^j}{\partial y^b} \delta^{ab} \, d\tau \, d\sigma. \tag{5.118}$$

When the spatial metric δ_{ij} is replaced by an arbitrary metric g_{ij} and the surface metric δ^{ab} by the reference metric m^{ab}, the area element corresponds to the integrand of the Polyakov action.

Alternatively, we can obtain the area element from a cross product of the two independent vectors $\partial \vec{x}/\partial y^1$ and $\partial \vec{x}/\partial y^2$ tangent to the surface:

$$\mathrm{d}A = \frac{1}{2} \sum_{i=1}^{3} \sum_{j=1}^{3} \sum_{k=1}^{3} \sum_{a=1}^{2} \sum_{b=1}^{2} \epsilon_{ijk} n^k \frac{\partial x^i}{\partial y^a} \frac{\partial x^j}{\partial y^b} \epsilon^{ab} \, d\tau \, d\sigma \tag{5.119}$$

where $\vec{n}$ is a unit vector normal to the surface, and ϵ^{ab} and ϵ_{ijk} are the completely antisymmetric tensors on the surface and in space, respectively. Here, the background field in space, $b_{ij} = \epsilon_{ijk} n^k$, is an antisymmetric two-tensor instead of a symmetric two-tensor δ_{ij} or $g_{\alpha\beta}$ in the first version of the area element.

The Polyakov action, making use only of a symmetric background tensor $g_{\alpha\beta}$ in spacetime, therefore relies on a restrictive assumption that the corresponding $b_{\alpha\beta} = 0$ vanishes. A general string action (in d spacetime dimensions) would include both options,

$$S[x^\alpha] = -\frac{1}{2\alpha'} \int_{-\infty}^{\infty} \int_{0}^{1} \left(\sqrt{-\det m} \sum_{a=0}^{1} \sum_{b=0}^{1} \sum_{\alpha=0}^{d} \sum_{\beta=0}^{d} m^{ab} g_{\alpha\beta} + \epsilon^{ab} b_{\alpha\beta} \right)$$

$$\frac{\partial x^\alpha}{\partial y^a} \frac{\partial x^\beta}{\partial y^b} d\tau d\sigma. \tag{5.120}$$

(The tensor ϵ^{ab}, defined to take the value $\epsilon^{01} = 0$ in any surface coordinate system, has density weight one, which can be shown in the same way as we derived the density weight -1 of ϵ_{ab} in Equation (5.55). Therefore, the second term need not be multiplied by $\sqrt{-\det m}$.) Like $g_{\alpha\beta}$, $b_{\alpha\beta}$ does not have units if we assume that the spacetime coordinates x^α are defined with length units, as we will do in what follows for compact dimensions.

This new action implies that x^α has momentum

$$P_\alpha = \frac{\delta S}{\delta(\partial x^\alpha/\partial\tau)} = \frac{1}{\alpha'}\sum_{\beta=0}^{d}\left(\frac{\sqrt{q}}{N}g_{\alpha\beta}\left(\frac{\partial x^\beta}{\partial\tau} - M\frac{\partial x^\beta}{\partial\sigma}\right) + b_{\alpha\beta}\frac{\partial x^\beta}{\partial\sigma}\right). \tag{5.121}$$

As in the case of an electric charge moving in a background vector potential, Equation (5.39), the canonical momentum differs from the kinematical momentum

$$p_\alpha^{\text{kin}} = \frac{\delta S}{\delta(\partial x^\alpha/\partial\tau)} = \frac{\sqrt{q}}{\alpha'N}\sum_{\beta=0}^{d}g_{\alpha\beta}\left(\frac{\partial x^\beta}{\partial\tau} - M\frac{\partial x^\beta}{\partial\sigma}\right) \tag{5.122}$$

determined by lapse and shift. Also as in the electric example (5.40), the Hamiltonian (constraint) derived from Equation (5.120) is independent of $b_{\alpha\beta}$ when expressed in terms of the kinematical momentum:

$$\text{H}[1] = \frac{1}{2}\int_0^1\sum_{\alpha=0}^{d}\sum_{\beta=0}^{d}\left(\alpha'p_\alpha^{\text{kin}}p_\beta^{\text{kin}}g^{\alpha\beta} + \frac{1}{\alpha'}\frac{\partial x^\alpha}{\partial\sigma}\frac{\partial x^\beta}{\partial\sigma}g_{\alpha\beta}\right)\frac{\text{d}\sigma}{\sqrt{q}} = 0. \tag{5.123}$$

Therefore, if we expand only the kinematical momentum in terms of modes, as in Equation (5.89), we can use the general expression of a mass spectrum as previously derived from the constraint.

In canonical language, the antisymmetric field $b_{\alpha\beta}$ appears mainly in the modified canonical momentum. The canonical momentum determines the quantized values (5.113) in the presence of compact extra dimensions because it is represented on wave functions as a derivative by the corresponding spacetime coordinate. However, the kinematical momentum appears in the constraint (5.123), and therefore in the mass spectrum derived from it. The latter is therefore modified in the presence of a non-zero antisymmetric field.

In order to derive the complete mass spectrum, we assume that $g_{\alpha\beta}$ and $b_{\alpha\beta}$ are constant, generalizing our previous derivation in flat spacetime. We use the quantized canonical momentum components

$$P_\alpha = N_\alpha\frac{\hbar}{R} \tag{5.124}$$

where N_α may be non-zero for any compact dimension α with circumference $2\pi R$. As before, see Equation (5.99), we have

$$P_\alpha = \int_0^1 P_\alpha(\sigma)\text{d}\sigma = \sqrt{C}p_{0\alpha}^{\text{kin}} + \frac{1}{\alpha'}\sum_{\beta=0}^{d}b_{\alpha\beta}\oint\text{d}x^\beta = P_\alpha^{\text{kin}} + \frac{2\pi R}{\alpha'}\sum_{\beta=0}^{d}b_{\alpha\beta}M^\beta, \tag{5.125}$$

introducing the integer winding numbers M^β along compact dimensions of circumference $2\pi R$.

Combining Equations (5.124) and (5.125), we obtain

$$P_\alpha^{\text{kin}} = N_\alpha \frac{\hbar}{R} - \frac{2\pi R}{\alpha'} \sum_{\beta=0}^{d} b_{\alpha\beta} M^\beta. \tag{5.126}$$

As in Equation (5.114), the mass spectrum acquires a contribution of

$$\frac{1}{c^2} \sum_{\alpha=0}^{d} \sum_{\beta=0}^{d} P_\alpha^{\text{kin}} P_\beta^{\text{kin}} g^{\alpha\beta}$$

$$= \sum_{\alpha=0}^{d} \sum_{\beta=0}^{d} g^{\alpha\beta} \left(N_\alpha \frac{\hbar}{cR} - \frac{2\pi R}{c\alpha'} \sum_{\beta=0}^{d} b_{\alpha\gamma} M^\gamma \right) \left(N_\alpha \frac{\hbar}{cR} - \frac{2\pi R}{c\alpha'} \sum_{\beta=0}^{d} b_{\alpha\delta} M^\delta \right) \tag{5.127}$$

$$= m_s^2 \sum_{\alpha=0}^{d} \sum_{\beta=0}^{d} g^{\alpha\beta} \left(N_\alpha \frac{\ell_s}{R} - \frac{2\pi R}{\ell_s} \sum_{\beta=0}^{d} b_{\alpha\gamma} M^\gamma \right) \left(N_\alpha \frac{\ell_s}{R} - \frac{2\pi R}{\ell_s} \sum_{\beta=0}^{d} b_{\alpha\delta} M^\delta \right)$$

from the spatial momentum in the energy $E^2 = m^2 c^4 + |\vec{P}|^2 c^2$, which is perceived as mass if the compact extra dimensions are too small to be resolved. Only the spatial components of the momentum P_α should be included in the relativistic expression for the energy, because the energy E itself is the zero component P_0 of a relativistic four-momentum. Our expression (5.127) does not directly take this requirement into account. However, because the sole time dimension in a higher-dimensional space-time is non-compact and only small compact dimensions contribute non-zero N^α and M^β, Equation (5.127) does not include time components of the four-momentum.

The winding modes also contribute to the second term of the constraint (5.123), implying a mass contribution of

$$\frac{1}{c^2 \alpha'^2} \sum_{\alpha=0}^{d} \sum_{\beta=0}^{d} g_{\alpha\beta} \frac{\partial x^\alpha}{\partial \sigma} \frac{\partial x^\beta}{\partial \sigma} = \frac{4\pi^2 R^2 m_s^2}{\ell_s^2} \sum_{\alpha=0}^{d} \sum_{\beta=0}^{d} g_{\alpha\beta} M^\alpha M^\beta. \tag{5.128}$$

Before we combine this term with the relativistic kinetic energy (5.127), we rescale coordinates by $x^\alpha \mapsto (\ell_s/R) x^\alpha$, mapping the coordinate circumference $2\pi R$ of compact dimensions to the string scale, $2\pi \ell_s$. The background fields are correspondingly transformed as two-tensors, $g^{\alpha\beta} \mapsto (\ell_s^2/R^2) g^{\alpha\beta}$ and $b_{\alpha\beta} \mapsto (R^2/\ell_s^2) b_{\alpha\beta}$. (The proper geometrical lengths of a compact dimension α, given by $\int \sqrt{g_{\alpha\alpha}} \, dx^\alpha$, remain unchanged by this transformation.)

All factors of ℓ_s/R then disappear from the mass spectrum, which after the transformation is given by

$$m_{j, N, M}^2 = m_s^2 \left(\sum_{\alpha=0}^{d} \sum_{\beta=0}^{d} g^{\alpha\beta} \left(N_\alpha - 2\pi \sum_{\beta=0}^{d} b_{\alpha\gamma} M^\gamma \right) \left(N_\alpha - 2\pi \sum_{\beta=0}^{d} b_{\alpha\delta} M^\delta \right) \right.$$

$$\left. + 4\pi^2 \sum_{\alpha=0}^{d} \sum_{\beta=0}^{d} g_{\alpha\beta} M^\alpha M^\beta + 4\pi \sum_{\alpha=2}^{d} \sum_{n=1}^{\infty} \left(j_{n\alpha} + \frac{1}{2} \right) n \right). \tag{5.129}$$

The first terms from compact dimensions can compactly be written as

$$m^2_{j,\,N,\,M} = m^2_s\left(\left(\begin{matrix}2\pi\vec{M}\\ \vec{N}\end{matrix}\right)^T \mathcal{G}\left(\begin{matrix}2\pi\vec{M}\\ \vec{N}\end{matrix}\right) + 4\pi\sum_{\alpha=2}^{d}\sum_{n=1}^{\infty}\left(j_{n\alpha} + \frac{1}{2}\right)n\right) \qquad (5.130)$$

where all momentum and winding quantum numbers have been combined into a single $2d$-dimensional vector. The coefficients define a symmetric $2d$-dimensional matrix

$$\mathcal{G} = \begin{pmatrix} g^{-1} & -g^{-1}b \\ bg^{-1} & g - bg^{-1}b \end{pmatrix}, \qquad (5.131)$$

using the inverse metric g^{-1} with components $g^{\alpha\beta}$ and the antisymmetric field b with components $b_{\alpha\beta}$ to define the d-dimensional blocks (Maharana & Schwarz 1993).

It follows from the definition of the blocks in $\mathcal{G}$ that this matrix is an element of the Lie group $O(d, d)$, defined as the set of all matrices that preserve the off-diagonal metric

$$\mathbb{D} = \begin{pmatrix} 0 & \mathbb{1} \\ \mathbb{1} & 0 \end{pmatrix} \qquad (5.132)$$

with the d-dimensional identity matrix $\mathbb{1}$:

$$\mathcal{G}^T\mathbb{D}\mathcal{G} = \mathbb{D}. \qquad (5.133)$$

In this formulation, called *double field theory*, a single T-duality transformation is replaced by the entire group $O(d, d)$, which acts on the vectors $(2\pi\vec{M}, \vec{N})$ in a doubled space of dimension $2d$ (Hull & Zwiebach 2009; Siegel 1993a, 1993b; Tseytlin 1990, 1991).

Based on the matrix $\mathcal{G}$ as a generalized metric, double field theory has been used to derive a new, non-Riemannian geometry. In this geometry, it is possible to combine the usual coordinate transformations of $g_{\alpha\beta}$ and $b_{\alpha\beta}$ together with gauge transformations of $b_{\alpha\beta}$, mapping $b_{\alpha\beta}$ to

$$b'_{\alpha\beta} = b_{\alpha\beta} + \frac{\partial\lambda_\alpha}{\partial\lambda_\beta} - \frac{\partial\lambda_\beta}{\partial\lambda_\alpha} \qquad (5.134)$$

with an arbitrary vector field λ_α. The latter transformation is itself a generalization of the common gauge transformation of vector potentials,

$$A'_\alpha = A_\alpha + \frac{\partial\lambda}{\partial x^\alpha}, \qquad (5.135)$$

to two-index tensors. The combined transformations are related to derivatives on the doubled space which, interestingly, are in general non-associative when they are applied at least three times in a row. Even though string theory is not background independent, it is therefore able to suggest new, non-classical spacetime structures (Hohm et al. 2010a, 2010b; Zwiebach 2011).

Non-commutative and Non-associative Structures

Rather than introducing non-associative derivatives on doubled space, it is easier to see how non-associative behavior may emerge for kinematical momenta in the presence of a *non-constant* antisymmetric field $b_{\alpha\beta}$. In contrast to our derivation of mass spectra, we now have to allow $b_{\alpha\beta}$ to depend on x^α in order to see this structure. Following the example of an electric charge moving in a background vector potential, (5.45); we first derive that the kinematical momenta (5.122) in general have non-zero brackets:

$$
\{P_\alpha^{\text{kin}}, P_\beta^{\text{kin}}\} = \int_0^1 \int_0^1 \{p_\alpha^{\text{kin}}(\sigma_1), p_\beta^{\text{kin}}(\sigma_2)\}\mathrm{d}\sigma_1\mathrm{d}\sigma_2
$$

$$
= \int_0^1 \int_0^1 \left\{ p_\alpha(\sigma_1) - \frac{1}{\alpha'}\sum_{\gamma=0}^d b_{\alpha\gamma}\frac{\partial x^\gamma}{\partial\sigma_1}, \; p_\beta(\sigma_2) - \frac{1}{\alpha'}\sum_{\gamma=0}^d b_{\beta\gamma}\frac{\partial x^\gamma}{\partial\sigma_2} \right\}\mathrm{d}\sigma_1\mathrm{d}\sigma_2
$$

$$
= \frac{1}{\alpha'}\int_0^1 \int_0^1 \sum_{\gamma=0}^d \left(\left(\frac{\partial b_{\beta\gamma}}{\partial x^\alpha(\sigma_1)}\frac{\partial x^\gamma}{\partial\sigma_2} - \frac{\partial b_{\alpha\gamma}}{\partial x^\beta(\sigma_2)}\frac{\partial x^\gamma}{\partial\sigma_1}\right)\delta(\sigma_1 - \sigma_2)\right.
\tag{5.136}
$$

$$
\left. - \left(b_{\alpha\beta}(x^\delta(\sigma_1))\frac{\partial\delta(\sigma_1 - \sigma_2)}{\partial\sigma_1} - b_{\beta\alpha}(x^\delta(\sigma_2))\frac{\partial\delta(\sigma_1 - \sigma_2)}{\partial\sigma_2}\right)\right)\mathrm{d}\sigma_1\mathrm{d}\sigma_2
$$

$$
= \frac{1}{\alpha'}\int_0^1 \sum_{\gamma=0}^d \left(\frac{\partial b_{\beta\gamma}}{\partial x^\alpha} - \frac{\partial b_{\alpha\gamma}}{\partial x^\beta} + 2\frac{\partial b_{\alpha\beta}}{\partial x^\gamma}\right)\frac{\partial x^\gamma}{\partial\sigma}\mathrm{d}\sigma.
$$

Momentum space is therefore non-commutative in the presence of a non-constant $b_{\alpha\beta}$ (Seiberg & Witten 1999; Herbst et al. 2001; Lüst 2010).

The last term in Equation (5.136) is zero because, using the chain rule, it is given by $\int_0^1 \mathrm{d}b_{\alpha\beta} = b_{\alpha\beta}(x^\gamma(1)) - b_{\alpha\beta}(x^\gamma(0))$, which vanishes for a closed string with $x^\gamma(1)$ and $x^\gamma(0)$ being the same event in spacetime. The integrand in Equation (5.136) can therefore be identified with the completely antisymmetric (or exterior) derivative

$$
H_{\alpha\beta\gamma} = \frac{\partial b_{\alpha\beta}}{\partial x^\gamma} + \frac{\partial b_{\gamma\alpha}}{\partial x^\beta} + \frac{\partial b_{\beta\gamma}}{\partial x^\alpha}
\tag{5.137}
$$

of the antisymmetric $b_{\alpha\beta}$. The field $H_{\alpha\beta\gamma}$ derived from $b_{\alpha\beta}$ is a higher-index version of the magnetic field

$$
B^i = \sum_{j=1}^3 \sum_{k=1}^3 \epsilon^{ijk}\frac{\partial A_j}{\partial x^k}
\tag{5.138}
$$

described by a vector potential A_i. The non-zero Poisson bracket (5.136) of kinematical momenta is therefore analogous to the non-zero Poisson bracket (5.45) for a charged particle in the presence of a magnetic field.

Using the new field $H_{\alpha\beta\gamma}$, we can write the Poisson bracket (5.136) as

$$\{P_\alpha^{\text{kin}}, P_\beta^{\text{kin}}\} = \frac{1}{\alpha'} \oint H_{\alpha\beta\gamma} \mathrm{d}x^\gamma \tag{5.139}$$

integrating over a closed string winding around a compact dimension. The result can be interpreted as the string average of a magnetic field.

The double Poisson bracket, analogous to Equation (5.46), follows directly from Equation (5.136):

$$\{\{P_\alpha^{\text{kin}}, P_\beta^{\text{kin}}\}, P_\gamma^{\text{kin}}\} = \frac{1}{\alpha'} \int_0^1 \left(\frac{\partial H_{\alpha\beta\delta}}{\partial x^\gamma} - \frac{\partial H_{\alpha\beta\gamma}}{\partial x^\delta} \right) \frac{\partial x^\delta}{\partial \sigma} \mathrm{d}\sigma. \tag{5.140}$$

As in Equation (5.136), the second term integrates to zero. Antisymmetrization of $\{\{P_\alpha^{\text{kin}}, P_\beta^{\text{kin}}\}, P_\gamma^{\text{kin}}\}$, using the fact that $H_{\alpha\beta\gamma}$ is already antisymmetric, then yields the Jacobiator

$$\{\{P_\alpha^{\text{kin}}, P_\beta^{\text{kin}}\}, P_\gamma^{\text{kin}}\}_{\text{antisymmetric}} = \frac{1}{3\alpha'} \int_0^1 \left(\frac{\partial H_{\alpha\beta\delta}}{\partial x^\gamma} + \frac{\partial H_{\gamma\alpha\delta}}{\partial x^\beta} + \frac{\partial H_{\beta\gamma\delta}}{\partial x^\alpha} \right) \frac{\partial x^\delta}{\partial \sigma} \mathrm{d}\sigma. \tag{5.141}$$

Now adding a suitable multiple of the vanishing $\int_0^1 (\partial H_{\alpha\beta\gamma}/\partial x^\delta)(\partial x^\delta/\partial \sigma)\mathrm{d}\sigma$ back into the expression, the Jacobiator equals

$$\{\{P_\alpha^{\text{kin}}, P_\beta^{\text{kin}}\}, P_\gamma^{\text{kin}}\}_{\text{antisymmetric}} = \frac{1}{3\alpha'} \oint (\mathrm{d}H)_{\alpha\beta\gamma\delta} \mathrm{d}x^\delta \tag{5.142}$$

with the antisymmetric, exterior derivative

$$(\mathrm{d}H)_{\alpha\beta\gamma\delta} = \frac{\partial H_{\alpha\beta\delta}}{\partial x^\gamma} + \frac{\partial H_{\gamma\alpha\delta}}{\partial x^\beta} + \frac{\partial H_{\beta\gamma\delta}}{\partial x^\alpha} - \frac{\partial H_{\alpha\beta\gamma}}{\partial x^\delta}. \tag{5.143}$$

This expression is analogous to the divergence of the magnetic field, or the magnetic monopole density. Since Gauss' theorem equates a spatial integral of the monopole density in some region with the magnetic flux through the surface, the appearance of $\mathrm{d}H$ in Equation (5.142) is referred to as a *geometric flux*.

For any $H_{\alpha\beta\gamma}$ derived from $b_{\alpha\beta}$ as in Equation (5.137), $\mathrm{d}H_{\alpha\beta\gamma\delta} = 0$ vanishes identically. The brackets derived here are therefore indeed Poisson brackets because they obey the Jacobi identity, which requires a vanishing Jacobiator. However, just as it is possible to assume that magnetic monopoles might lead to a non-zero div $\vec{B}$, it is possible to construct background spacetimes for moving strings with non-trivial fields and topologies, such that the analog of $\mathrm{d}H$ is non-zero. Such string backgrounds imply non-associative structures when quantized (Cornalba & Schiappa 2002; Blumenhagen & Plauschinn 2011; Blumenhagen et al. 2011; Mylonas et al. 2012; Lüst 2012; Bakas & Lüst 2014).

5.2.6 Canonical Quantum Gravity

Starting with Wheeler (1968; DeWitt 1967), several different proposals have been made to quantize gravity in a canonical manner, turning the classical phase space and constraints into operators acting on a Hilbert space. The variety of proposals within this approach is a consequence of technical problems in the quantization of fields that describe spacetime rather than being supported on a pre-existing (background) spacetime: Background independent quantum gravity aims to provide a quantum field theory *of* spacetime, not *on* spacetime as used in particle physics.

Algebras

In particular, the standard quantization rule that turns momentum components into derivative operators by configuration variables is subtle when the configuration variables are functions, such as the components of a metric, and provide an infinite number of degrees of freedom. While it is possible to write formal expressions such as canonical commutator relations $[\hat{x}_i, \hat{p}_j] = i\hbar\delta_{i,j}$ if i and j have infinite, continuous ranges, such as x as a label of independent field values $\psi(x)$, the resulting algebras and their quantum states are hard to control. The continuous-label version of the Kronecker delta $\delta_{i,j}$ is Dirac's delta "function," $\delta(x - y)$ (rather, a distribution). However, it takes an infinite value if $x = y$ and therefore does not define an algebra with finite coefficients.

In standard quantum field theory, the problem of defining suitable operators for fields and their momenta is solved by "smearing" their classical expressions: One replaces field values $\psi(x)$ with partially integrated fields, such as

$$\psi[f] = \int_{-\infty}^{\infty} f(x)\psi(x) \tag{5.144}$$

on a flat one-dimensional space, where $f(x)$ is an arbitrary function with some general properties, such as continuity and differentiability. The infinite continuous range of the field label x is then replaced by a function space, which is infinite-dimensional and therefore does not simply remove infinitely large label spaces. However, commutators that include at least one smeared field, such as $\psi[f]$, result in integrated delta functions, and therefore produce finite values as long as the smearing function $f(x)$ is finite for any x.

If space is not flat, or maybe not even geometrical as in canonical quantum gravity where the metric itself has to be turned into an operator, an integration such as Equation (5.144) would be coordinate dependent if no factor of $\sqrt{\det q}$ is included, with the spatial metric q_{ij}. It is then more convenient to smear the momentum of the field, because field momenta, as seen in the canonical string example (5.71); include a factor of $\sqrt{\det q}$ and are therefore densities of weight one.

For instance, if we start with a scalar field $\phi(x)$ and its momentum $p_\phi(x)$ on some spatial manifold with points x, the usual substitution rule would result in a momentum operator

$$\hat{p}_\phi(x) = \frac{\hbar}{i}\frac{\delta}{\delta\phi(x)} \tag{5.145}$$

with a functional derivative $\delta/\delta\phi(x)$. The commutator of the field and its momentum,

$$[\hat{\phi}(x), \hat{p}_\phi(y)] = -\frac{\hbar}{i}\frac{\delta\phi(x)}{\delta\phi(y)} = i\hbar\delta(x - y) \tag{5.146}$$

is not a finite function.

However, if we choose as our basic operators the previous $\phi(x)$ together with

$$p_\phi[F] = \int_{-\infty}^{\infty} F(y)p_\phi(y)\mathrm{d}y, \tag{5.147}$$

we have the finite commutator

$$[\hat{\phi}(x), \hat{p}_\phi[F]] = i\hbar \int_{-\infty}^{\infty} F(y)\delta(x - y)\mathrm{d}y = i\hbar F(x). \tag{5.148}$$

Since we can choose any finite function for F, such as the plane waves $F_k(x) = \exp(ikx)$ for all real numbers k, we can calculate the complete $p_\phi(x)$ if we know all $p_\phi[F]$, for instance by inverse Fourier transformation. We can therefore describe the complete phase space of a scalar field theory by using $\phi(x)$ and $p_\phi[F]$ and quantize by choosing a representation of their well-defined algebra defined by Equation (5.148).

If our scalar field is supported on a curved manifold, the same procedure can be used without modifications. We do need an additional factor of the measure in integrations, $\sqrt{\det q}$ if q is the metric on our space. However, this factor is already contained in p_ϕ if it is derived for a scalar field on a curved manifold, as already seen in examples. The metric factor appears in any canonical formulation of a covariant theory because the Lagrangian of such a field theory, $L = \int \sqrt{\det q}\,\mathcal{L}\mathrm{d}x$ with the Lagrangian density $\mathcal{L}$, is spatially integrated and must contain the same measure factor, which will then also appear in $p_\phi(x) = \delta L/\delta\dot{\phi}(x)$.

If the field to be quantized is the metric q_{ij} itself, however, the procedure is problematic, for several reasons. If we include the measure factor in the smeared metric, we are dealing with non-linear objects such as $\int \sqrt{q}\,q\mathrm{d}x$ for which it is impossible to use representation theory of linear algebras.

If we smear the metric's momentum, we do not need to include the measure factor, relying on the density weight of field momenta. However, in three spatial dimensions, the metric q_{ij} and its momentum p^{ij} are tensors with indices. Since the tensor transformation law (1.26) is local, it is not respected by smeared fields in the general case of non-linear coordinate transformations:

$$\det(\partial y^k / \partial y^{k'}) \sum_{i=1}^{3} \sum_{j=1}^{3} \int_{-\infty}^{\infty} \int_{-\infty}^{\infty} \int_{-\infty}^{\infty} F(x) \frac{\partial x^{i'}}{\partial x^i} \frac{\partial x^{j'}}{\partial x^j} p^{ij} \mathrm{d}^3 x$$

$$\neq \det(\partial y^k / \partial y^{k'}) \sum_{i=1}^{3} \sum_{j=1}^{3} \frac{\partial x^{i'}}{\partial x^i} \frac{\partial x^{j'}}{\partial x^j} \int_{-\infty}^{\infty} \int_{-\infty}^{\infty} \int_{-\infty}^{\infty} F(x) p^{ij} \mathrm{d}^3 x.$$

(5.149)

Therefore, $\int_{-\infty}^{\infty} \int_{-\infty}^{\infty} \int_{-\infty}^{\infty} F(x) p^{ij} \mathrm{d}^3 x$, unlike p^{ij}, does not transform like a tensor density.

Before tensor fields can be smeared covariantly, they should be turned into suitable scalar functions by multiplication with vectors or other tensors instead of functions $F(x)$. Smearing tensors, such as P_{ij} with an index structure dual to the momentum p^{ij} and Q^{ij} (with density weight one) dual to the metric q_{ij}, would be subject to coordinate transformations just like the smeared tensors, given by the metric or its momentum. The tensorially smeared expressions

$$q[Q^{ij}] = \int_{-\infty}^{\infty} \int_{-\infty}^{\infty} \int_{-\infty}^{\infty} Q^{ij} q_{ij} \mathrm{d}^3 x \quad \text{and} \quad p[P_{ij}] = \int_{-\infty}^{\infty} \int_{-\infty}^{\infty} \int_{-\infty}^{\infty} P_{ij} p^{ij} \mathrm{d}^3 x \quad (5.150)$$

are coordinate independent and linear in the canonical fields. It is then possible that representations of the well-defined commutator

$$[\hat{q}[Q^{ij}], \hat{p}[P_{ij}]] = i\hbar \int_{-\infty}^{\infty} \int_{-\infty}^{\infty} \int_{-\infty}^{\infty} Q^{ij} P_{ij} \mathrm{d}^3 x \qquad (5.151)$$

can be found. Constructing such a representation of an infinite-dimensional algebra may still be challenging, but it can be aided by suitable choices of the smearing tensors.

Representations

The metric does not lend itself easily to an introduction of suitable smearing tensors for a well-defined algebra to be obtained, mainly because the space of two-index smearing functions Q^{ij} with density weight one is not very intuitive. Canonical quantizations of metric variables, in an approach usually referred to as Wheeler–DeWitt quantization, have therefore remained formal: They work with functional derivatives of wave functions with respect to metric components without turning them into well-defined operators in an infinite-dimensional algebra.

Even at this formal level, different versions exist of canonical quantizations of metrics, depending on the representation proposed. The most direct approach represents momenta of the metric (related to extrinsic curvature) as functional derivatives by the metric components (Wheeler 1968; DeWitt 1967). However, for a well-defined geometry the metric is subject to the condition that it be non-degenerate, that is, that its determinant be non-zero. If this condition is included in the formulation of a classical phase space of general relativity, the phase space has a boundary. A derivative by the metric, however, is exponentiated to a translation on the space of metrics, which in general does not respect the boundary as shown in the one-dimensional minisuperspace case in Equation (4.44). Affine quantum gravity (Klauder 2003, 2006) is a canonical approach that quantizes the bounded phase space, much like the quantum-cosmological representation in Section 4.3.1 that respects positivity of the scale factor. This approach has, in fact, mainly been evaluated in cosmological models, as in Bergeron et al. (2015), and not much is known about the behavior of a full theory of quantum gravity in this setting.

Loop quantum gravity (Rovelli 2004; Thiemann 2007; Ashtekar & Lewandowski 2004; Nicolai et al. 2005; Alexandrov & Roche 2011), which is also a canonical approach, has simplified the general representation problem by replacing the metric with new functions of different tensorial types, for which it is easier to find well-defined smearing tensors. In fact, instead of merely introducing smearing tensors, the approach makes use of geometrical structures, such as loops in space or two-dimensional surfaces (Rovelli & Smolin 1990), which provide smearing tensors through their tangent and normal vectors, respectively. The smeared fields are then no longer subject to coordinate transformations, but rather to the simpler transformations obtained by deforming loops and surfaces with diffeomorphisms. Moreover, the intuitive geometrical meaning of loops and surfaces makes it easier to control the algebras that result from this kind of smearing. Before the inception of loop quantum gravity, this crucial idea had already been conceived in Gambini & Trias (1980) and applied to a quantization of the electromagnetic field, which like gravity is also described by tensorial objects.

Fluxes
Preparing metric variables for smearing, in a first step one replaces the spatial metric q_{ij} with a *densitized triad* E_I^i, related to the inverse metric q^{ij} by

$$q^{ij} \det q = \sum_{I=1}^{3} E_I^i E_I^j. \tag{5.152}$$

(The choice of index labels used here follows the other chapters of this book. It differs slightly from common conventions in loop quantum gravity.) A *triad*, already seen in Equation (4.45), is a set of three independent vectors at each point which, by requiring it to be orthonormal, indirectly determines the spatial metric. The three vectors are labeled by the index $I = 1$, 2, 3 in Equation (5.152). A densitized triad is obtained from a triad by providing it with a density weight one, multiplying it with $\sqrt{\det q}$ where q_{ij} is the spatial metric determined by the triad. The defining relation used here, Equation (5.152), is then indeed consistent with orthonormality of the triad: It implies

$$3 = \sum_{i=1}^{3}\sum_{j=1}^{3} q_{ij} q^{ij} = \sum_{I=1}^{3}\sum_{i=1}^{3}\sum_{j=1}^{3} q_{ij} \frac{E_I^i}{\sqrt{\det q}} \frac{E_I^j}{\sqrt{\det q}}. \tag{5.153}$$

The right-hand side also equals three when interpreted as summing the normalization conditions for the three triad vectors with components $E_I^i/\sqrt{\det q}$.

A triad or a densitized triad can be left-handed or right-handed, depending on the spatial orientation which is indicated by the sign of $\det(E_i^j)$. What appears as a phase-space boundary in a metric formulation, given by degenerate metrics with $\det(q_{ij}) = 0$, is only a surface separating the two orientations of space in a triad formulation. Since there is no need to restrict triads to a fixed orientation, the problem addressed in affine quantum gravity never appears in a triad formulation such as loop quantum gravity.

Like p_ϕ in a scalar field theory, the densitized triad already contains the measure factor $\sqrt{\det q}$ required for spatial integrations. Its spatial index i can be eliminated if we smear the field by integrating it over two-dimensional surfaces, such that there is a distinguished co-vector n_i that can be multiplied with E_i^i. Using completely antisymmetric tensors, the co-vector is obtained as

$$n_i = \frac{1}{2} \sum_{j=1}^{3} \sum_{k=1}^{3} \sum_{a=1}^{2} \sum_{b=1}^{2} \epsilon_{ijk}\epsilon^{ab}\frac{\partial x^j}{\partial y^a}\frac{\partial x^k}{\partial y^b} \tag{5.154}$$

from a parameterization $x^i(y^a)$ of the surface. This co-vector, unlike the more common vector $n^i = \sum_{j=1}^{3} q^{ij}n_j$, does not depend on the metric of space, and is therefore independent of E_I^i in a densitized-triad formulation. (The norm of this co-vector, computed using the spatial metric q^{ij}, equals the area element (5.119) used to motivate the field $b_{\alpha\beta}$ in worldsheet formulations of string theory.)

The resulting smeared objects,

$$F_I[S] = \sum_{i=1}^{3} \int_S E_I^i n_i d\sigma \tag{5.155}$$

are called *fluxes* in the present context. We do not include a measure factor on the surface because n_i, through ϵ^{ab} in Equation (5.154), has density weight one on the surface. The spatial density weight of the densitized triad is removed by a density weight -1 of ϵ_{ijk} in Equation (5.154); see Equation (5.55). Fluxes are therefore coordinate independent, and they are linear in E_I^i. They can therefore be used to define a linear algebra of smeared fields. The remaining index I of fluxes is not tensorial and therefore does not imply coordinate transformations of the smeared objects. It is, however, subject to an independent transformation of triad rotations, which we will discuss soon.

There is no explicit smearing function in Equation (5.155) that could be used to define an inverse transformation, reconstructing the field E_I^i from the fluxes $F_I[S]$. But because the infinite integration range in Equation (5.144) has been replaced with finite ranges defined by arbitrary surfaces S, the complete field E_I^i can still be computed if all the fluxes are known. In order to reconstruct $E_I^i(x)$ at a given point x, one just needs to evaluate the $F_I[S]$ on three families of infinitesimally small surfaces approaching the point x in a limit, such that surfaces in a given family have normals pointing in a fixed coordinate direction.

Holonomies

The densitized triad (divided by $8\pi G/c^3$) is the momentum of the co-vector version of *extrinsic curvature* K_{ij}, defined as one half the derivative of the spatial metric in a spacetime direction normal to a spatial slice used to set up the canonical formulation. Extrinsic curvature therefore plays the role of a velocity of the metric, and it is not surprising that it appears in momenta. The co-vector version relevant for triad variables is related to K_{ij} by

$$K_{iI} = \sum_{j=1}^{3} \frac{K_{ij}E_I^j}{\sqrt{\det q}}. \tag{5.156}$$

A co-vector can be integrated in a well-defined way along one-dimensional curves C, which come with a distinguished vector field $t^i = dx^i/d\lambda$ given by the tangent vector where λ parameterizes the curve:

$$K_I[C] = \sum_{i=1}^{3} \int_C K_{iI} t^i d\lambda. \tag{5.157}$$

Also here, knowing $K_I[C]$ for all curves C in space is sufficient to reconstruct a unique K_{iI}. If we combine fluxes with this integrated extrinsic curvature, the curve and surface integration can, generically, be combined to a three-dimensional integration, removing infinities from delta functions as in Equation (5.148).

A final subtlety arises because the components of E_i^I and K_{iI}, respectively, can for fixed i be transformed into one another by applying the rotation group: The spatial geometry, defined by a line element or a metric tensor, does not imply a unique triad. We can always rotate the triplet of vectors in a triad without changing their orthonormality properties. These rotations can be performed locally, with different angles at different spatial points; they form a local gauge group. In general, local operations are incompatible with non-local integrations as required for smearing. There are two possible solutions, considered here for K_{iI}, to be specific. First, we may introduce a second smearing function for the I-index, $f^I(x)$ also subject to local rotations, such that

$$K[C, f] = \sum_{I=1}^{3} \sum_{i=1}^{3} \int_C f^I K_{iI} t^i d\lambda. \tag{5.158}$$

Since both f^I and K_{iI} are rotated locally, the product is invariant and can be integrated without coordinate or rotation problems.

Alternatively, it is possible to turn the tensor K_{iI} into a vector potential, $A_i^I = \Gamma_i^I + \gamma K_i^I$, where Γ_i^I is a *spin connection* derived from the densitized triad, generalizing the Christoffel symbol (2.26) derived from a metric. The parameter γ is an arbitrary but fixed non-zero number, the so-called Barbero–Immirzi parameter (Immirzi 1997; Barbero 1995). Like the vector potential A_i of electromagnetism which can be transformed to $A_i + \partial g/\partial x^i$ with an arbitrary function g while maintaining the same magnetic field, see also Equation (5.135), the gravitational vector potential, also referred to as the Ashtekar–Barbero connection (Ashtekar 1987; Barbero 1995), is subject to gauge transformations. They map A_i^I to

$$A_i^{I'} = \sum_{J=1}^{3} \sum_{K=1}^{3} \epsilon_{I'JK} g^J A_i^K + \frac{\partial g^{I'}}{\partial x^i} \tag{5.159}$$

for any triple of functions, g^I. The completely antisymmetric tensor $\epsilon_{I'JK}$ here refers to the basic commutator relations $[R_I, R_J] = \sum_{K=1}^{3} \epsilon_{IJK} R_K$ satisfied by generators R_I of the rotation group, also known as angular-momentum operators (divided by $i\hbar$) in quantum mechanics. Since rotation matrices, in general, do not commute with one another, the local gauge group is non-commutative, or non-Abelian.

A vector potential, subject to rotations acting on the index I (or, more generally, subject to non-Abelian gauge transformations) can be turned into a well-defined smeared expression if it is not only integrated but also exponentiated. The combined procedure defines the *parallel transport*

$$h[C] = \mathcal{P} \exp\left(\sum_{I=1}^{3} \sum_{i=1}^{3} \int_C \tau_I A_i^I t^i d\lambda \right) \tag{5.160}$$

on the curve C. Here τ_I are matrix representations of the infinitesimal rotations R_I, such as $\tau_I = \frac{1}{2} i\sigma_I$ with Pauli matrices σ_I. These matrices, like R_I do not commute with one another. The definition of their matrix exponential through a power series therefore requires a specific choice of ordering. The symbol $\mathcal{P}$ indicates *path ordering*, which orders factors of $\sum_{I=1}^{3} \tau_I A_i^I(x^j(\lambda))$ according to their position on the curve $x^j(\lambda)$. (See also Equation (5.163) below for an explicit definition of path ordering.)

Exponentiation is important for well-defined local rotation properties, as can be seen more easily by comparison with the example of an Abelian transformation group such as the one-dimensional restriction of three-dimensional rotations obtained by fixing a rotation axis. This one-dimensional group can conveniently be written in terms of complex numbers, where $e^{i\varphi}$ represents a rotation around the fixed axis by an angle φ. The restricted group therefore has a single infinitesimal generator, $i = \lim_{\varphi \to 0} de^{i\varphi}/d\varphi$. In this case, the curve integral of a vector potential can easily be transformed by inserting $A_i' = A_i + \partial g/\partial x^i$:

$$\sum_{j=1}^{3} \int_C A_j' dx^j = \int_C \left(\sum_{j=1}^{3} A_j dx^j + dg \right) = \sum_{j=1}^{3} \int_C A_j dx^j + g(C_1) - g(C_0) \tag{5.161}$$

where we have denoted the initial and final points of the curve by C_0 and C_1, respectively. An Abelian parallel transport, exponentiating the curve integral, therefore transforms according to

$$\exp\left(\sum_{j=1}^{3}\int_C A_j'\mathrm{d}x^j\right) = \exp(g(C_1))\exp\left(\sum_{j=1}^{3}\int_C A_j\mathrm{d}x^j\right)\exp(g(C_0))^{-1}. \quad (5.162)$$

For a non-Abelian group, path ordering complicates this argument. However, it can still be used if we first decompose the full curve integral into smaller, approximately straight pieces:

$$\mathcal{P}\exp\left(\sum_{i=1}^{3}\sum_{I=1}^{3}\int_C \tau_I A_i^I t^i \mathrm{d}\lambda\right) = \lim_{N\to\infty}\exp\left(\sum_{n=1}^{N}\sum_{i=1}^{3}\sum_{I=1}^{3}\int_{x_n}^{x_{n+1}} \tau_I A_i^I t^i \mathrm{d}\lambda\right)$$

$$= \lim_{N\to\infty}\prod_{n=1}^{N}\exp\left(\sum_{i=1}^{3}\sum_{I=1}^{3}\int_{x_n}^{x_{n+1}} \tau_I A_i^I t^i \mathrm{d}\lambda\right). \quad (5.163)$$

(This decomposition can be seen as an explicit definition of path ordering because it correctly lines up the non-commuting contributions $\sum_{i=1}^{3}\sum_{I=1}^{3}\tau_I A_i^I t^i$ along the path.) In each infinitesimal contribution, we use

$$\exp\left(\sum_{i=1}^{3}\sum_{I=1}^{3}\int_{x_n}^{x_{n+1}} \tau_I A_i^I t^i \mathrm{d}\lambda\right) \sim 1 + \sum_{i=1}^{3}\sum_{I=1}^{3}\int_{x_n}^{x_{n+1}} \tau_I A_i^I t^i \mathrm{d}\lambda \quad (5.164)$$

in the limit of $N\to\infty$, such that $x_{n+1}\to x_n$. In this limit, Equation (5.164) is linear in the non-commuting objects $\tau_I A_i^I$, and is therefore free of ordering ambiguities.

In each such expression, we use an analog of Equation (5.161), and then rewrite the results as exponential functions, transforming as

$$\exp\left(\sum_{i=1}^{3}\sum_{I=1}^{3}\int_{x_n}^{x_{n+1}} \tau_I A_i'^I t^i \mathrm{d}\lambda\right)$$

$$= \exp(g(x_{n+1}))\exp\left(\sum_{i=1}^{3}\sum_{I=1}^{3}\int_{x_n}^{x_{n+1}} \tau_I A_i^I t^i \mathrm{d}\lambda\right)\exp(g(x_n))^{-1} \quad (5.165)$$

where g now represents a non-Abelian matrix $g = \sum_{I=1}^{3}g^I\tau_I$ with coefficients as in Equation (5.159). In the final product of the transformed Equation (5.163), all intermediate $\exp(g(x_n))$ with x_n not an endpoint of C cancel out, and the whole exponentiated integral, or the parallel transport, transforms multiplicatively:

$$h'[C] = \exp(g(C_1))h[C]\exp(g(C_0))^{-1}. \quad (5.166)$$

Without exponentiation, $g(C(\lambda))$ on intermediate points would not have canceled out, implying complicated non-local gauge relationships.

In particular, the matrix trace of the exponentiated integral is invariant if $g(x_1) = g(x_N)$, which is always the case if the curve C is a closed loop, $C_0 = C_1$. Since the trace of a matrix product remains unchanged if the factors are exchanged by cyclic permutations,

$$\mathrm{tr}h'[C] = \mathrm{tr}(\exp(g(C_1))h[C]\exp(g(C_0))^{-1}) = \mathrm{tr}(h[C]\exp(g(C_0))^{-1}\exp(g(C_1)))$$
$$= \mathrm{tr}(h[C]) \tag{5.167}$$

is gauge invariant. The parallel transport around the closed loop is called the *holonomy* of the loop, and its trace is called a *Wilson loop*. Applied to the non-Abelian vector potential A_i^I of gravity, we have the loops of loop quantum gravity.

Holonomy-flux Representation

Loop quantum gravity constructs a representation of holonomy and flux operators acting on states defined as wave functions $\psi(A_i^I)$ that depend on A_i^I. Holonomies are turned into basic multiplication operators,

$$\hat{h}[C]\psi(A_i^I) = \mathcal{P}\exp\left(\sum_{i=1}^{3}\sum_{I=1}^{3}\int_C \tau_I A_i^I t^i \mathrm{d}\lambda\right)\psi(A_i^I). \tag{5.168}$$

A simple state is one that does not depend on A_i^I at all. (Such a state is normalizable in loop quantum gravity because the compactness of the rotation group allows one to construct a compact, but still infinite-dimensional, configuration space for the A_i^I; Ashtekar et al. 1995.) Acting with a finite number of holonomy operators, $\hat{h}[C_1]$, ..., $\hat{h}[C_2]$, generates non-trivial A_i^I-dependence of ψ along the curves $C_1, \ldots, C_n$ used in these holonomies. The curves can collectively be described by a graph or a network in space.

If we apply a holonomy operator for the same curve multiple times, different dependences on A_i^I are generated in each step. The links of the graph should therefore be labeled by additional quantum numbers in order to determine how often the corresponding holonomy has been used. Because holonomies take values in the rotation group, quantum numbers of the rotation group, just like spin and angular momentum in quantum mechanics, determine the independent ways of the dependence. Together with a spin quantum number on each link, the network is turned into a *spin network*. (There are additional quantum numbers at the vertices where two or more links intersect, telling us how the matrix elements of rotation-group elements in their spin representations are multiplied.) However, in contrast to spin in quantum mechanics, the spin quantum numbers in loop quantum gravity are not related to physical spinning motion because they refer to the gauge symmetry of triad rotations rather than physical rotations of space. For the same reason, and also because they do not carry electric charge, spin networks do not imply new interactions of space with a magnetic field.

The rotation group has a unique normalized integration measure invariant under rotations. This measure can be extended to any pair of spin networks to define an inner product on wave functions $\psi(A_i^I)$. Spin-network states that have different spin quantum numbers on the same link are orthogonal, just like two states with different spin quantum numbers are orthogonal in quantum mechanics. Moreover, if one spin-network state does not include a link that is part of another spin network state with non-zero spin, the two states are orthogonal. To show this without explicit calculation, note that an absent link is, in its A_i^I-dependence, equivalent to a link with spin zero, which is orthogonal to any state with non-zero spin. Spin-network states therefore define a convenient orthogonal basis.

In a representation based on wave functions $\psi(A_i^I)$, fluxes $\hat{F}[S]$ are angular-momentum-like derivative operators, integrated over surfaces. Acting with a flux operator $\hat{F}[S]$ on a spin-network state, the result is non-zero only if the surface S intersects the network at least once because there is then an $A_i^I(x)$ on which the state depends and whose derivative appears in $\hat{F}[S]$ with the same x. Each such contribution depends on the spin quantum number of the link intersecting S in the usual angular-momentum manner. For multiple intersection points, the contributions add up. Flux operators therefore have discrete spectra, and since they quantize the densitized triad which, in turn, represents the spatial metric, spatial geometry is discrete in loop quantum gravity (Rovelli & Smolin 1995; Ashtekar & Lewandowski 1997, 1998; Loll 1995).

As already seen, algebraic properties of these operations can more easily be demonstrated using the example of holonomies fluxes for the Abelian group of rotations around a fixed axis, which are of the form

$$h[C] = \exp\left(i \int_C \sum_{j=1}^{3} A_j \mathrm{d}x^j\right). \tag{5.169}$$

Mathematically, a flux in this example is nothing but the standard electric flux $F[S] = \int_S \sum_{i=1}^{3} E^i n_i \mathrm{d}^2 y$ through the surface S. However, we still use gravitational fields, in which the densitized triad is the momentum of $c^3 A_i/(8\pi\gamma G)$. (We drop the label I because it is fixed when we consider only a one-dimensional subgroup of the three-dimensional rotation group.) Therefore, replacing $E^i(x^j)$ with $8\pi\gamma G\hbar/(ic^3) = 8\pi\gamma\ell_P^2/i$ times a functional derivative with respect to $A_i(x^j)$, we obtain the flux operator

$$\hat{F}[S] = \frac{8\pi\gamma\ell_P^2}{i} \int_S \sum_{i=1}^{3} n_j \frac{\delta}{\delta A_j(\vec{x}(y))} \mathrm{d}^2 y \tag{5.170}$$

acting on states $\psi(A_i(\vec{x}))$. In the functional derivative, $\vec{x}(y)$ denotes the embedding map of the surface (with coordinates y^α) into space (with coordinates x^i).

The commutator of a flux and a holonomy is given by

$$[\hat{F}[S], \hat{h}[C]] = \frac{8\pi\gamma\ell_P^2}{i} \int_S \sum_{i=1}^{3} n_j \frac{\delta h[C]}{\delta A_j(\vec{x}(y))} \mathrm{d}^2 y$$

$$= 8\pi\gamma\ell_P^2 \left(\int_S \int_C \delta(\vec{z} - \vec{x}(y)) \sum_{j=1}^{3} n_j(\vec{x}(y)) \mathrm{d}z^j \mathrm{d}^2 y\right) \hat{h}[C]. \tag{5.171}$$

The parenthesis in the second line is non-zero only if the surface S and the curve C have intersection points where C is transversal to S. The non-zero contribution from the delta function at $\vec{z} = \vec{x}(y)$ is then included in the integration range. Any points where C is tangential to S are removed by a vanishing dot product of the normal vector n_j with the tangent vector dx^j. In all non-zero contributions, the two-dimensional integration over S and the one-dimensional integration over C then combine to a three-dimensional integration $\sum_{j=1}^{3} n_j(\vec{x}(y))dx^j d^2y = d^3x$, using Equation (5.154), which removes the delta function. The result is an integer because it is the integrated delta function, $\iiint \delta(\vec{z} - \vec{x}(y))d^3z = 1$, times the number of all intersection points where such non-zero contributions are produced; it is called the *intersection number* of the surface S and the curve C:

$$[\hat{F}[S], \hat{h}[C]] = 8\pi\gamma\ell_P^2 \text{Int}(S, C)\hat{h}[C]. \tag{5.172}$$

The intersection number may be negative, depending on the alignment of the tangent vector of the curve with respect to the normal of the surface.

Spatial Geometry

Because holonomy operators are used to generate a basis given by spin-network states, starting with a constant state $\psi_0(A_i^I)$ that does not actually depend on A_i^I, the commutator (5.172) can be used to derive the spectrum of flux operators. On ψ_0, we have $\hat{F}[S]\psi_0 = 0$. Any other state $\psi(A_i^I)$ can be written as $\psi = \hat{h}[C]\psi_0$ with a curve C that describes the entire spin network, including overlapping pieces of the curve C where a basic holonomy operator has been applied multiple times. Any such state is an eigenstate of $\hat{F}[S]$ with eigenvalue given by

$$\hat{F}[S]\psi = \hat{F}[S]\hat{h}[C]\psi_0 = \hat{h}[C]\hat{F}[S]\psi_0 + [\hat{F}[S], \hat{h}[C]]\psi_0 = 8\pi\gamma\ell_P^2 \text{Int}(S, C)\psi. \tag{5.173}$$

Flux operators in this representation have discrete spectra, given by integer multiplies of $8\pi\gamma\ell_P^2$. The spacing depends on the Planck length mainly for reasons of units. Its numerical value is determined by the Barbero–Immirzi parameter γ, which is arbitrary in the classical theory while it affects quantum properties. It therefore represents a quantization ambiguity.

Loop quantum gravity shows how it may be possible to construct geometry through quantum excitations. There is no background geometry to begin with, only a manifold with a topology and differential structure in which spin-network states and fluxes are set up. These structures are represented by the unexcited state, ψ_0. This state is not like the vacuum (or "empty space") of particle physics, because it does not even have space in the sense of geometry. It is emptier than empty space, or "emptiest space" because it contains only the bare minimum of what we might consider space. Conceptually, the state ψ_0 is therefore unlike the harmonic ground state, defined by zero excitation levels j_{na} in the mass formula (5.108) of strings moving in a flat spacetime. Here, we have a clear indication of the more challenging nature of background-independent quantizations, such as loop quantum gravity, compared with background-dependent ones.

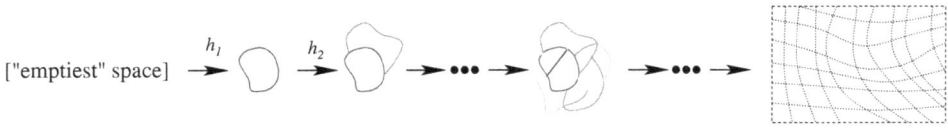

Figure 5.14. Illustration of the states of loop quantum gravity. The basic state in this background-independent theory is emptiest space, in which no matter and not even geometry is excited. This state has a simple mathematical description, a constant wave function ψ_0, but defies any intuitive classical description. Successive actions of holonomy operators generate geometry localized along one-dimensional excitations that form a spin network, with links that carry independent quantum numbers (colors). It is hoped that suitable highly excited states might approximate continuum space that could be used in a semiclassical approximation.

As states are excited by acting with holonomy operators, illustrated in Figure 5.14; fluxes obtain non-zero values. A non-zero metric is therefore represented, but only along one-dimensional flux lines. It is conceivable that highly excited states with many links might approximate a continuum geometry, but a mathematical treatment of such many-body states in a highly interacting theory is challenging. It therefore remains unclear whether loop quantum gravity contains suitable states that could approximate continuum space, even simple ones such as flat Euclidean space.

The representation is invariant under spatial diffeomorphisms if one lets them act on spin-network states simply by moving the networks as well as surfaces of flux operators. Intersection numbers of links and surfaces and their spin representations do not change when this operation is performed, demonstrating the invariance. With this symmetry, there are strong uniqueness statements about the possible quantum representations, eliminating some ambiguities (Lewandowski et al. 2006; Fleischhack 2006, 2009). (The Barbero–Immirzi parameter γ, however, remains undetermined by this result.)

The construction of a mathematically well-defined and, given the assumptions, largely unique representation of spatial geometry, together with potential physical implications of the discreteness, is considered a success of the theory. However, several questions remain open even before we try to address the complicated dynamics of the theory, related to quantum generators of hypersurface deformations in spacetime. If we look closely at the degrees of freedom represented by spin-network states, we notice a certain redundancy that is not completely resolved by imposing the required constraints. States in quantum physics represent phase-space degrees of freedom, on which the wave function may depend directly (as A_i^I here) or which are determined by it through derivatives (for momenta such as E_i^i). Spin-network states depend on A_i^I, which are used as configuration variables of the gravitational phase space.

In addition, spin-network states also depend on the positions of their vertices and links. This dependence is clearly related to coordinates, and it is removed by imposing spatial diffeomorphism invariance through its action on the spatial network structure. However, in addition to the action of diffeomorphisms on points in space, there is the diffeomorphism constraint D[M]. Classically, as seen in our explicit examples (2.184) for spherically symmetric models and Equation (2.243) for perturbative inhomogeneity, this constraint is a function on phase space which, when imposed, restricts the possible relationships between vector potentials and densitized triads. The way the constraint is imposed at the quantum level, by contrast, does not imply any relationship between densitized triads (represented through spin quantum numbers) and vector potentials (appearing in the dependence of the wave function). Therefore, it seems that the implementation of spatial diffeomorphisms fails to implement the required phase-space restrictions. It only removes auxiliary degrees of freedom, given by the positions of vertices, which have been introduced to set up quantum states but do not appear in a classical formulation. For this reason, it remains unclear whether loop quantum gravity correctly represents spatial geometries. It is even less clear whether it correctly represents gravitational degrees of freedom in *spacetime* because the quantum dynamics of the theory remains poorly understood.

Dynamics
The discrete geometry of loop quantum gravity is, at first, only a geometry of space: The canonical formalism implies that $A_i^I(x)$ refers only to spatial positions x, and $\psi(A_i^I)$ determines probabilities for "measurements" of the vector potential at a fixed time. (Because constraints required for hypersurface deformations have not been imposed at this stage, these "measurements" do not refer to a physical process but merely suggest a mathematical analogy with the usual interpretation of quantum mechanics.) Spin networks as well as surfaces in fluxes are therefore spatial objects that do not extend into the time direction.

If we had an absolute time, a Hamilton operator would tell us how a state evolves and changes as time goes on. In a relativistic theory, the Hamiltonian is replaced by a constraint $H[N] = 0$ for any spatial function N. In addition to time evolution there are transformations that tell us how a state should change when an observer's perspective is modified, for instance by a boost. As shown in Section 2.3, these transformations correspond to hypersurface deformations in spacetime which, when quantized, may be similar to the classical behavior or receive quantum corrections.

In a background-independent theory such as loop quantum gravity, the structure of spacetime is not predetermined but should be derived from the theory. The main task of any canonical approach to quantum gravity is then to construct a representation of the constraints $H[N]$ and $D[M]$ of general relativity as operators, such that their commutators are closed, and consistent with the classical behavior under hypersurface deformations at low curvature. It has been known for some time that this problem is a very difficult one (Komar 1979a, 1979b), mainly because the bracket (2.154) of two Hamiltonian constraints results in a *metric-dependent* function (2.155) inserted in a diffeomorphism constraint. When quantized, one has to find a specific ordering of all non-commuting operators that appear in $H[N]$ and $D[M]$, including Equation (2.155), such that the commutator of two $H[N]$ is indeed some operator of the form $D[M]$. At present, it is not known whether such an ordering exists without any symmetry restrictions. (Suitable orderings have been found in spherically symmetric models; Brahma 2015.)

In loop quantum gravity, using the new mathematical machinery given by its well-defined holonomy-flux representation, it is possible to define operator versions of H[N]. The resulting expressions are complicated because they represent a rather involved classical one, given by

$$
H[N] = \frac{c^4}{16\pi G} \int N \sum_{i=1}^{3} \sum_{j=1}^{3} \sum_{I=1}^{3} \sum_{J=1}^{3} \sum_{K=1}^{3} \frac{E_I^i E_J^j}{\sqrt{|\det E|}} \epsilon^{IJK}
$$

$$
\left(F_{ij}^K + (1 + \gamma^2) \sum_{M=1}^{3} \sum_{N=1}^{3} \epsilon_{KMN} K_i^M K_j^N \right) d^3x
$$

(5.174)

in triad variables, where

$$
F_{ij}^K = \frac{\partial A_j^K}{\partial x^i} - \frac{\partial A_i^K}{\partial x^j} - \sum_{I=1}^{3} \sum_{J=1}^{3} \epsilon_{IJK} A_i^I A_j^J
$$

(5.175)

is the non-Abelian magnetic field of the vector potential A_i^I. In Equation (5.174), the basic variables are the densitized triad E_I^i and the vector potential A_i^I because they are directly defined as operators in the holonomy-flux representation. Written in terms of these fields, the expression is non-polynomial because $K_i^M = \gamma^{-1}(A_i^M - \Gamma_i^M)$ depends non-polynomially on E_I^i through the spin connection Γ_i^M, in a way similar to the Christoffel symbol.

Choosing the imaginary value $\gamma = i$ eliminates the complicated contribution through K_i^M, as originally pointed out by Ashtekar (1987), but the resulting complex theory does not describe the correct physics. It is interesting to note that Euclidean gravity is obtained by replacing $1 + \gamma^2$ in Equation (5.174) with $1 - \gamma^2$. The constraint can then be simplified by choosing the real value $\gamma = 1$. As in other approaches to quantum gravity, Euclidean signature therefore appears simpler than Lorentzian signature, even though the reason in loop quantum gravity is unrelated to transition amplitudes derived from path integrals.

A quantization of H[N] requires one to write the dependence on A_i^I and E_I^i through holonomies and fluxes. Since flux operators have discrete spectra containing zero, an initial concern is that there is no well-defined operator for the inverse $1/\sqrt{|\det E|}$ in this representation. A well-defined operator for H[N] nevertheless exists, as found by Thiemann (1998), because one can use the classical identity

$$
\frac{c^3}{8\pi\gamma G} \frac{E_I^i E_J^j}{\sqrt{|\det E|}} \epsilon^{IJK} = \epsilon^{ijk} \{A_k^K, V\} \, \text{sgn}(\det E)
$$

(5.176)

where $V = \int \sqrt{|\det E|}\, d^3x$ is the volume of some finite region in space, restricted only by the condition that it contain the point where the left-hand side of Equation (5.176) is to be evaluated. (This identity relies on determinant equations such as Equation (5.54).) A well-defined quantization is therefore given by the commutator of a holonomy operator with the volume operator, which latter can be derived from flux operators.

The existence of well-defined operators for H[N], and similarly for matter Hamiltonians (Thiemann 1998), suggests a certain finiteness of quantum field theories on discrete space(time) as constructed in loop quantum gravity. However, this finiteness is much weaker than renormalizability because it does not rely on the specific dynamics, and it has not been explored yet in possible derivations of scattering amplitudes or perturbative interactions. In fact, such an analysis is currently impossible in loop quantum gravity because it is unknown how to construct a suitable superposition of spin-network states that could approximate flat space, or any other continuum geometry. Compared with causal dynamical triangulations in which stable large-scale continuum states have been found, or string theory which has solved the instability problem caused by tachyons by incorporating supersymmetry, loop quantum gravity has so far been unable to check this important condition.

Turning the constraint H[N] into an operator in loop quantum gravity requires several modifications of the classical expression, mainly because non-linear holonomies are used for an expression which classically is quadratic in the vector potential. In addition, any specific operator requires a choice of ordering of non-commuting factors in H[N] that depend on E_i^i and A_i^I, respectively. Some of these modifications, in particular the application of holonomies, may be viewed as regularization required to express the constraint as an operator in the holonomy-flux representation. Seen in this view, such modifications are not uncommon in quantum field theories. However, $\hat{H}[N]$ is not just a single operator but a family of operators, defined for all spatial functions N. In addition to operator properties, there are further consistency conditions that require covariance in the form of a well-defined behavior of the commutators $[\hat{H}[N_1], \hat{H}[N_2]]$ which in a consistent theory must be proportional to some operator $\hat{D}[M]$ and at least in a classical limit should resemble Equation (2.154) with Equation (2.155). These commutator relations turn out to be very complicated, and so far it is not clear whether they can be realized. It is therefore unknown whether loop quantum gravity provides an acceptable quantization of spacetime, not just of space, and whether it can be covariant. Moreover, even if it may be covariant, its spacetime structure remains unknown until a quantum version of Equations (2.154) and (2.155) has been derived.

There has been some progress on this important question (Laddha & Varadarajan 2011; Henderson et al. 2013a, 2013b; Tomlin & Varadarajan 2013; Laddha 2014), but the analysis remains incomplete. Moreover, loop quantum gravity encounters an additional difficulty given by a large number of quantization ambiguities. Since modifications of H[N] in preparation of using holonomies are not unique, there is a large number of possible quantum constraints, even if the basic holonomy-flux representation may be largely unique. Also their commutators could then be modified compared with the classical structure of hypersurface deformations, as long as they form a closed set. The closure condition can be used to rule out many possible quantizations, but at present it is not known whether any consistent version exists. The covariance problem therefore remains unsolved in loop quantum gravity.

Addressing the same problem in symmetric model systems, such as spherically symmetric ones, has shown that it may be possible to close the commutators for holonomy-modified constraints, but only if the commutator structure is also modified (Reyes 2009; Bojowald et al. 2014; Brahma 2015). Even if there may be a well-defined spacetime structure in loop quantum gravity, there are indications that it is non-classical and does not agree with Riemannian geometry. The resulting implications for quantum cosmology will be discussed in the final Chapter 6.

Other Approaches

We have not yet exhausted the set of approaches to quantum gravity. Here, we briefly describe two remaining ones in which implications for spacetime and possible physical effects are currently less clear. Although they are not canonical in their basic formulation, they are conceptually close to the canonical approach of loop quantum gravity.

Spin-foam Models.

An attempt to simplify the dynamics of canonical quantum gravity, in particular loop quantum gravity, has been made by the introduction of spin-foam models (Reisenberger & Rovelli 1997). Spin foams are two-dimensional objects in space-time, with intersecting surfaces labeled by spin quantum numbers. They may be viewed as evolving spin networks, in which the dynamical creation of new vertices leads to surfaces branching off and forming a foam-like structure. Formally, spin foam models are often presented as discrete path-integral versions of the canonical dynamics given by loop quantum gravity.

One initial motivation for trying to control the dynamics of discrete quantum gravity by using spin-foam models came from a comparison of the spacetime and canonical formulations, respectively, of general relativity. The spacetime action (2.48) looks deceptively simple, given by the integral of a single function, the Ricci scalar. The Hamiltonian constraint of the canonical formulation, such as Equation (5.174), is much more involved, and subject to complicated consistency conditions through the hypersurface-deformation structure it has to obey. The latter implements covariance which is also a condition on the spacetime version. However, in spacetime formulations it is usually easy to tell whether proposed actions are covariant just by looking at the tensorial nature of objects they depend on. If all indices are paired up correctly, only spacetime scalars appear in the action and the theory is covariant.

Unfortunately, the apparent simplicity of spacetime action principles is deceptive when quantum effects are to be introduced. Since quantum gravity is a quantum theory of spacetime, the spacetime structure cannot be presupposed and may well be non-classical. It is therefore unclear how Riemannian properties of tensor calculus, used in standard considerations of curvature invariants, could be applied *before* one has completed the definition of the theory. There may be significant quantum corrections not only in the dynamics of quantum gravity, but also in its spacetime structure. Formally, the Ricci scalar that defines the classical action might look like a simple expression when interpreted as an intuitive expression of curvature, but written as a combination of partial derivatives of the metric, it is non-polynomial and therefore anything but simple. The resulting metric theory is strongly interacting and should be expected to receive large quantum corrections.

An alternative to guessing the spacetime structure before quantization would be, in a path integral formulation, to propose a theory and then check whether it is covariant by analyzing invariance properties of the integration measure. In spin-foam models, both the action and the measure must remain covariant after discretization, applied to make them integrable over paths determined by spin foams. Transformation properties of the path-integral measure are much harder to analyze than closure conditions on constraints for which effective approximations are available, as we will see in the next chapter. Defining a consistent theory of quantum gravity in spin-foam form is therefore more difficult than in a canonical formulation, in spite of the initial expectation that the dynamics might be simplified. In fact, general properties of constrained quantization indicate that the measure usually used in spin-foam models violates covariance (Bojowald & Perez 2010; Alexandrov 2008).

A systematic but still ongoing program has therefore been launched to study whether it is possible to restore covariance through renormalization, approaching a continuum limit (Bahr 2014; Dittrich et al. 2016; Bahr & Steinhaus 2016; Dittrich et al. 2016; Bahr & Steinhaus 2017; Bahr et al. 2018). So far, most calculations in spin-foam models are restricted to discretizations with small numbers of building blocks, making it difficult to see how continuum properties can emerge.

Group-field Theory.

Group-field theory (Calcagni et al. 2012) is defined as an application of quantum field theory to functions supported on a group manifold rather than spacetime. Spacetime and gravity therefore enter the theory rather indirectly, initially through an observation that the structure of spin foams can be viewed as Feynman graphs of a specific group-field theory (De Pietri et al. 2000). In this case, the group on which fields are defined is given by several copies of the rotation group, as used in spin networks. The whole perturbation series of all Feynman diagrams then amounts to a sum over spin foams, or a version of a discretized spacetime path integral. Reformulating the expansion as a group-field theory does not, in itself, improve convergence properties, but it has led to several interesting connections (Carrozza et al. 2014a, 2014b) with abstract renormalization in tensor models which are also used in condensed-matter physics.

In terms of gravitational applications, group-field theory has mainly been used in a cosmological setting in which minisuperspace models could be reinterpreted as collective excitations or condensate states in an underlying inhomogeneous theory (Gielen et al. 2013, 2014; Gielen 2014; de Cesare et al. 2016). However, even though inhomogeneity can in principle be included (Gielen & Oriti 2014; Gielen 2015), the question of covariance and spacetime structure remains unexplored. Stability is also an open question as it has not been shown yet that condensate states remain condensed in the full dynamics of a group-field theory.

5.3 Appreciation

The different approaches to quantum gravity are distinguished from one another by the principles they use and emphasize, as well as the mathematical methods they introduce to solve formal problems. Each approach claims certain successes in addressing questions that may be unique to this approach, or shared with other approaches but, if so, usually in different ways and with different ranges of applicability. It is therefore difficult to compare approaches or to rank them in terms of their promise.

The appreciation proposed here focuses on successes in providing partial or complete solutions to general problems that every approach to quantum gravity has to face, such as the problem of covariance or the semiclassical limit, rather than solutions to specific problems that are intrinsic to some approach but may not appear at all in other approaches. An example of the latter is given by attempts to formulate spin-foam models such that they resemble certain features expected for the dynamics of loop quantum gravity.

1. String theory is unique in that it is the only approach to quantum gravity that has solved the covariance problem and has the correct semiclassical limit. In fact, it is often criticized for producing "too many" classical limits because its higher dimensions as well as supersymmetric content can be reduced to possible four-dimensional low-energy behaviors in myriad ways. However, string theory has not merely postulated supersymmetry or extra dimensions, but was led to them by implementing consistency and covariance. No other approach has been able to provide complete solutions to these general requirements. It is therefore unknown what additional ingredients other approaches may have to face. A fair comparison should therefore not focus on the necessity of such ingredients.

2. Causal dynamical triangulations have not completely solved the covariance problem, but they have produced several interesting results about the semiclassical and continuum limit of a discrete approach to spacetime structure. Open questions remain in how to implement Lorentzian rather than Euclidean spacetime signature.

3. Non-commutative geometry has provided several elegant mathematical formulations of non-classical spacetime structures with potential gravitational dynamics that is covariant in a modified (non-commutative) way. Both the covariance problem and the semiclassical limit can be addressed, but questions remain in how to formulate consistent dynamics, in particular in Lorentzian rather than Euclidean signature. Non-associative geometry may at some time reach a similar status, but at present it is rather new and not much developed.

4. Loop quantum gravity is the leading canonical approach. Its successes, mainly of kinematical nature (describing space rather than spacetime), culminate in uniqueness statements about representations of basic operators for holonomies and fluxes. However, it remains unclear how relevant these results are for a formulation of the dynamics, in particular for physical

applications close to continuum spacetime geometries that would correspond to highly excited states in the representation of loop quantum gravity. The dynamics is not only complicated, but also, according to current understanding, introduces a large number of quantization ambiguities that significantly weaken the advantage of having uniqueness results about kinematical representations of basic operators. Both the covariance problem and the semiclassical limit remain largely unresolved. Most of the physical effects proposed in loop quantum cosmology (as well as black hole models) are based on an unfortunate over-reliance on minisuperspace models, as we will see in the next chapter.

References

Agishtein, M. E., & Migdal, A. A. 1992, NuPhB, 385, 395

Alexandrov, S. 2008, PhRvD, 78, 044033

Alexandrov, S., & Roche, P. 2011, PhR, 506, 41

Alvarez-Gaume, L., Meyer, F., & Vazquez-Mozo, M. 2006, NuPhB, 753, 92

Ambjørn, J., Coumbe, D. N., Gizbert-Studnicki, J., & Jurkiewicz, J. 2015, JHEP, 08, 033

Ambjørn, J., Drogosz, Z., Gizbert-Studnicki, J., et al. 2016, PhRvD, 94, 044010

Ambjørn, J., Gizbert-Studnicki, J., Görlich, A., et al. 2017, EPJC, 77, 152

Ambjørn, J., Gizbert-Studnicki, J., Görlich, A., Grosvener, K., & Jurkiewicz, J. 2017, NuPhB, 922, 226

Ambjørn, J., Gizbert-Studnicki, J., Görlich, A., & Jurkiewicz, J. 2014, JHEP, 1406, 034

Ambjørn, J., Görlich, A., Jurkiewicz, J., et al. 2011, NuPhB, 849, 144

Ambjørn, J., Görlich, A., Jurkiewicz, J., & Loll, R. 2008, PhRvL, 100, 091304

Ambjørn, J., Görlich, A., Jurkiewicz, J., & Loll, R. 2012, PhR, 519, 127

Ambjørn, J., & Jurkiewicz, J. 1992, PhLB, 278, 42

Ambjørn, J., Jurkiewicz, J., & Loll, R. 2004, PhRvL, 93, 131301

Ambjørn, J., Jurkiewicz, J., & Loll, R. 2005, PhRvD, 72, 064014

Ambjørn, J., Jurkiewicz, J., & Loll, R. 2011, PhRvL, 107, 211303

Ambjørn, J., & Loll, R. 1998, NuPhB, 536, 407

Arnold, V. I. 1997, Mathematical Methods of Classical Mechanics (Berlin: Springer)

Arnowitt, R., Deser, S., & Misner, C. W. 1962, in Gravitation: An Introduction to Current Research (New York: Wiley)

Arnowitt, R., Deser, S., & Misner, C. W. 2008, GReGr, 40, 1997

Aschieri, P., Blohmann, C., Dimitrijevic, M., et al. 2005, CQGra, 22, 3511

Aschieri, P., Dimitrijevic, M., Meyer, F., & Wess, J. 2006, CQGra, 23, 1883

Ashtekar, A. 1987, PhRvD, 36, 1587

Ashtekar, A., & Lewandowski, J. 1997, CQGra, 14, A55

Ashtekar, A., & Lewandowski, J. 1998, AdTMP, 1, 388

Ashtekar, A., & Lewandowski, J. 2004, CQGra, 21, R53

Ashtekar, A., Lewandowski, J., Marolf, D., Mourão, J., & Thiemann, T. 1995, JMaPh, 36, 6456

Bahr, B. 2014, arXiv: 1407.7746

Bahr, B., Rabuffo, G., & Steinhaus, S. 2018, PhRvD, 98, 106026

Bahr, B., & Steinhaus, S. 2016, PhRvL, 117, 141302

Bahr, B., & Steinhaus, S. 2017, PhRvD, 95, 126006

Bakas, I., & Lüst, D. 2014, JHEP, 01, 171

Barbero, G. J. F. 1995, PhRvD, 51, 5507

Baytaş, B., Bojowald, M., Crowe, S., & Mielczarek, J. 2020, JCAP, 01, 019

Bergeron, H., Czuchry, E., Gazeau, J.-P., Malkiewicz, P., & Piechocki, W. 2015, PhRvD, 92, 124018

Bialas, P., Burda, Z., Krzywicki, A., & Petersson, B. 1996, NuPhB, 472, 293

Blumenhagen, R., Deser, A., Lüst, D., Plauschinn, E., & Rennecke, F. 2011, JPhA, 44, 385401

Blumenhagen, R., & Fuchs, M. 2016, JHEP, 07, 019

Blumenhagen, R., Fuchs, M., Hassler, F., Lüst, D., & Sun, R. 2014, JHEP, 1404, 141

Blumenhagen, R., & Plauschinn, E. 2011, JPhA, 44, 015401

Bojowald, M., Brahma, S., Büyükçam, U., Guglielmon, J., & van Kuppefeld, M. 2018, PhRvL, 121, 201602

Bojowald, M., Brahma, S., Büyükçam, U., & Strobl, T. 2017, JHEP, 04, 028

Bojowald, M., Paily, G. M., & Reyes, J. D. 2014, PhRvD, 90, 025025

Bojowald, M., & Perez, A. 2010, GReGr, 42, 877

Bonanno, A., Eichhorn, A., Gies, H., et al. 2004, arXiv: 2004.06810

Brahma, S. 2015, PhRvD, 91, 124003

Buchert, T. 2011, CQGra, 28, 164007

Burgess, C. P. 2004, LRR, 7, 5

Calcagni, G. 2010, PhRvL, 104, 251301

Calcagni, G. 2013, JCAP, 12, 041

Calcagni, G., Gielen, S., & Oriti, D. 2012, CQGra, 29, 105005

Candelas, P., Horowitz, G. T., Strominger, A., & Witten, E. 1985, NuPhB, 258, 46

Carrozza, S., Oriti, D., & Rivasseau, V. 2014a, CMaPh, 327, 603

Carrozza, S., Oriti, D., & Rivasseau, V. 2014b, CMaPh, 330, 581

Catterall, S., Kogut, J. B., & Renken, R. 1994, PhLB, 328, 277

Connes, A. 1994, Non-commutative Geometry (Boston, MA: Academic)

Connes, A. 1996, CRAS, 323, 1231

Cornalba, L., & Schiappa, R. 2002, CMaPh, 225, 33

Coumbe, D. N., Gizbert-Studnicki, J., & Jurkiewicz, J. 2016, JHEP, 02, 144

de Bakker, B. V. 1996, PhLB, 389, 238

de Cesare, M., Pithis, A. G. A., & Sakellariadou, M. 2016, PhRvD, 94, 064051

De Pietri, R., Freidel, L., Krasnov, K., & Rovelli, C. 2000, NuPhB, 574, 785

DeWitt, B. S. 1967, PhRv, 160, 1113

Dimitrijević Ćirić, M., Giotopoulos, G., Radovanović, V., & Szabo, R. J. 2020, arXiv: 2005.00454

Dirac, P. A. M. 1931, RSPSA, 133, 1

Dittrich, B., Mizera, S., & Steinhaus, S. 2016, NJPh, 18, 053009

Dittrich, B., Schnetter, E., Seth, C. J., & Steinhaus, S. 2016, PhRvD, 94, 124050

Donoghue, J. F. 1994a, PhRvD, 50, 3874

Donoghue, J. F. 1994b, PhRvL, 72, 2996

Donoghue, J. F. 2020, FrP, 8, 56

Doplicher, S., Fredenhagen, K., & Roberts, J. E. 1995, CMaPh, 172, 187

Eliezer, C. J., & Roy, S. K. 1962, PCPS, 58, 401

Feldbrugge, J., Lehners, J.-L., & Turok, N. 2017, PhRvL, 119, 171301

Feldbrugge, J., Lehners, J.-L., & Turok, N. 2018, PhRvD, 97, 023509

Fleischhack, C. 2006, PhRvL, 97, 061302

Fleischhack, C. 2009, CMaPh, 285, 67

Gambini, R., & Trias, A. 1980, PhRvD, 22, 1380

Gasperini, M., & Veneziano, G. 1993, APh, 1, 317

Gasperini, M., & Veneziano, G. 2003, PhR, 373, 1

Gielen, S. 2014, CQGra, 31, 155009

Gielen, S. 2015, PhRvD, 91, 043526

Gielen, S., & Oriti, D. 2014, NJPh, 16, 123004

Gielen, S., Oriti, D., & Sindoni, L. 2013, PhRvL, 111, 031301

Gielen, S., Oriti, D., & Sindoni, L. 2014, JHEP, 1406, 013

Gingras, M. J. P. 2009, Sci, 326, 375

Günaydin, M., Piron, C., & Ruegg, H. 1978, CMaPh, 61, 69

Günaydin, M., & Zumino, B. 1986, in Symp. Old and New Problems in Fundamental Physics, held in Honor of G. C. Wick, ed. L. A. Radicati di Bronzolo (Pisa: Publications of the Scuola Normale Superiore), 43

Henderson, A., Laddha, A., & Tomlin, C. 2013a, PhRvD, 88, 044029

Henderson, A., Laddha, A., & Tomlin, C. 2013b, PhRvD, 88, 044028

Herbst, M., Kling, A., & Kreuzer, M. 2001, JHEP, 0109, 014

Hohm, O., Hull, C., & Zwiebach, B. 2010a, JHEP, 1007, 016

Hohm, O., Hull, C., & Zwiebach, B. 2010b, JHEP, 1008, 008

Hořava, P. 2009, PhRvD, 79, 084008

Hull, C., & Zwiebach, B. 2009, JHEP, 0909, 099

Immirzi, G. 1997, CQGra, 14, L177

Jackiw, R. 1985, PhRvL, 54, 159

Jackiw, R. 2004, IJMPA, 19S1, 137

Jordan, S., & Loll, R. 2013a, PhLB, 724, 155

Jordan, S., & Loll, R. 2013b, PhRvD, 88, 044055

Klauder, J. 2003, IJMPD, 12, 1769

Klauder, J. 2006, IJGMMP, 3, 81

Klimcik, C., & Strobl, T. 2002, IGP, 43, 341

Knorr, B., & Saueressig, F. 2018, PhRvL, 121, 161304

Komar, A. 1979a, PhRvD, 19, 2908

Komar, A. 1979b, PhRvD, 20, 830

Kwapisz, J. H., & Meissner, K. A. 2005, arXiv: 2005.03559

Laddha, A. 2014, arXiv: 1401.0931

Laddha, A., & Varadarajan, M. 2011, CQGra, 28, 195010

Lewandowski, J., Okołów, A., Sahlmann, H., & Thiemann, T. 2006, CMaPh, 267, 703

Loll, R. 1995, PhRvL, 75, 3048

Lüst, D. 2010, JHEP, 1012, 084

Lüst, D. 2012, arXiv: 1205.0100

Lüst, D., & Theisen, S. 1989, Lectures on String Theory (Berlin: Springer)

Maharana, J., & Schwarz, J. H. 1993, NuPhB, 390, 3

Mylonas, D., Schupp, P., & Szabo, R. J. 2012, JHEP, 09, 012

Mylonas, D., Schupp, P., & Szabo, R. J. 2014, JMaPh, 55, 122301

Nicolai, H., Peeters, K., & Zamaklar, M. 2005, CQGra, 22, R193

Niedermaier, M., & Reuter, M. 2006, LRR, 9, 5

Pachner, U. 1978, ArMa, 30, 89

Palmer, R. F., & Taylor, J. G. 1968, Natur, 219, 1033

Park, J.-S. 2000, arXiv: hep-th/0012141

Polchinski, J. 1998, String Theory, Vol. I and II (Cambridge: Cambridge University Press)

Regge, T. 1961, NCim, 19, 558

Reisenberger, M., & Rovelli, C. 1997, PhRvD, 56, 3490

Reyes, J. D. 2009, PhD thesis, The Pennsylvania State University

Rovelli, C. 2004, Quantum Gravity (Cambridge: Cambridge University Press)

Rovelli, C., & Smolin, L. 1990, NuPhB, 331, 80

Rovelli, C., & Smolin, L. 1995, NuPhB, 442, 593 [Erratum: 1995, NuPhB, 456, 753]

Sakellariadou, M. 2012, arXiv: 1204.5772

Sathiapalan, B. 1987, PhRvL, 58, 1597

Seeley, R. T. 1967, Proc. of Symposia in Pure Mathematics, 10, 288

Seiberg, N., & Witten, E. 1999, JHEP, 09, 032

Severa, P., & Weinstein, A. 2001, PThPS, 144, 145

Siegel, W. 1993a, PhRvD, 48, 2826

Siegel, W. 1993b, PhRvD, 47, 5453

Steinacker, H. 2007, JHEP, 07, 049

Steinacker, H. 2010, CQGra, 27, 133001

Thiemann, T. 1998, CQGra, 15, 839

Thiemann, T. 1998, CQGra, 15, 1281

Thiemann, T. 2007, Introduction to Modern Canonical Quantum General Relativity (Cambridge: Cambridge University Press)

Tomlin, C., & Varadarajan, M. 2013, PhRvD, 87, 044039

Tseytlin, A. A. 1990, PhLB, 242, 163

Tseytlin, A. A. 1991, PhRvL, 66, 545

Vant-Hull, L. L. 1968, PhRv, 173, 1412

Veneziano, G. 1991, PhLB, 265, 287

Veneziano, G. 2000, in The Primordial Universe—L'univers primordial ed. P. Binétruy, R. Schaeffer, J. Silk, & F. David Les Houches—Ecole d'Ete de Physique Theorique, Vol. 71 (Berlin-Heidelberg: Springer)

Weinberg, S. 1979, in General Relativity: An Einstein Centenary Survey (Cambridge: Cambridge University Press) 790

Wetterich, C. 1993, PhLB, 301, 90

Wheeler, J. A. 1968, in Battelle Rencontres (New York: Benjamin) 242

Zwiebach, B. 2012, Strings and Fundamental Physics, ed. M. Baumgartl, I. Brunner, & M. Haack, Lecture Notes in Physics 851 (Berlin-Heidelberg: Springer), 265

Chapter 6

Quantum Cosmology of Inhomogeneous Spacetimes

Quantum cosmology faces its biggest challenges in addressing inhomogeneity, which exposes it to the main consistency conditions as well as potential observational pressure in its description of structure formation. Moreover, in all approaches to quantum gravity, the formulation or derivation of a sufficiently unambiguous cosmological sector with a reliable connection to the full quantum theory remains incomplete, but to varying degrees.

String theory is the only approach to quantum gravity in which the important consistency condition of covariance has already been fully addressed. Moreover, several qualitative features of potential solutions are known, although it is often not clear whether they are completely consistent in a full quantum description (Vafa 2005). There are two main questions left for cosmological investigations: Controlling the large number of possible spacetime models generated at low energies by breaking supersymmetry and reducing extra dimensions; and providing ingredients that seem crucial for current cosmology, such as a positive cosmological constant. (Evaluations of string theory appear simpler in the presence of a negative cosmological constant.) It is interesting to note that both main problems for string-based cosmological model-building are consequences of the successful solution to the covariance condition, which in string theory requires supersymmetry and extra dimensions (or a large number of internal degrees of freedom).

In discrete approaches to quantum gravity, by contrast, consistency conditions remain in the foreground because at present it is not known whether they can be fully covariant. The structure of spacetime, or even whether there is any meaningful sense to "spacetime" in such theories, therefore remains unclear, making it difficult to construct fully justified cosmological models. This challenge is particular severe in loop quantum gravity, which is not only discrete but also canonical such that the structure of spacetime, encoded in symmetry transformations, has to be derived from a complicated interacting theory. While there has been a large number of

doi:10.1088/2514-3433/ab9c98ch6

phenomenological activities in this approach, they remain subject to consistency conditions which in most cases have not been addressed. In cases in which the covariance condition has been addressed, it has shown the possibility of strong deviations from classical spacetime structure which are in conflict with what has been assumed in most phenomenological models.

From a foundational perspective, it seems most crucial to analyze the formal ingredients relating to spacetime structure before one gets carried away too much by unfounded phenomenology. In this final chapter, we therefore first present several examples of unexpected shortcomings of attempted circumventions of the full covariance problem in discrete, canonical quantum gravity. The second part of the chapter will show a few examples in the wide range of potential physical scenarios that have been proposed as cosmological models based on various versions of quantum gravity.

6.1 Homogeneity in Inhomogeneous Models

The importance of foundational questions in cosmological models of canonical discrete approaches can best be illustrated by a discussion of "instructive failures:" These are attempts to avoid the consistency issues of inhomogeneous spacetime models by reducing them to homogeneous models, or to separate descriptions of a homogeneous background on the one hand, and perturbations on such a background on the other hand. Such constructions, based on minisuperspace models, are rather common in canonical approaches because the severe requirements of covariance in inhomogeneous settings are completely trivialized in homogeneous minisuperspace models: Complicated relationships between space and time transformations, summarized by Equations (2.154) and (2.155), are then reduced to commuting reparameterizations of the time coordinate.

There are two different ways in which homogeneity can be used to describe special inhomogeneous situations. First, one may exploit accidental symmetries in known classical solutions that imply homogeneity in some regions of an otherwise inhomogeneous spacetime. The most well-known example, as seen in Section 1.5.4, is the black hole interior of the Schwarzschild solution (1.154) within the horizon, $r < R_S$, where the same vector field that implies staticity in the exterior turns spacelike. Together with three-dimensional rotational symmetry of the entire solution, the interior then has a four-dimensional symmetry of spacelike transformations on suitable spacelike slices. This symmetry acts transitively on the spatial coordinates (x, ϑ, φ) of the Kantowski–Sachs geometry (1.156) because rotations can be used to transform a given (ϑ_1, φ_1) to any other (ϑ_2, φ_2), while translations in the x-direction of the interior geometry can map a given x_1 to any other x_2. A transitive action of a symmetry group on space implies spatial homogeneity.

If one assumes that this general behavior of symmetries is also realized, at least approximately, in quantum versions of the solution, even while the dynamics may be modified by quantum corrections, one may use homogeneous minisuperspace quantizations in order to understand the interior region of non-rotating black holes. Such models might reveal indications as to how quantum gravity could resolve black hole singularities. However, questions of how such restricted models can be realized in a full theory of quantum gravity, or how interior quantum solutions might be connected to an inhomogeneous exterior would require more advanced methods, which are currently out of reach in canonical quantum gravity or loop-related approaches such as spin foams.

Second, cosmological structure formation in general relativity can be described by a perturbative approach, in which the dynamics of an expanding homogeneous background is used as a time-dependent stage for evolving inhomogeneous modes. As long as inhomogeneity remains small, back-reaction can be ignored, such that a simple homogeneous model is again sufficient to describe the background. A homogeneous model can be quantized in different ways, as described in Chapter 4, and then coupled to a separate quantization of inhomogeneous modes. In order to avoid complicated issues of quantum gravity in inhomogeneous settings, such as covariance questions, one may use more standard methods from quantum field theory on a curved background. Since the background would be dynamical in this situation, the field theory of inhomogeneous modes has time-dependent coefficients for a given background solution. If the background is homogeneous and quantized by minisuperspace methods, these time-dependent coefficients, such as mode frequencies, likely differ from their classical values. Indirectly, quantum-gravity effects in minisuperspace models could therefore be transferred to inhomogeneous quantizations of perturbative modes.

While these homogeneity-based approaches can be made viable from a purely formal perspective, implementing internal consistency conditions such as unitary quantum evolution, they fail to describe spacetime for rather subtle but still important reasons. In broad terms, this failure is related to the fact that all these approaches use certain properties of symmetries in classical solutions and then *assume* that they are still valid after quantization. However, such assumptions are usually unjustified, in particular in discrete approaches to quantum gravity. It is not easy, but instructive, to pinpoint where exactly these assumptions fail, which we do in the present section. Knowing where assumptions fail can then lead to less biased procedures in which they are replaced with dedicated derivations within a given approach to quantum gravity.

6.1.1 Schwarzschild Interior

The Schwarzschild solution

$$\mathrm{d}s^2 = -\left(c^2 - \frac{2Gm}{r}\right)\mathrm{d}t^2 + \frac{1}{1 - 2Gm/(c^2 r)}\mathrm{d}r^2 + r^2(\mathrm{d}\vartheta^2 + \sin^2\vartheta\,\mathrm{d}\varphi^2) \qquad (6.1)$$

is valid as long as $r \neq R_\mathrm{S} = 2Gm/c^2$. This condition separates spacetime covered by the solution into two regions, the exterior with $r > 2Gm/c^2$ and the interior with $r < 2Gm/c^2$. In the exterior, any vector field pointing just in the t-direction is timelike, which implies staticity of the solution because the metric does not depend on t. In the interior, by contrast, while changing t still leaves the metric unchanged and therefore amounts to a symmetry, any vector field in the t-direction is spacelike.

The symmetry therefore no longer implies staticity. Instead, this spacelike symmetry can be combined with the three independent rotations of the spherically symmetric geometry to form a four-dimensional algebra of infinitesimal symmetry transformations on spacelike slices given by $r = $ const. (As elaborated in more detail in Section 1.5.4, we may use r as a time coordinate in the interior, where $g_{rr} < 0$ and therefore any vector field that points only in the r-direction is timelike.)

The four-dimensional symmetry algebra can be used to generate transformations that map any point $(t, r, \vartheta, \varphi)$ in the interior to any other point in the same region with the same value of r. The action of the corresponding symmetry group is therefore transitive, such that the spacelike slices given by $r = $ const are homogeneous. Since a three-dimensional group would be sufficient to imply spatial homogeneity, there must be an additional rotational axis that amounts to the fourth independent generator. With this additional symmetry, it turns out, the homogeneous geometry is not contained in the Bianchi classification of Section 1.3.3, but rather defines a separate model, named after Kantowski & Sachs (1966).

Demonstrating that the interior of a non-rotating black hole solution in a given approach to quantum gravity is spatially homogeneous, as in the simple classical solution, is a challenging problem. (In fact, it is likely that no spacetime region in quantum gravity can be exactly homogeneous because quantum fluctuations of inhomogeneous modes of the metric should be restricted by uncertainty relations, such that they are generically non-zero.) Instead of solving this problem, one might *assume* that there is such a homogeneous quantum solution, or at least a homogeneous approximation of some quantum solution, and describe it as a minisuperspace quantization of the classically reduced model using methods as in Chapter 4.

Such models are obtained by quantization after restricting the Schwarzschild dynamics to the symmetric form of the classical interior region. Given the anisotropy, we cannot use quantized Friedmann models, but instead include an additional independent scale factor in a generic line element of the form

$$ds^2 = -N(t)^2 c^2 dt^2 + a(t)^2 dx^2 + b(t)^2 (d\vartheta^2 + \sin^2 \vartheta d\varphi^2). \tag{6.2}$$

This line element has the same symmetry properties as in Equation (1.156), written with a lapse function N, while $a(t)$ replaces $L(t, r)$ and $b(t)$ replaces $S(t, r)$.

Because this line element is spherically symmetric, it is a special case of Equation (1.119) with r-independent coefficients, as well as $M = 0$. The canonical structure is therefore determined by a restriction of Equation (1.140) to $M = 0$,

$$p_a = -\frac{b}{N} \frac{\partial b}{\partial t}, \quad p_b = -\frac{1}{N} \frac{\partial (ab)}{\partial t}. \tag{6.3}$$

The canonical Hamiltonian then follows from Equation (2.225) with $V(S) = -2/S$:

$$\mathrm{H}[N] = N \left(-\frac{p_a p_b}{b} + \frac{a p_a^2}{2b^2} - \frac{a}{2} \right). \tag{6.4}$$

It implies the equations of motion

$$\frac{\mathrm{d}a}{\mathrm{d}t} = \frac{\partial \mathrm{H}[N]}{\partial p_a} = N\left(-\frac{p_b}{b} + \frac{ap_a}{b^2}\right) \tag{6.5}$$

$$\frac{\mathrm{d}b}{\mathrm{d}t} = \frac{\partial \mathrm{H}[N]}{\partial p_b} = -N\frac{p_a}{b} \tag{6.6}$$

$$\frac{\mathrm{d}p_a}{\mathrm{d}t} = -\frac{\partial \mathrm{H}[N]}{\partial a} = \frac{1}{2}N\left(1 - \frac{p_a^2}{b^2}\right) \tag{6.7}$$

$$\frac{\mathrm{d}p_b}{\mathrm{d}t} = -\frac{\partial \mathrm{H}[N]}{\partial b} = -N\left(\frac{p_a p_b}{b^2} - \frac{ap_a^2}{b^3}\right). \tag{6.8}$$

If we introduce a time coordinate t such that

$$N(t) = \frac{1}{\sqrt{2Gm/(c^2 t) - 1}}, \tag{6.9}$$

we can use the Schwarzschild solution to read off classical solutions for the homogeneous variables $a(t)$ and $b(t)$:

$$a(t) = \sqrt{\frac{2Gm}{c^2 t} - 1}, \quad b(t) = t. \tag{6.10}$$

These functions indeed solve the canonical equations of motion and the constraint $\mathrm{H}[N] = 0$, with

$$p_a(t) = -t\sqrt{\frac{2Gm}{c^2 t} - 1}, \quad p_b(t) = 1 - \frac{Gm}{c^2 t}. \tag{6.11}$$

They reach a singularity ($t = 0$) at a phase-space point where $b(0) = 0$ and $a(0) \to \infty$. It is more convenient to express the approach to the black hole singularity in terms of new variables $q_1 = b^2$ and $q_2 = ab$, which are in fact components of a densitized triad with the required symmetry, restricting the general expression introduced in the context of loop quantum gravity, Equation (5.152). These variables both vanish at the singularity, making the black hole singularity look more like the Big Bang singularity with vanishing scale factor.

Any quantum effects that may make non-singular solutions possible in minisuperspace models for cosmology can also be applied to resolve the black hole singularity in a homogeneous black hole interior. In this new context, the implications are perhaps more significant because non-singular evolution then not only tells us more about a distant past before the Big Bang which we cannot see directly, but may also indicate that black holes are not perpetual final states of stellar collapse. If there is spacetime beyond the spacelike singularity of a non-rotating black hole, complete gravitational collapse is not inevitable. The black hole horizon is no longer a veil that cannot be lifted but might open up, perhaps after the black hole has slowly evaporated through Hawking radiation. (Currently, and for a long time to come, astrophysical black holes gain more mass by absorbing photons from the cosmic microwave background than they lose from evaporating Hawking photons. Hawking evaporation can therefore help to reveal a potentially non-singular black hole interior only long after the universe has cooled down much more due to expansion.)

However, even if it may be possible to construct non-singular quantum models for the homogeneous Schwarzschild interior, it is not guaranteed that we, as external observers that never cross the horizon, will be allowed to have a future peek inside. Future observations of the interior will be possible only if the interior remains connected to *our* exterior after evaporation. There are two assumptions in this statement. First, given only the interior quantization, we still have to make sure that it is connected to an inhomogeneous exterior region even before evaporation, which is not an easy task in a quantum setting. Second, two broad outcomes can in general be assumed for a non-singular quantum spacetime after evaporation: The interior may remain connected to the original exterior, in which case the black hole will open up in a unique surrounding spacetime. Or, it may be connected to a new exterior that is no longer causally accessible from the original exterior. In this case, the black hole would open up into a baby universe, its interior remaining invisible from a viewpoint in the parent universe. Such questions can be addressed only if we have access to quantum solutions in the exterior, which is spatially inhomogeneous and therefore sensitive to the challenging question of how to formulate covariant quantum spacetime models.

6.1.2 Timelike Homogeneity

The Schwarzschild exterior is not spatially homogeneous. However, it has a timelike symmetry in the t-direction, such that there is a symmetry group acting transitively on slices with $r = $ const, as in the case of the interior except that these slices are not spacelike. Nevertheless, the classical Hamiltonian formalism can formally be performed even if "time" (or, more precisely, the coordinate in a direction normal to the chosen three-dimensional slices) is spacelike. The resulting Hamiltonian description can then be quantized in much the same way as in the standard spacelike case, in particular if the slices are homogeneous and subtleties of the unconventional causal picture can be ignored. An application of such constructions to an understanding of black hole exteriors has been suggested by Ashtekar et al. (2018), using homogeneous models even in the inhomogeneous region of Schwarzschild spacetime.

Formally, it is easy to adjust the classical description by replacing the previous metric component a as well as the lapse function N with i times the original variables, while b remains unchanged. This transformation,

$$A = ia, \quad p_A = -ip_a, \quad n = iN \tag{6.12}$$

indeed transforms the line element (6.2) into the line element

$$ds^2 = n(t)^2c^2dt^2 - A(t)^2dx^2 + b(t)^2(d\vartheta^2 + \sin^2\vartheta d\varphi^2) \qquad (6.13)$$

in which slices of constant t are timelike. In the Schwarzschild interior, where Equation (6.2) applied, the new line element does not have homogeneous constant-t slices because the metric depends on t. However, if we apply the same transformation in the exterior, where metric coefficients depend on x, as for instance in Equation (1.154), the partial complexification of variables implies a homogeneous line element

$$ds^2 = n(x)^2c^2dt^2 - A(x)^2dx^2 + b(x)^2(d\vartheta^2 + \sin^2\vartheta d\varphi^2) \qquad (6.14)$$

on timelike slices of constant x.

The transformation (6.12) of the pair (a, p_a) includes the definition of a new momentum so as to make the transformation canonical, $\{A, p_A\} = \{a, p_a\} = 1$. Hamilton's equations for A and p_A will therefore have the usual expressions in terms of partial derivatives of the Hamiltonian. Similarly, minisuperspace quantization can be applied as we are used to from the spacelike case because the basic commutator of $\hat{A}$ and $\hat{p}_A$ has the standard form, $[\hat{A}, \hat{p}_A] = i\hbar$. A quantum description of the exterior can therefore be obtained, at least in one slicing distinguished by its homogeneity.

However, the same symmetry that guarantees homogeneity of timelike slices implies that there should, in any covariant theory, also be a spherically symmetric, spacelike slicing with static solutions. The latter would be inhomogeneous and more difficult to quantize or solve, but fortunately the possible versions of covariant theories for vacuum spherically symmetric spacetime are well understood, and rather highly restricted: If they are local with standard momenta, they must be of the form of a (1+1)-dimensional dilaton gravity action in which, as shown in Section 2.3.4, only the dilaton potential $V(S)$ is free to choose. Since the Hamiltonian timelike slicing also (implicitly) assumes locality because the number of classical metric degrees of freedom is not enhanced, any covariant quantum correction that modifies the timelike homogeneous model must, after transforming it to the spherically symmetric slicing, correspond to a modified dilaton potential. The required dependence on the dilaton potential on only one of the metric components makes this test of slicing independence rather constraining.

Interestingly, the main effect commonly used in loop quantum cosmology does *not* pass the test. Here, the classical Hamiltonian (6.4) is modified according to an anisotropic version of Equation (4.58), where the periodic function of p_V/p_0 is replaced by a suitable combination of periodic functions of p_A and p_b. The precise form is subject to quantization ambiguities, but for our purposes it suffices to assume a generic expression for a modified Hamiltonian of the form

$$H[n] = n\left(-\frac{p_A p_b}{b} + \frac{A p_A^2}{2b^2} + \frac{A}{2} + \delta h(A, b, p_A, p_b)\right), \qquad (6.15)$$

where δ is a parameter (analogous to $1/p_V$ in Equation (4.58)) that goes to zero in the classical limit. The function $h(A, b, p_A, p_b)$ models holonomy modifications of loop quantum gravity and need not be specified in detail here. We will only use the general property that it is non-polynomial in p_A and p_b if it is indeed derived from holonomies, in which p_A and p_b are homogeneous versions of the vector potential A_i^I. Compared with Equation (6.4), the version (6.15) for timelike slices differs only by the sign of the momentum-independent term, $A/2$.

While the geometrical meaning of A and b is the same as in the classical model, determined by their appearance in the line element (6.14), the relationship between momenta p_A and p_b and first-order derivatives of A and b, respectively, is modified by the new term δh. Because h is in general not quadratic in momenta, this modification obscures derivative orders in expressions that depend on configuration variables and momenta. It is therefore advantageous to make time derivatives explicit by computing Hamilton's equations of motion for $\dot{A}$ and $\dot{b}$ and inverting them. Instead of Equations (6.5) and (6.6), we now have

$$\frac{\mathrm{d}A}{\mathrm{d}t} = \frac{\partial \mathrm{H}[n]}{\partial p_A} = n\left(-\frac{p_b}{b} + \frac{Ap_A}{b^2} + \delta\frac{\partial h}{\partial p_A}\right) \tag{6.16}$$

$$\frac{\mathrm{d}b}{\mathrm{d}t} = \frac{\partial \mathrm{H}[n]}{\partial p_b} = n\left(-\frac{p_A}{b} + \delta\frac{\partial h}{\partial p_b}\right). \tag{6.17}$$

We have to invert the equations of motion in order to obtain the momenta in terms of time derivatives of A and b, which we can then insert in the Hamiltonian (6.15). Given the non-polynomial form of h, and therefore of $\partial h/\partial p_A$ and $\partial h/\partial p_b$, such an inversion cannot be performed in general, in particular if we do not know the precise form of h. Fortunately, it suffices to find a perturbative inversion which is valid up to first order in δ, as this order will tell us whether the leading quantum corrections can be covariant. Since an effective theory, expanded in some parameter δ, is covariant if and only if it is covariant to all orders in δ, a first-order calculation is sufficient to reveal problems with covariance. Such a perturbative inversion of the equations of motion yields

$$p_A = -\frac{b}{n}\frac{\mathrm{d}b}{\mathrm{d}t} + \delta b\frac{\partial h}{\partial p_b} \tag{6.18}$$

$$p_b = -\frac{1}{n}\left(b\frac{\mathrm{d}A}{\mathrm{d}t} + A\frac{\mathrm{d}b}{\mathrm{d}t}\right) + \delta\left(A\frac{\partial h}{\partial p_b} + b\frac{\partial h}{\partial p_A}\right). \tag{6.19}$$

The right-hand sides still depend on p_A and p_b through h, but only in terms multiplied by δ. To first order in δ, those momenta that appear in h can therefore be identified with their classical expressions in terms of time derivatives.

The modified Hamiltonian then takes the form

$$
\begin{aligned}
\mathrm{H}[n] = &- n\left(\frac{b}{n^2}\frac{\mathrm{d}A}{\mathrm{d}t}\frac{\mathrm{d}b}{\mathrm{d}t} + \frac{1}{2}A\left(\frac{1}{N^2}\left(\frac{\mathrm{d}b}{\mathrm{d}t} \right)^2 - 1 \right) \right) \\
&+ n\delta\left(-p_A\frac{\partial h}{\partial p_A} - p_b\frac{\partial h}{\partial p_b} + h \right).
\end{aligned}
\tag{6.20}
$$

There is a non-zero correction of first order in δ unless h is linear in momenta. This exception is never realized in models of loop quantum cosmology.

If the model is covariant, the homogeneous line element (6.14) on timelike slices should be transformable to a static spherically symmetric one,

$$
\mathrm{d}s^2 = -K(X)^2c^2\mathrm{d}T^2 + L(X)^2\mathrm{d}X^2 + S(X)^2(\mathrm{d}\vartheta^2 + \sin^2\vartheta\mathrm{d}\varphi^2)
\tag{6.21}
$$

subject only to the condition of zero shift because we will be interested in static solutions that can correspond to homogeneous timelike slices. Comparing Equation (6.21) and (6.14), given by

$$
\mathrm{d}s^2 = n(x)^2c^2\mathrm{d}t^2 - A(x)^2\mathrm{d}x^2 + b(x)^2(\mathrm{d}\vartheta^2 + \sin^2\vartheta\mathrm{d}\varphi^2),
\tag{6.22}
$$

we can read off a unique transformation of coordinates and variables that could accomplish a mapping between the two slicings: In the homogeneous slicing, the "time" coordinate ct is normal to timelike homogeneous slices, and therefore corresponds to the spatial coordinate X along spherically symmetric slices. The homogeneous coordinate x in Equation (6.2) for this slicing corresponds to the actual time direction, given by T in the spherically symmetric slicing. Coordinates therefore transform according to $X = ct$, $cT = x$; see Figure 6.1.

By comparing the corresponding coefficients in the line elements, we find that the two slicings can be related by a transformation of variables given by

$$
A = K, \quad b = S, \quad n = L.
\tag{6.23}
$$

The constraint equation $\mathrm{H}[n] = 0$ imposed in the homogeneous slicing must then be satisfied also after a transformation to the spherically symmetric slicing, if both slicings are to be realized in a single spacetime solution. (However, the condition $\mathrm{H}[n] = 0$ in the homogeneous slicing need not be identical to the corresponding constraint in the spherically symmetric slicing. Since the spherically symmetric slicing is required to be static, there are additional conditions that the constraint in the homogeneous slicing could, and does as we will see, correspond to.)

In order to transform equations of motion or constraints, we use Equation (6.23) and replace any time derivative in equations of the timelike homogeneous model, where t is the coordinate normal to the slices, with a spatial derivative by X in the spherically symmetric model. The constraint (6.20) is therefore transformed to the expression

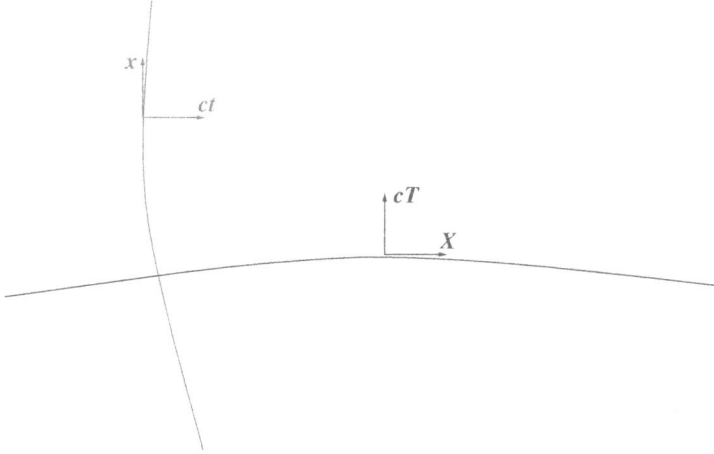

Figure 6.1. A homogeneous timelike slicing (coordinates (ct, x) in the line element (6.14), red) and a spherically symmetric spacelike slicing (coordinates (cT, X) in a line element (6.21), blue) exist in any spacetime region of a static spherically symmetric solution. The covariance of quantum corrections derived in homogeneous cosmological models can therefore be tested by trying to find a suitable transformation between quantum-corrected equations.

$$\mathrm{H}[L] = -\frac{S}{L}\frac{\mathrm{d}K}{\mathrm{d}X}\frac{\mathrm{d}S}{\mathrm{d}X} - \frac{1}{2}KL\left(\frac{1}{L^2}\left(\frac{\mathrm{d}S}{\mathrm{d}X}\right)^2 - 1\right)$$
$$+ L\delta\left(-p_A\frac{\partial h}{\partial p_A} - p_b\frac{\partial h}{\partial p_b} + h\right) \tag{6.24}$$

where

$$p_A = -\frac{S}{L}\frac{\mathrm{d}S}{\mathrm{d}X} + O(\delta) \tag{6.25}$$

$$p_b = -\frac{1}{L}\left(S\frac{\mathrm{d}K}{\mathrm{d}X} + K\frac{\mathrm{d}S}{\mathrm{d}X}\right) + O(\delta) \tag{6.26}$$

from Equations (6.18) and (6.19).

The derivative $\mathrm{d}K/\mathrm{d}X$ in Equation (6.24) seems unusual because K is the lapse function of the spherically symmetric slicing with line element (6.21). The expression $\mathrm{H}[L]$ given in Equation (6.24) therefore cannot be the Hamiltonian of the spherically symmetric slicing, which would depend on K but not on its partial derivatives. Nevertheless, we need $\mathrm{H}[L] = 0$ in the spherically symmetric slicing if it is to describe the same spacetime region in which $\mathrm{H}[n] = 0$ on homogeneous timelike slicings.

A second condition in the spherically symmetric slicing is given by the fact that it should be static. Therefore, the momenta of L and S must vanish, which gives us a new equation derived from the equation of motion (1.144) in general spherically symmetric models. (We need only one further condition, and therefore will not use Equation (1.143) in the present context. This equation is more complicated but would give equivalent results.) With $M = 0$ and $p_L = 0$, Equation (1.144) turns into

$$0 = -\frac{S}{KL^2}\frac{dS}{dX}\frac{dK}{dX} - \frac{1}{2L^2}\left(\frac{dS}{dX}\right)^2 - \frac{1}{4}SV(S) \tag{6.27}$$

with a general dilaton potential $V(S)$. We can therefore eliminate dK/dX from Equation (6.24), such that

$$H[L] = \frac{1}{2}KL\left(1 + \frac{1}{2}SV(S)\right) + O(\delta). \tag{6.28}$$

This expression vanishes on spherically symmetric slices if and only if the dilaton potential is the function determined by classical spherically symmetric models, $V(S) = -2/S$. (Dilaton models are covariant as field theories with one spatial dimension, x, for any dilaton potential. However, in general they do not have static solutions of Kantowski–Sachs form and therefore do not show the same relationship with homogeneous timelike slicings as seen in the case of spherically symmetric general relativity.)

For non-zero δ, the Hamiltonian $H[n]$ has additional terms compared with Equation (6.28), unless h is linear in p_A and p_b. These terms depend on p_b, which according to Equation (6.26) is a function of the spherically symmetric lapse function, K, and its derivative dK/dX. Using the staticity condition (6.27), while dK/dX is algebraically related to L, S and the derivatives of S, the lapse function K itself can be determined only by solving a differential equation. The variable K for static solutions, as required for a comparison with homogeneous timelike slices, is therefore non-local in X.

Quantum corrections of order δ are more complicated than the classical terms, in which the K-dependence in Equation (6.28) was reduced to a single factor, which could be eliminated by setting the parenthesis in Equation (6.28) equal to zero, restricting only the S-dependence of the dilaton potential. With δ-corrections, it is not possible to eliminate all terms in the analog of Equation (6.28) by absorbing them in the dilaton potential, because the latter is required to be a local function only of S. For any δ-corrections from non-polynomial h, as in models of loop quantum gravity, it is therefore impossible to ensure $H[L] = 0$ on spherically symmetric slices even though $H[n] = 0$ on homogeneous timelike slices. These quantum corrections violate slicing independence, and covariance in the classical sense.

While the discussion here was placed mainly in the context of black hole models, in which static spherically symmetric slicings are used most often, it has an important lesson for quantum cosmological models. Static solutions in spherical symmetry provide a systematic tool by which proposed modifications in homogeneous minisuperspace models can be tested for covariance. Even if one intends to apply a certain modification only in a cosmological context, it should also be applicable to homogeneous timelike slices in the Schwarzschild exterior if it can be part of a general prescription derived from quantum gravity. Any such modification should therefore be subject to the covariance condition derived from slicing independence in the Schwarzschild exterior. As seen here, this condition is very restrictive, ruling out any non-polynomial dependence on momenta of metric coefficients in modified Hamiltonians.

There are two possible ways to evade this conclusion. First, one might try to find a consistent *non-local* minisuperspace model, which would not be as strongly restricted by our covariance condition because the dilaton potential describes local covariant theories while non-local theories could provide further options. However, the set of non-local theories remains rather uncontrolled and poorly understood. Moreover, non-local minisuperspace models would have to deal with infinitely many momenta in order to describe time derivatives of arbitrary orders in a Hamiltonian for a non-local theory in time.

Second, one could try to find consistent *non-classical* spacetime structures which are covariant in some generalized sense (maintaining the same number of local gauge transformations as found in general relativity) while they do not implement classical slicing independence. This option, which is indeed possible, will be discussed further below.

6.1.3 Separate Quantization of Background and Perturbations

Since any homogeneity assumption eliminates most consistency conditions from covariance, it is tempting to set up an inhomogeneous description by combining a quantum minisuperspace model as a background on which quantized inhomogeneous modes evolve. If inhomogeneity is small, its back-reaction on the background can be ignored in an approximation. Quantum corrections in the minisuperspace dynamics, for instance according to Equation (4.58), may then change parameters in the mode Hamiltonian or the equations of motion it generates, such as the frequencies (1.90) in the classical perturbation Equation (1.88), which could have an influence on the formation of structure. Several such constructions have been proposed in canonical quantum gravity, for instance the "hybrid" approach (Martín-Benito et al. 2008) or the "dressed-metric" approach (Agulló et al. 2013), based in Ashtekar et al. (2009), or an alternative version developed by Dapor et al. (2013).

At first sight, it seems that covariance questions can be avoided in this setting even though inhomogeneity is included: Instead of plain metric perturbations, we can use Bardeen potentials (1.73) or, better, the curvature perturbations (1.75) in the presence of a matter scalar field in order to eliminate the dependence of modes on small coordinate changes. (As we have seen in Section 1.4.2, the Bardeen potentials, unlike curvature perturbations, are not completely invariant under small homogeneous transformations in the time direction.) The remaining transformations of large homogeneous coordinate changes, amounting to time reparameterizations, could be taken care of by applying deparameterization to the background dynamics; see Section 4.2.2.

Unfortunately, however, such constructions are problematic not only because deparameterization is imperfect, as discussed in Section 4.2.2. The separate treatment of small transformations that may be inhomogeneous and large homogeneous ones misrepresents the interrelated nature of these transformations in a covariant theory (Bojowald 2020). If the components ξ^α of a vector field, specifying the direction in which we perform an infinitesimal coordinate transformation, depend on time, as they generically do, the transformation does not commute with a general time reparameterization $t' = t + f(t)$ of the background time coordinate $t = x^0$: For a generic function $\phi(x^\alpha)$ on spacetime, we have

$$\sum_{\alpha=0}^{3} \xi^\alpha \frac{\partial \phi(t', x^1, x^2, x^3)}{\partial x^\alpha} - \left(\sum_{\alpha=0}^{3} \xi^\alpha \frac{\partial \phi(x^0, x^1, x^2, x^3)}{\partial x^\alpha} \right) \Bigg|_{x^0 = t'} \quad (6.29)$$

$$= \xi^0 \frac{\mathrm{d}f(t)}{\mathrm{d}t} \frac{\partial \phi}{\partial x^0} - \sum_{\alpha=0}^{3} f(t) \frac{\partial \xi^\alpha}{\partial t} \frac{\partial \phi}{\partial x^\alpha} \neq 0.$$

The first term in the second line is a consequence of the chain rule applied to $\mathrm{d}t'/\mathrm{d}x^0$ in the first term in the first line, while the second term in the second line is implied by a first-order Taylor expansion of $\xi^\alpha(t')$ in the second term of the first line. (Higher-order terms can be ignored in infinitesimal transformations.)

Applied to perturbative inhomogeneity, the components ξ^α should be small in order to implement small inhomogeneous coordinate transformations that maintain the perturbative order of fields. The background transformation $f(t)$, however, is not subject to this condition. The non-zero terms in Equation (6.29) are therefore of the same order as ξ^α and cannot be ignored in a theory of first-order perturbative inhomogeneity. The non-zero commutator (6.29) of coordinate transformations for the background and for inhomogeneity shows that these two symmetries cannot be implemented separately in a quantization, as would be done in an attempt to combine inhomogeneous curvature perturbations with deparameterization of the background.

In the canonical formulation of Section 2.3, the analog of a non-zero commutator (6.29) of infinitesimal coordinate transformations is given by non-zero brackets of hypersurface deformations. For perturbative inhomogeneity, the general bracket (2.154) with Equation (2.155), applied to a background transformation (spatially constant $\bar{N}$) and a small inhomogeneous transformation (small δN) in the normal direction in analogy with Equation (6.29), is reduced to

$$[\mathrm{H}[\bar{N}], \mathrm{H}[\delta N]] = \mathrm{D}[a^{-2} \bar{N} \nabla \delta N]. \quad (6.30)$$

This commutator is non-zero even if no explicit time dependence of the background function $\bar{N}$ is assumed: A spatially inhomogeneous normal deformation of a spatial slice by δN changes the normal directions on the slice, such that performing a background deformation by $\bar{N}$ before the inhomogeneous deformation differs from the same background deformation applied after the inhomogeneous deformation; see Figure 6.2.

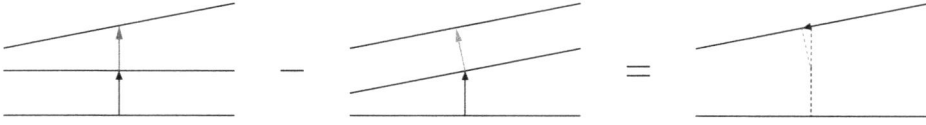

Figure 6.2. Non-zero commutator of a background deformation and a small inhomogeneous perturbation (here, linear). Since the inhomogeneous deformation changes the normal directions on the slice, a constant background deformation before the inhomogeneous deformation differs from the same background deformation applied after the inhomogeneous deformation.

Imposing invariance separately for background and perturbations, as in combinations of deparameterized background models with perturbative inhomogeneity, is inconsistent with the relation (6.30). Even though such an approach apparently implements the required transformations that originate from coordinate changes, it does not respect their algebraic structure. The algebraic structure is, however, important in order to make sure that the metric constructed from background and perturbations transforms such that it defines an invariant line element

$$\mathrm{d}s^2 = \sum_{\alpha=0}^{3} \sum_{\beta=0}^{3} g_{\alpha\beta} \mathrm{d}x^\alpha \mathrm{d}x^\beta \tag{6.31}$$

where the $\mathrm{d}x^\alpha$ transform by standard coordinate changes. If a quantum corrected $g_{\alpha\beta}$ has modified transformations, as implicitly implied by a separate treatment of background and perturbations, the resulting line element cannot be invariant. Any geometry based on such an object would be meaningless because distance measurements would depend on one's choice of coordinates.

6.2 Consistent Spacetime Structures

While canonical treatments are useful for various quantization methods, they sometimes make it difficult to recognize the relevant spacetime structure. Nevertheless, this important property is under full control even in canonical quantizations if one makes sure that hypersurface-deformation brackets are represented correctly. It remains unclear whether a full representation, without any assumption about symmetries or semiclassical behavior, is possible in the presence of quantum corrections. Nevertheless, several models indicate how the structure of spacetime may be changed by quantum effects such as spatial discreteness.

6.2.1 Closed Brackets for Spherically Symmetric Spacetimes

A simple model that illustrates the essential features, given by a single scalar field $\phi(x)$ with momentum $p(x)$ on a one-dimensional space with coordinate x, subject to a dynamics generated by the Hamiltonian (2.280), has already been discussed in Section 2.3.6. In this model, it is possible to maintain closed brackets even if the quadratic classical dependence on the momentum is replaced by a non-polynomial function. However, the bracket (2.154) is then modified as in Equation (2.279).

In particular, its sign can change for certain values of the momentum, indicating that Lorentzian spacetime is replaced by Euclidean-signature space in certain regions.

In this model, while the bracket $\{H[N_1], H[N_2]\}$ is closed, the bracket $[H[N], D[w]]$ is not a combination of H or D even with quadratic momentum dependence. This single-field model therefore cannot be fully covariant, a failure which can be repaired by introducing a second field and choosing suitable functional dependence of $H[N]$ on the two fields and their momenta. The resulting structure is a generalized version of what we have already seen in spherically symmetric models. In spite of the more complicated nature, one can show that the same modified brackets of the form (2.279) can be obtained.

The Hamiltonian

$$H[N] = \int dx N \left(p^2 - \frac{1}{4}\left(\frac{d\phi}{dx}\right)^2 - \frac{1}{2}\phi\frac{d^2\phi}{dx^2} \right) \tag{6.32}$$

of the toy model (2.280) does not describe a covariant theory because its integrand does not have density weight one, as required for a well-defined spatial integration. As determined by the constraint

$$D[w] = -\int dx w \phi \frac{dp}{dx} \tag{6.33}$$

of the same model, ϕ has density weight one, while p has density weight zero; see the transformations (2.281) generated by the constraint. The integrand of $H[N]$ therefore is a sum of terms with different density weights, none of which equals one. (In one spatial dimension, taking a derivative by x increases the density weight by one. The three terms in $H[N]$ therefore have density weight zero for the kinetic term and four for each of the "curvature" terms containing spatial derivatives.)

Modified Constraints
In order to construct a covariant system, we need to introduce a second field such that suitable ratios of the fields with their momenta or spatial derivatives in the numerator have density weight one. An example of such a system is given by the spherically symmetric Hamiltonian (2.225), in which the field L has density weight one and therefore can play the role of ϕ in the toy model, while p_L, like p in the toy model, has density weight zero. The second field, S, does not have a density weight, but its momentum p_S does have weight one. All combinations added up in the constraint

$$H[N] = \int_{-\infty}^{\infty} N \left(-\frac{p_L p_S}{S} + \frac{L p_L^2}{2S^2} + \frac{1}{2}\frac{(S')^2}{L} + \frac{SS''}{L} - \frac{SS'L'}{L^2} + \frac{1}{4}LSV(S) \right) dx \tag{6.34}$$

therefore have density weight one. According to our derivation of this Hamiltonian in Section 2.3.4, $H[N]$ together with

$$D[\epsilon] = \int_{-\infty}^{\infty} \epsilon (p_S S' - L p_L') \, dx \tag{6.35}$$

indeed have closed brackets of the required form, (2.154) with (2.155), (2.158), and (2.156).

In the toy model, we were able to modify the quadratic dependence on p and maintain closed brackets, albeit with modified structure functions (2.155). A similar modification with closely related structure function is possible in spherically symmetric gravity. Also here, we use motivations from loop quantum gravity which suggest that the classical quadratic dependence on momenta, or extrinsic curvature contained in the gravitational vector potential, is replaced by non-polynomial functions if the Hamiltonian is expressed in the holonomy-flux representation. Holonomies, in general, should be applied to both momentum components, p_L and p_S in spherically symmetric models. However, since they have different density weights, they allow different well-defined integrations. While p_S has density weight one and can directly be integrated over one-dimensional curves, $\int_{x_1}^{x_2} p_S \, dx$, as required for the exponent of a holonomy, p_L has density weight zero. Such a function cannot be integrated without an additional measure factor, but it can simply be exponentiated in a pointwise fashion, that is by using the exponentiated field $h(x) = \exp(i p_L(x))$ instead of the original field, $p_L(x)$. These exponentiated fields are referred to as *point holonomies* (Thiemann 1998) because they share with holonomies the property that they can be turned into well-defined operators in a holonomy-flux representation, even though they are not integrated.

In a spherically symmetric model with holonomy modifications, p_S would appear in non-local expressions integrated one-dimensionally, while p_L appears in local point holonomies. A derivative expansion of non-local features would, to leading order, not include any modifications of the dependence on p_S, while the local holonomy modification of p_L remains. We are therefore led to a generic modified Hamiltonian of the form

$$H[N] = \int_{-\infty}^{\infty} N \left(-\frac{f_2(p_L) p_S}{S} + \frac{L f_1(p_L)}{2S^2} + \frac{1}{2} \frac{(S')^2}{L} + \frac{SS''}{L} - \frac{SS'L'}{L^2} \right. $$
$$\left. + \frac{1}{4} L S V(S) \right) dx \tag{6.36}$$

with two free functions, f_1 and f_2, that could be used to model holonomy modifications. These functions are restricted by the condition that $H[N]$ should still have closed brackets after the modification.

Computing these brackets is very similar to our detailed derivations in Section 2.3.4. In particular, basic Poisson brackets of fields and momenta, applied to the expression $H[N]$ which contains spatial derivatives of S and L, produce derivatives of delta functions of up to second order. Eliminating these delta functions therefore requires several integrations by parts. The final result,

$$\{H[N_1],\ H[N_2]\} = \int_{-\infty}^{\infty} (N_1 N_2' - N_1' N_2)$$
$$\times \left(\frac{S'}{LS}\left(f_2(p_L) - \frac{1}{2}\frac{df_1}{dp_L} \right) + \frac{1}{L^2}\frac{df_2}{dp_L}(p_S S' - L p_L') \right) dx, \tag{6.37}$$

should equal $D[L^{-2}\beta(p_L)(N_1 N_2' - N_1' N_2)]$ for a covariant system, with some function $\beta(p_L)$ that approaches $\beta \to 1$ for small curvature, $p_L \to 0$. (Since the modification does not change the density weight of the integrand of $H[N]$, the bracket $\{H[N],\ D[\epsilon]\}$ is not affected.) This condition is fulfilled, provided

$$f_2(p_L) = \frac{1}{2}\frac{df_1}{dp_L} \tag{6.38}$$

and

$$\beta(p_L) = \frac{df_2}{dp_L}. \tag{6.39}$$

Only one function, f_1, then remains free. The combination of the two conditions,

$$\beta(p_L) = \frac{1}{2}\frac{d^2 f_1}{dp_L^2}, \tag{6.40}$$

shows that β is negative around any local maximum of $f_1(p_L)$, in which case we have signature change from Lorentzian to Euclidean.

The modified Hamiltonian does not introduce additional degrees of freedom and remains of first order in time derivatives, unlike most higher-curvature theories. It implies signature change whenever p_L is such that $\beta < 0$, which happens around local maxima of the function f_1. Although the theory is not of higher-curvature form, it is interesting to note that some higher-curvature effects can also imply signature change, for instance Einstein-dilaton Gauss–Bonnet gravity (Ripley & Pretorius 2019, 2020). In this case, however, only the dilaton field is subject to signature change, while the metric is always Lorentzian in this classical spacetime structure. Signature change of the dilaton in Einstein-dilaton Gauss–Bonnet gravity is therefore not a fundamental geometrical effect but rather an instability of the dilaton dynamics.

Hidden Signature Change
Signature change has been shown to be a generic consequence of holonomy modifications in models of loop quantum gravity (Barrau et al. 2015; Bojowald et al. 2020), which imply bounded modification functions such as f_1 with local maxima. In fact, signature change follows from the same general property of these modifications that has led to claims that the Big Bang singularity may be replaced by a bounce in loop quantum cosmology, based on modified Friedmann equations such

as Equation (4.58), which transfer bounded holonomy modifications to bounded energy densities. However, the modified spacetime structure implied by signature change in consistent models of holonomy modifications, drastically alters the bounce picture: Spacetime around the Big Bang is replaced by a sliver of four-dimensional space of Euclidean signature. These questions will be discussed more in Section 6.2.3.

There have been attempts to introduce holonomy modifications without having to face the consequence of modified spacetime structures. However, in all cases, signature change reappears when a detailed analysis is made. For instance, Gambini & Pullin (2013) made the interesting observation that, somewhat akin to the Virasoro algebra (5.81) and (5.82) on the string worldsheet, the constraints of spherically symmetric models can be rearranged by linear combinations so as to remove the complicated structure function (2.155): If we keep $D[\epsilon]$ and replace $H[N]$ with

$$H[2PS'/L] + D[2Pp_L/(SL)]=$$
$$\int_{-\infty}^{\infty} P\frac{d}{dx}\left(-\frac{p_L^2}{S} + \frac{S(S')^2}{L^2} + \frac{1}{2}\int SV(S)dS\right)dx, \tag{6.41}$$

p_S can be eliminated, and the remaining integrand is given by the new multiplier function P times a total derivative. (For spherically symmetric models, $SV(S)$ with the dilaton potential $V(S) = -2/S$ can easily be integrated: $\int SV(S)dS = -2S$.) The new constraint is equivalent to the constraint

$$C[Q] = \int_{-\infty}^{\infty} Q\left(-\frac{p_L^2}{S} + \frac{S(S')^2}{L^2} + \frac{1}{2}\int SV(S)dS + C_0\right)dx \tag{6.42}$$

where an arbitrary constant C_0 appears after integrating the total x-derivative in the original expression. (The original constraint (6.41) also, implicitly, implies an undetermined constant because it does not impose local field equations for constant P.)

The new multiplier, Q, has density weight minus one because, compared with Equation (6.41), a derivative has been eliminated in the integrand. We therefore have the Poisson bracket

$$\{C[Q], D[\epsilon]\} = -C[(\epsilon Q)'] \tag{6.43}$$

reflecting the density weight by replacing the usual $\epsilon Q'$ with $(\epsilon Q)' = \epsilon Q' + Q\epsilon'$ on the right-hand side. More importantly, the constraints $C[Q]$ commute with each other,

$$\{C[Q_1], C[Q_2]\} = 0. \tag{6.44}$$

This result follows immediately from the antisymmetry of the bracket together with the fact that C only depends on p_L, L, S, and S', but not on p_S or spatial derivatives of L. Antisymmetry implies that non-zero terms can be produced from Poisson brackets only of the form $Q_1'Q_2 - Q_1Q_2'$, which requires derivatives of delta functions. However, no such derivatives appear if the only momentum variable that appears in the constraint, p_L, is conjugate to a field, L, that does not appear with spatial derivatives in the constraint.

The brackets (6.43) and (6.44) rely on general features that do not require a specific dependence on the momentum p_L, which moreover does not have a density weight. The quadratic dependence p_L^2 in C[Q] can therefore be replaced by any function $f_1(p_L)$. The brackets remain not only closed but also unchanged compared with the classical result (6.44). This result is in apparent conflict with our previous derivation, (6.37), which showed that the spacetime structure is necessarily affected by holonomy modifications.

However, after reformulating the constrained system by introducing C[Q], spacetime structure has been obscured because we no longer see the relationship with classical hypersurface-deformation brackets (2.154). In order to make the spacetime structure manifest after we modified the constraint C[Q] to

$$C[Q] = \int_{-\infty}^{\infty} Q\left(-\frac{f_1(p_L)}{S} + \frac{S(S')^2}{L^2} + \frac{1}{2}\int SV(S)dS + C_0\right)dx, \qquad (6.45)$$

we should retrace our steps that led us from H[N] and D[ϵ] to D[ϵ] and C[Q]. In particular, as in Equation (6.41), a linear combination of some modified $\tilde{H}[N]$ and D[ϵ] should be equal to a multiplier function P times the x-derivative of the integrand of C[Q]:

$$\tilde{H}[2PS'/L] + D[2Pf_3(p_L)/(SL)]$$

$$= \int_{-\infty}^{\infty} P\frac{d}{dx}\left(-\frac{f_1(p_L)}{S} + \frac{S(S')^2}{L^2} + \frac{1}{2}\int SV(S)dS\right)dx$$

$$\qquad\qquad (6.46)$$

$$= \int_{-\infty}^{\infty} 2P\left(-\frac{p_L'}{2S}\frac{df_1}{dp_L} + \frac{S'}{L}\left(\frac{Lf_1(p_L)}{2S^2} + \frac{(S')^2}{2L} + \frac{SS''}{L} - \frac{SS'L'}{L^2} + \frac{LSV(S)}{4}\right)\right)dx.$$

The multiplier in D[M] used in this equation may have to be adjusted compared with classical case, as indicated by the function $f_3(p_L)$.

Solving this equation for $\tilde{H}[N]$ (which, incidentally, requires $f_3(p_L) = \frac{1}{2}df_1/fp_L = f_2(p_L)$), we find that it is exactly of the form (6.36) in which the consistency condition (6.38) has already been implemented. Computing the Poisson bracket $\{\tilde{H}[N_1], \tilde{H}[N_2]\}$ therefore must lead to the previous results, Equation (6.37), with the same modification function (6.40). In particular, modifying or quantizing the simplified constraint C[Q] implies the same kind of signature change as seen for the original Hamiltonian H[N] (Bojowald et al. 2015).

6.2.2 Effective Line Element

Implications of the modified bracket on spacetime structure can be seen directly if one reconstructs the form of line elements compatible with the modified transformations, using the methods of canonical gravity. Classically, any transformation generated by the constraints of hypersurface deformations,

$$\delta F = \{F, \mathrm{H}[\epsilon^0] + \mathrm{D}[\vec{\epsilon}]\},\qquad(6.47)$$

is equivalent to a derivative in the direction of a spacetime vector field with components ξ^α, where

$$\xi^0 = \frac{\epsilon^0}{N},\quad \vec{\xi} = \epsilon^i - \frac{\vec{M}}{N}\epsilon^0\qquad(6.48)$$

using the background lapse and shift functions, N and $\vec{M}$, of the spacetime on which the derivative is performed. These relations merely rewrite the reference directions normal and tangential to spatial slices, used in ϵ^0 and $\vec{\epsilon}$, as coordinate directions:

$$\epsilon^0 n^\alpha + \epsilon^i s_i^\alpha = \frac{\epsilon^0}{N}t^\alpha + \left(\epsilon^i - \frac{N^i}{N}\epsilon^0\right)s_i^\alpha = \xi^0 t^\alpha + \xi^i s_i^\alpha.\qquad(6.49)$$

A general coordinate transformation may also change the background functions N and $\vec{M}$, defined such that

$$ds^2 = -N^2 dt^2 + \sum_{i=1}^{3}\sum_{j=1}^{3} q_{ij}(dx^i + M^i dt)(dx^j + M^j dt)\qquad(6.50)$$

with the spatial metric q_{ij}. This behavior is important in the present context because it depends on the brackets of constraints and therefore changes if the brackets are modified. In general, if the brackets of constraints C_A are of the form

$$\{C_A, C_B\} = \sum_D F_{AB}^D C_D\qquad(6.51)$$

with suitable functions F_{AB}^D, a transformation of the metric should be combined with the transformation

$$\delta_\epsilon N^A = \frac{\partial \epsilon^A}{\partial t} + \sum_{BC} N^B \epsilon^C F_{BC}^A\qquad(6.52)$$

of the background functions. (The notation here is somewhat formal, as the summations may include integrations over spatial coordinates.) This condition ensures that time evolution in the form of constrained Hamilton's equations,

$$\frac{dF}{dt} = \{F, \mathrm{H}[N] + \mathrm{D}[\vec{M}]\},\qquad(6.53)$$

is compatible with gauge transformations (Bojowald 2010).

For the brackets of classical canonical gravity, the general relation (6.51) implies

$$\delta_\epsilon N = \frac{\partial \epsilon^0}{\partial t} + \vec{\epsilon}\cdot\nabla N - \vec{M}\cdot\nabla\epsilon^0\qquad(6.54)$$

and

$$\delta_\epsilon \vec{M} = \frac{\partial \vec{\epsilon}}{\partial t} + (\vec{\epsilon} \cdot \nabla)\vec{M} - (\vec{M} \cdot \nabla)\vec{\epsilon} - \mathbf{q}^{-1}(N\nabla\epsilon^0 - \epsilon^0\nabla N), \qquad (6.55)$$

where $\mathbf{q}^{-1}$ represents the inverse metric as a matrix with components q^{ij}. However, if the bracket (2.279) is modified, the latter relation should be changed to

$$\delta_\epsilon \vec{M} = \frac{\partial \vec{\epsilon}}{\partial t} + (\vec{\epsilon} \cdot \nabla)\vec{M} - (\vec{M} \cdot \nabla)\vec{\epsilon} - \beta\mathbf{q}^{-1}(N\nabla\epsilon^0 - \epsilon^0\nabla N). \qquad (6.56)$$

The coefficients of the line element (6.50) therefore transform differently, in a way that no longer matches the coordinate transformations of dt and dx^i. In order to obtain an invariant expression, the form (6.50) must be modified.

Still working with spherically symmetric models, we have two components, q_{xx} and $q_{\varphi\varphi}$, of the spatial metric in

$$ds^2 = -N(x, t)^2 dt^2 + q_{xx}(x, t)(dx + M(x, t)dt)^2 + q_{\varphi\varphi}(x, t)(d\vartheta^2 + \sin^2\vartheta d\varphi^2). \qquad (6.57)$$

If we apply a small coordinate transformation $t' = t + \xi^0$ and $x' = x + \xi^x$, where

$$\xi^0 = \frac{\epsilon^0}{N}, \qquad \vec{\xi} = \vec{\epsilon} - \frac{\vec{M}}{N}\epsilon^0 \qquad (6.58)$$

in terms of gauge parameters, the infinitesimal coordinate differences transform as

$$dx' = d(x + \epsilon^x - (N^x/N)\epsilon^0) = dx + \frac{\partial(\epsilon^x - (M/N)\epsilon^0)}{\partial t}dt$$
$$+ \frac{\partial(\epsilon^x - (M/N)\epsilon^0)}{\partial x}dx \qquad (6.59)$$

and

$$dt' = d(t + \epsilon^0/N) = dt + \frac{\partial(\epsilon^0/N)}{\partial t}dt + \frac{\partial(\epsilon^0/N)}{\partial x}dx. \qquad (6.60)$$

The derivation of these transformations is analogous to methods used in the construction of Bardeen variables in Section 1.4.2. Inserting these transformations in the line element, we obtain the corresponding change

$$\delta q_{xx} = \frac{\epsilon^0}{N}\frac{\partial q_{xx}}{\partial t} + \left(\epsilon^x - \frac{M}{N}\epsilon^0\right)\frac{\partial q_{xx}}{\partial x} + 2q_{xx}\frac{\partial}{\partial x}\left(\epsilon^x - \frac{M}{N}\epsilon^0\right)$$
$$+ 2q_{xx}M\frac{\partial}{\partial x}\left(\frac{\epsilon^0}{N}\right) \qquad (6.61)$$

of the spatial metric, along with

$$\delta M = \frac{\partial\epsilon^x}{\partial t} + \epsilon^x\frac{\partial M}{\partial x} - M\frac{\partial\epsilon^x}{\partial x} - \frac{1}{q_{xx}}\left(N\frac{\partial\epsilon^0}{\partial x} - \epsilon^0\frac{\partial N}{\partial x}\right) \qquad (6.62)$$

and

$$\delta N = \frac{\partial \epsilon^0}{\partial t} + \frac{\partial N}{\partial x} \epsilon^x - M \frac{\partial \epsilon^0}{\partial x}. \tag{6.63}$$

In particular, Equation (6.62) is not compatible with the transformation (6.56) that results from modified brackets.

Applying the same methods (Bojowald et al. 2018) shows that the modified bracket is compatible with standard coordinate changes of t and x in an effective line element

$$ds^2 = -\beta(t)N(x, t)^2 dt^2 + q_{xx}(x, t)(dx + M(x, t)dt)^2 + q_{\varphi\varphi}(x, t)$$
$$(d\vartheta^2 + \sin^2 \vartheta d\varphi^2), \tag{6.64}$$

using the same function β as in Equation (2.279), and assuming that β depends only on time. (This assumption applies in situations of perturbative inhomogeneity, in which modifications β result only from the background dynamics. If the assumption is violated, it remains unknown whether a more general version of an effective line element still exists, or whether the spacetime structure is entirely non-Riemannian.) When the sign of β changes, it is now clear that we obtain a positive definite line element, as expected for four-dimensional space of Euclidean signature.

6.2.3 Implications of Signature Change

Signature change is an unexpected and counter-intuitive phenomenon. It is an implication not only of certain discrete spatial structures in a consistent spacetime (Bojowald & Paily 2012; Mielczarek 2014; Bojowald & Mielczarek 2015), but has also been found, in a variety of technical realizations, in minisuperspace models (Martin 1994), string theory (Perry & Teo 1993), matrix models (Chaney et al. 2015; Chaney & Stern 2017; Steinacker 2018; Stern & Xu 2018), and causal dynamical triangulations (Ambjørn et al. 2015). While signature change is possible also in classical solutions, in which case it has to be arranged by hand, patching together solutions of Lorentzian and Euclidean gravity along a hypersurface of zero extrinsic curvature, it is always singular in this case (Ellis et al. 1992; Hayward 1993; Dray et al. 1995; Alty & Fewster 1996; Dray et al. 1997). Modified spacetime structures, by contrast, allow for non-singular signature change, for instance when the function β in Equation (2.279) smoothly goes through zero.

In the present context, setting aside details of signature change, the general outcome that a non-classical geometrical structure is realized when the quadratic classical curvature dependence of the Hamiltonian is replaced by a non-quadratic function, evades the slicing problem observed in Section 6.1.2: It is possible to have non-quadratic modifications in a covariant theory, but the covariance relations implied by closed Poisson brackets of the constraints are modified and do not correspond to classical general covariance. The corresponding non-classical spacetime structure is not compatible with the usual form of slicing independence, which is why we obtained a contradiction in Section 6.1.2. Slicing independence is obtained only after transforming to a new effective line element such as Equation (6.64), if one exists. However, the precise transformation remains unknown in the general situation in which the spacetime dependence of β is not only through time.

Conceptually, signature change also helps to explain why models of loop quantum cosmology can be non-singular with standard matter sources, even though they do not violate the usual energy conditions applied in singularity theorems of general relativity. Holonomy modifications in loop quantum cosmology do modify the equations of motion, but the most general singularity theorems are insensitive to the specific dynamics of general relativity and only use broadly applicable geometrical properties such as the geodesic deviation equation. It is therefore not obvious how precisely loop quantum cosmology may be able to evade these theorems.

Signature change presents one possible answer to this question, the only one known so far. In this view, the mathematical outcome of singularity theorems remains unchanged, as it should. These theorems show that, given certain assumptions on initial conditions such as the existence of universal expansion or the presence of trapped surfaces, timelike geodesics cannot last for an infinite amount of time. In the usual spacetime interpretation, this mathematical result is taken to show that spacetime has a boundary at some finite time in the past or future beyond which general relativity does not provide an extension. This finite boundary presents a singularity, not necessarily in terms of diverging densities or curvatures, but as a limitation of the spacetime description derived from the theory.

Signature change solves the singularity problem almost by cutting the Gordian knot; it removes the conceptual problem posed by the required inextendability of timelike geodesics by removing time itself: The assumptions and the outcome of singularity theorems remain mathematically valid, but timelike geodesics end just because time exists only in a limited region within a large universe subject to signature change. As a manifold, the universe does have an extension beyond the point where timelike geodesics end, but only in a four-dimensional spacelike manner. This extended universe is still geometrical, but the geometry is not Riemannian throughout. Only in certain subregions, might an effective line element exist which can map the geometry restricted to such regions to Riemannian form. With this generalized interpretation, singularity theorems do not imply boundaries to the whole manifold of the universe. However, there are still boundaries to causal behavior because there is no deterministic evolution through a region with Euclidean signature.

In particular, a well-posed system of partial differential equations in a geometry with signature change requires some final conditions and cannot be obtained purely as an initial-value problem. This general conclusion follows from the fact that a Euclidean-signature metric implies elliptic partial differential equations for test fields, rather than hyperbolic wave equations. The same general outcome can be seen also in the context of geodesics. If we assume that an effective line element of the form (6.64) can be used (as is the case for instance in the Schwarzschild interior), its time-time component approaches zero in a region where $\beta \to 0$, close to signature change. All light cones in such a region, determined as usual by the condition

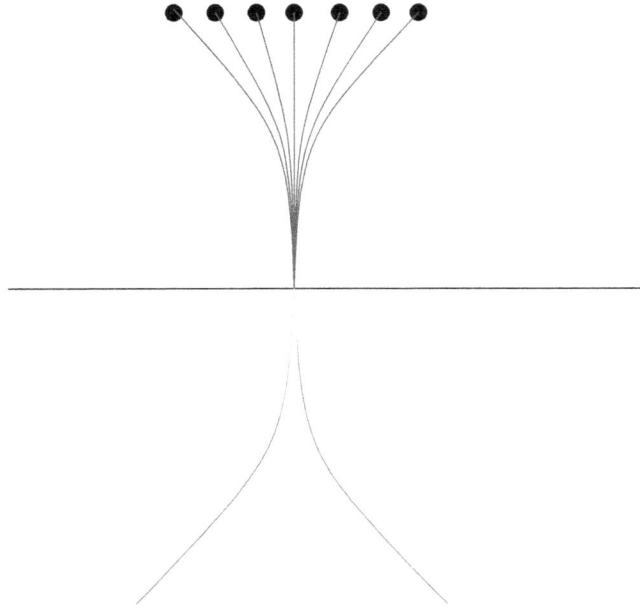

Figure 6.3. Illustration of a family of geodesics in a geometry with signature change, such that Lorentzian signature is realized in the bottom half and Euclidean signature in the top half. In the Lorentzian region, light cones (red) exist but collapse to a unique direction at the transition hypersurface. All timelike geodesics (yellow) therefore approach the same direction. Similarly, all spacelike geodesics in the Euclidean region that approach the tip of the red light cone on the transition hypersurface approach this point in the same asymptotic direction. Any blue spacelike geodesic could therefore be used to extend a given yellow timelike geodesic in a differentiable manner. Unique extensions of timelike geodesics into the Euclidean region as spacelike geodesics (blue) are possible only if a second endpoint is specified (black circles), selecting a specific blue geodesic. The required final position amounts to future data required for a well-posed problem.

$ds^2 = 0$, have tiny spatial extensions which collapse to zero at $\beta = 0$. Therefore, all timelike geodesics (as well as lightlike ones) approach a unique direction at the hypersurface of signature change, as shown in Figure 6.3.

There are no light cones in the Euclidean region, but an analysis of the geodesic equation shows that also there, geodesics that approach the hypersurface of signature change do so along the normal direction asymptotically close to the hypersurface. Intuitively, this happens because geodesics in a space of Euclidean signature have the shortest distance among all curves connecting two different points. If t is a coordinate normal to the hypersurface of signature change (using the usual letter for the time coordinate even when there is no time in the Euclidean region), the metric component g_{tt} goes through zero where signature change takes place. Therefore, the line element in the Euclidean region is the smallest in the direction of t, which is the direction that geodesics have to follow.

We demonstrate this property formally in two dimensions, using a general line element

$$ds^2 = th_1(t, x)dt^2 + 2\sqrt{|t|}\, h_2(t, x)dtdx + g_{xx}(t, x)dx^2 \tag{6.65}$$

with positive coefficients $h_1 > 0$ and $g_{xx} > 0$, such that $d := h_1 g_{xx} - h_2^2 > 0$. Therefore,

$$\det g = th_1 g_{xx} - |t|h_2^2 = -|t|(h_1 g_{xx} + h_2^2) = -|t|(d + 2h_2^2) < 0 \tag{6.66}$$

for $t < 0$ (the Lorentzian region) while

$$\det g = th_1 g_{xx} - th_2^2 = t(h_1 g_{xx} - h_2^2) > 0 \tag{6.67}$$

for $t > 0$ (the Euclidean region).

Along spacelike or timelike worldlines, we have

$$\frac{ds^2}{d\lambda^2} = th_1\left(\frac{dt}{d\lambda}\right)^2 + 2\sqrt{|t|}\, h_2 \frac{dt}{d\lambda}\frac{dx}{d\lambda} + g_{xx}\left(\frac{dx}{d\lambda}\right)^2 = \pm 1 \tag{6.68}$$

with the upper sign for $t > 0$ (spacelike worldlines in the Euclidean region, where the curve parameter λ is then proper distance) and the lower sign for $t < 0$ (timelike worldlines in the Lorentzian region, where λ is then proper time). Thus, near $t = 0$ the component $dt/d\lambda$ of the tangent vector dominates because the equation can be solved for

$$\frac{dt}{d\lambda} = -\frac{1}{\sqrt{|t|}\, h_1}\left(\text{sgn}(t)h_2\frac{dx}{d\lambda} \pm \sqrt{(h_2^2 - \text{sgn}(t)h_1 g_{xx})\left(\frac{dx}{d\lambda}\right)^2 \pm \text{sgn}(t)h_1}\right) \tag{6.69}$$

$$= O(|t|^{-1/2})$$

in terms of $dx/d\lambda = O(1)$. The dominant contributions to the geodesic equation are then

$$\frac{d^2t}{d\lambda^2} \sim -\frac{1}{2t}\left(\frac{dt}{d\lambda}\right)^2(1 + O(t)), \quad \frac{d^2x}{d\lambda^2} \sim \left(\frac{dt}{d\lambda}\right)^2 O(\sqrt{|t|}). \tag{6.70}$$

Therefore, $x(\lambda)$ is approximately constant, while the first equation, written as

$$0 \sim 2t\frac{d^2t}{d\lambda^2} + \left(\frac{dt}{d\lambda}\right)^2 = 2\sqrt{|t|}\, \text{sgn}(t)\frac{d}{d\lambda}\left(\sqrt{|t|}\frac{dt}{d\lambda}\right) \tag{6.71}$$

implies that $\sqrt{|t|}\, dt/d\lambda$ is approximately constant. Along geodesics, the slope

$$\frac{dt}{dx} = \frac{dt/d\lambda}{dx/d\lambda} \sim \frac{\text{const}}{\sqrt{|t|}} \tag{6.72}$$

therefore approaches infinity near $t = 0$ in both the Lorentzian and the Euclidean region, as shown in Figure 6.3.

No timelike geodesics exist in the Euclidean region, but we can try to extend a given timelike geodesic arriving at the hypersurface of signature change from the Lorentzian side as a spacelike geodesic extending into the Euclidean region.

However, because all timelike geodesics arriving at a given point on the hypersurface of signature change do so in the same direction, the extension cannot be unique. In addition to specifying initial values to determine which timelike geodesic we are interested in, we have to specify future conditions in the Euclidean region in order to determine an extension by a spacelike geodesic. Because future conditions are required, evolution cannot be deterministic.

The relevant geometry therefore implies a mixture of different, previously irreconcilable scenarios for the interior of black holes. Since quantum modifications are used in the dynamics, it may be possible that the interior does not reach infinite curvature or density of collapsing matter but rather bounces or extends otherwise through the classical singularity. However, interpreted in the corresponding space-time structure, this transition is not deterministic and requires future data. This latter feature is reminiscent of proposals that tried to address the singularity problem in black hole models by imposing a future condition for wave functions at the singularity (Horowitz & Maldacena 2004; Bouhmadi-López et al. 2019), as some kind of generalization of DeWitt's condition in cosmological models.

Although there is a consistent realization of spacetime in such models, their analysis remains difficult (Bojowald 2015; Ben Achour et al. 2018a, 2018b; Aruga et al. 2019). In particular, the question of how a non-singular interior might be connected to an exterior spacetime, opening up into the original universe or splitting off into a baby universe, remains open. In order to address this question one would like to have access to an effective line element, but β cannot be just time dependent in the exterior as required for Equation (6.64). The non-Riemannian geometry in the exterior therefore remains hard to address.

6.2.4 Perturbative Inhomogeneity

Deformed hypersurface-deformation brackets have been found also for perturbative inhomogeneity in cosmological models (Bojowald et al. 2008; Cailleteau et al. 2012, 2014; Cuttell & Sakellariadou 2014; Cuttell 2019), in a form very similar to those in spherically symmetric models (Barrau et al. 2015). In particular, there is a multiplier β as in Equation (6.30) which becomes negative at Planckian curvature, indicating signature change.

Equations of motion for matter fields, such as a scalar in models of inflation, are relevant in cosmology. These are partial differential equations of a type that depends on the signature of spacetime—hyperbolic in Lorentzian signature and elliptic in Euclidean signature. The conclusion that this behavior is directly related to the sign of β can be obtained from a reconstruction of consistent equations of motion from the (deformed) brackets, following the methods of Kuchař (1976a, 1976b, 1976c).

It is now convenient to Legendre transform the brackets for Hamiltonians, such as $H(\phi, p_\phi)$ for a scalar field ϕ with momentum p_ϕ, to Lagrangian form, introducing the "velocity" $v(x) = \partial H / \partial p_\phi(x) = \{\phi, H[N]\}/N$. The Lagrangian (density) is then defined as $L = p_\phi v - H$, such that

$$\frac{\partial H}{\partial \phi}\bigg|_{p_\phi \text{ constant}} = -\frac{\partial L}{\partial \phi}\bigg|_{v \text{ constant}} \tag{6.73}$$

and

$$p_\phi(x) = \frac{\partial L}{\partial v(x)}. \tag{6.74}$$

The bracket of two smeared Hamiltonian constraints then equals

$$\{H[N], H[M]\} = \int d^3x \int d^3y\, N(y)\frac{\delta H(y)}{\delta\phi(x)}v(x)M(x) - (N \leftrightarrow M) \tag{6.75}$$

$$= -\int d^3x \int d^3y\, \frac{\delta L(y)}{\delta\phi(x)}v(x)N(y)M(x) - (N \leftrightarrow M) \tag{6.76}$$

$$= D[\beta(N\nabla M - M\nabla N)], \tag{6.77}$$

introducing the diffeomorphism constraint $D[\vec{\epsilon}] = \int \vec{\epsilon} \cdot p_\phi \nabla\phi d^3x$ in the last step, where we assume brackets of the form (2.279). Comparing the last two lines,

$$\frac{\delta L(x)}{\delta\phi(x')}v(x') + \beta(x)\vec{D}(x) \cdot \nabla_x\delta(x, x') - (x \leftrightarrow x') = 0 \tag{6.78}$$

where $\vec{D}(x) = p_\phi(x)\nabla\phi(x)$.

Using the standard geometrical structure of space, the Lagrangian density must be of the form $\mathcal{L} = \sqrt{\det q}\, L(\phi, v, \psi)$ with a Lagrange function L that can depend pointwise on ϕ, its velocity v normal to spatial slices, and the rotationally invariant combination

$$\psi = \sum_{i=1}^{3}\sum_{j=1}^{3} q^{ij}\frac{\partial\phi}{\partial x^i}\frac{\partial\phi}{\partial x^j} \tag{6.79}$$

of spatial derivatives. Comparing coefficients of $\nabla\phi$ in Equation (6.78), after an application of the chain rule

$$\frac{\delta L(x)}{\delta\phi(x')} = \frac{\partial L(x)}{\partial\phi(x)}\frac{\delta\phi(x)}{\delta\phi(x')} + 2\frac{\partial L(x)}{\partial\psi(x)}\nabla\phi(x) \cdot \nabla\delta(x, x'), \tag{6.80}$$

implies

$$\beta p_\phi + 2v\frac{\partial L}{\partial\psi} = \beta\frac{\partial L}{\partial v} + 2v\frac{\partial L}{\partial\psi} = 0. \tag{6.81}$$

Therefore, L must be of the form $L(\phi, \psi - v^2/\beta)$.

For $\beta > 0$, the combination $\psi - v^2/\beta$ has coefficients of opposite signs in its spatial and temporal derivatives, respectively, implying hyperbolic equations of motion from any Lagrangian consistent with a bracket (2.279). For $\beta < 0$, by contrast, spatial and temporal derivatives enter the Lagrangian with equal signs (as in Euclidean field theory) and imply elliptic partial differential equations for the field.

If β changes its sign dynamically, for instance if it depends on a varying curvature scale as in Equation (6.40), we have a mixed-type problem of partial differential equations that requires a combination of initial values and final values. For instance, a free scalar field on a cosmological background with Hubble parameter H might be subject to the mixed-type wave equation

$$-\frac{\partial^2 \phi}{\partial t^2} + \beta(H)\Delta\phi = 0 \tag{6.82}$$

(using the spatial Laplacian Δ) if spacetime is modified, with hypersurface deformations according to Equation (2.279).

A well-posed formulation, introduced by Tricomi (1968) and shown in two dimensions in Figure 6.4; imposes initial values on a single light ray C, or half the usual characteristic problem. Where the light ray approaches the hypersurface of signature change, it is smoothly connected to an arc A in the Euclidean region. If the arc reaches all the way to the end of a second light ray C' starting at the same event as C but moving in the opposite direction, the problem is well-posed: For smooth data of ϕ (but none of its derivatives) on the union $C \cup A$, there is a unique and smooth solution in the region bounded by C, A and C' which depends stably on the data. (Had one tried to evolve usual initial data into the Euclidean region, there might still be a unique solution but it would not depend stably on the data.)

Notice, however, that solutions generically have a root-like pole at the far end of A where it touches C'. This "cosmic boom" is an analog of the sonic boom in hydrodynamics which can be described by similar mixed-type equations. It implies that even small perturbations may lead to solutions with large derivatives. Even if one can extend a metric into the Euclidean region, there may therefore be curvature divergences. In contrast to the standard Big Bang singularity, such divergences would be located at isolated points rather than on an entire spatial slice. As in hydrodynamics, they would be singularities of the effective spacetime description while the underlying fundamental theory, given by a consistent set of canonical constraints H[N] and D[M], would be non-singular. Nevertheless, such singularities in an effective cosmological description complicate any physical analysis based on perturbative inhomogeneity.

These examples show that spacetime properties in models of quantum cosmology can be rather counter-intuitive. An analysis of possible evolution at large curvature therefore requires great care and good control on mathematical conditions such as the treatment of well-posed initial-boundary problems that are not required classically. None of these features could have been found had one stayed at a restricted level of minisuperspace models, or perturbations on minisuperspaces.

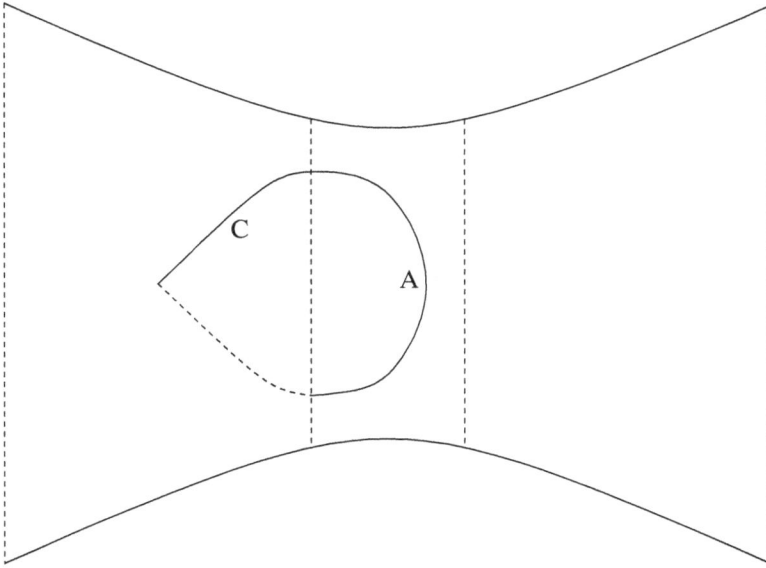

Figure 6.4. The two-dimensional Tricomi problem on a bouncing cosmological background. Initial data are posed on one light ray C in the Lorentzian region, smoothly connected to an arc A in the Euclidean region that ends at the endpoint of the second light ray starting at the same event as C.

6.3 Cosmology at Large Curvature

There are (at least) two general reasons why perturbative inhomogeneity cannot be sufficient to describe quantum spacetime very close to the Big Bang.

First, if we consider a contracting universe in order to understand the approach to a spacelike singularity, either in a potential pre-bounce universe or in backward evolution of our expanding universe, enhanced gravitational collapse in a dense universe implies large inhomogeneity. For instance, black hole formation is expected to happen frequently when the matter distribution gets confined to smaller and smaller space. And when we approach the Planck density, which after all equals more than one trillion solar masses in a proton-sized region, it is hard to imagine how further collapse could keep macroscopic regions at a sufficiently homogeneous distribution for perturbation theory to remain available. If we use perturbative inhomogeneity around a minisuperspace model in order to study the approach to a spacelike singularity, we therefore have to keep adjusting the size V_0 of homogeneous regions used to set up the minisuperspace model by a process of infrared renormalization (Bojowald 2019). This process must end when we reach Planckian size, at which time the infrared scale V_0 would merge with the ultraviolet.

Second, if inflation took place in the early universe, it can be consistent with observations only if it lasted for a sufficiently long time, stretching space by 60 e-folds, or a factor of at least $e^{60} = 1.14 \times 10^{26}$. Moreover, inflation must happen sufficiently early on in the history of the universe because high-energy processes of matter formation, such as baryogenesis, are still to take place after inflation ends.

While it is possible to construct inflation models in which these two conditions are met and the Planck scale is avoided even at the beginning of inflation, inflation is an inherently unstable process which is not always easy to stop once it gets started. Without fine-tuned initial conditions, inflation generically lasts longer than 60 e-folds, and if it overshoots this target by only about ten e-folds, some modes of inhomogeneity that are relevant for current observations of the cosmic microwave background would have had a wave length less than the Planck length at the beginning of inflation. It might still be possible to describe these modes as perturbative ones, but their equations of motion would be subject to quantum spacetime effects that, in general, require a derivation beyond perturbation theory. This *trans-Planckian problem* shows that the inflation scenario can be used to constrain models of quantum cosmology and the underlying approaches to quantum gravity (Martin & Brandenberger 2001; Brandenberger & Martin 2001; Niemeyer 2001; Bedroya & Vafa 2019; Bedroya et al. 2020).

6.3.1 Infrared Renormalization

Considering a collapsing universe, the BKL scenario, described in Section 1.6, suggests a new role for homogeneous dynamics asymptotically close to a spacelike singularity. Even if collapse has led to structure formation on all length scales, the dynamics at each spatial point in this scenario resembles the dynamics of a spatially homogeneous cosmological model.

However, this reference to homogeneity is local and does not imply that the entire space of a minisuperspace model is obtained. If we approach high curvature in forward or backward evolution, we may use homogeneous models, perhaps combined with perturbative inhomogeneity, but the size of regions described by such an approximation must progressively be reduced in order to keep track of gravitational collapse. If we fix a spatial coordinate system at some initial time, a useful parameter to describe this process is the coordinate volume of the region in which perturbative inhomogeneity is admissible as an approximation. Any modes that stretch over distances greater than $\sqrt[3]{V_0}$ must be described by non-perturbative inhomogeneity, or by a full theory of quantum gravity; see Figure 6.5.

As the comoving scale of inhomogeneity decreases, the value of V_0 should be reduced progressively, in a process of infrared renormalization described in Section 4.1.3. This process cannot be performed in a traditional minisuperspace model which works with a fixed set of operators and wave functions in which V_0 (appearing in the basic commutators) cannot be varied. Infrared renormalization therefore requires a new kind of effective theory in quantum cosmology, for instance based on algebraic equations with running V_0. Additional parameters that are usually believed to be constants in minisuperspace models, such as the magnitude of holonomy modifications in loop quantum cosmology given by p_0 in Equation (4.57), run along with V_0 and can therefore show different behaviors in low and high curvature regimes.

Also quantum corrections generically depend on V_0. This behavior initially seems puzzling in minisuperspace models because V_0 refers to the coordinate size of an

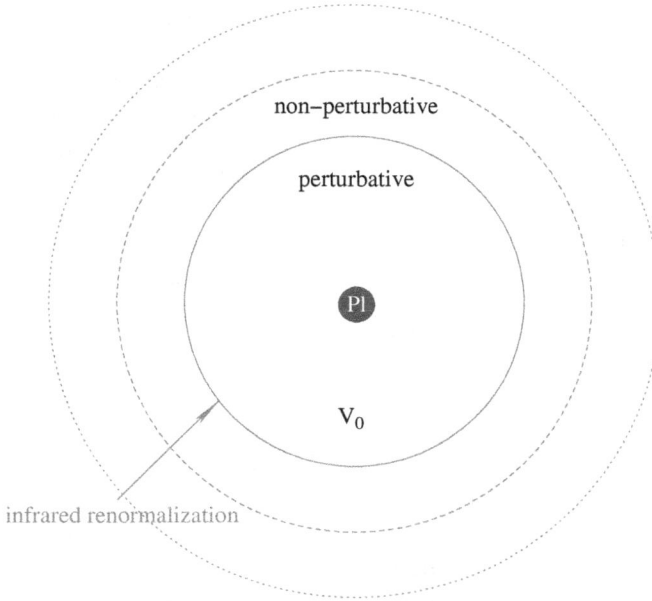

Figure 6.5. Infrared renormalization reduces the size V_0 of any region in which perturbative inhomogeneity can be used, tracking the scale of gravitational collapse in an approach to high curvature (red circles). A lower limit for the size is set by the Planck scale (blue circle).

arbitrary averaging region used to define homogeneous variables. And even if spatial coordinates have been fixed, the averaging region may be changed without affecting the classical dynamics. In classical cosmology, there is no obvious physical process that corresponds to changing V_0, suggesting that observables or their quantum corrections should be independent of V_0. However, quantum corrections generically depend on V_0 because it appears in the basic variables, in particular $V = V_0 a^3$, that define a quantum minisuperspace model. Uncertainty relations, such as

$$\Delta_\omega V \Delta_\omega H \geqslant \frac{2\pi G \hbar}{c^2} = 2\pi c \ell_{\rm Pl}^2 \tag{6.83}$$

in a state ω, based on the momentum $p_V = -H c^2/(4\pi G)$ of V according to Equation (1.16), by contrast, impose a lower bound given by fundamental parameters. Numerator and denominator in unitless relative fluctuations, $V^{-1}\Delta_\omega V$, must therefore depend differently on V_0 because the classical scaling, $V \propto V_0$ and H independent of V_0 realized for expectation values $\omega(\hat{V})\, \omega_\omega(\hat{H})$, cannot be applied to volume fluctuations, $\Delta_\omega V$ together with $\Delta_\omega H$, without violating the uncertainty bound.

If V_0 is macroscopic, which is a valid assumption at late times given large-scale homogeneity, relative volume fluctuations can be miniscule, implying largely classical behavior. For microscopic V_0, by contrast, large relative fluctuations and significant quantum effects are expected. Importantly, even the tiny Planckian value on the right-hand side of Equation (6.83) does not imply that fluctuation terms can always be ignored because the BKL scenario is asymptotic and does not provide any lower bound on the size of regions in which homogeneity may be assumed, close to a spacelike singularity.

From a more general perspective that includes the build-up of inhomogeneity, the value of V_0 does have physical meaning once spatial coordinates have been fixed. The time dependence of V_0 describes the varying scale of inhomogeneity, limiting the size of regions in which minisuperspace models can be used. If a quantum version of the BKL scenario is available, local quantum effects should depend only on the volume of the typical regions that may be considered homogeneous at any given time, but not on back-reaction effects from inhomogeneous modes on scales greater than $\sqrt[3]{V_0}$. The initially puzzling dependence of quantum corrections on V_0, seen in Section 4.1, therefore models the quantum interactions (or their absence) between different modes. As V_0 shrinks in a BKL-type scenario, more modes are pushed into the full inhomogeneous region and therefore, by the BKL assumption, no longer interact significantly with the homogeneous center.

A valid version of numerical quantum cosmology should include infrared renormalization by combining homogeneous or perturbative quantum cosmological models in the interior region of Figure 6.5 with numerical relativity in the exterior in order to track the scale of inhomogeneity. One would have to extract the typical value of V_0 from the exterior simulation and feed it into the running scale of the interior. At high curvature, a full exterior simulation of quantum gravity would have to be used. The complexity of such simulations has so far prevented detailed implementations. The approach to a spacelike singularity in minisuperspace-based quantum cosmology therefore remains unclear. This caveat applies in particular to bounce claims made in loop quantum cosmology.

The process of infrared renormalization, adjusting V_0 to the infrared scale of inhomogeneity, is valid as long as curvature does not grow too much. At Planckian curvature, the infrared scale V_0 of a minisuperspace or perturbed model is pushed into the ultraviolet, indicating the breakdown of the minisuperspace perspective even if a quantum version of the BKL scenario may be realized. At this scale, quantum cosmology in all its versions necessarily becomes highly speculative.

6.3.2 The Trans-Planckian Problem

The necessity of infrared renormalization implies that minisuperspace models, on their own, cannot provide a full picture of spacetime at large curvature. Such models eventually break down in an approach to a spacelike singularity because the infrared scale on which they are based will have to be pushed into the ultraviolet in order to maintain an approximation of spacetime by an evolving homogeneous region.

The inflation scenario, described in Section 1.4.4, implies the opposite transition, from ultraviolet to infrared, because modes with short, Planckian wavelength are physically expanded by a large factor in expanding spacetime, and may become relevant for late-time observations of the cosmic microwave background long after inflation has ended. The trans-Planckian problem (Martin & Brandenberger 2001; Brandenberger & Martin 2001; Niemeyer 2001) or, in a stronger version, the trans-Planckian censorship conjecture (Bedroya & Vafa 2019; Bedroya et al. 2020) states that such modes should in consistent inflation models be prevented from becoming relevant at late times, given the (at least at present) uncontrolled nature of strong quantum-gravity effects.

A prominent example of modifications that may be required if quantum gravity becomes necessary in early stages of inflationary structure formation is given by non-classical dispersion relations, which would change relationships such as Equation (1.90) that involve the frequency of inhomogeneous modes for a given wave number. The classical continuum relationship could, for instance, be affected by discrete spacetime (much like dispersion effects implied by solid-state structures), fractal dimensions as derived from causal dynamical triangulations (Mielczarek 2017), higher-curvature terms in covariant actions or in Hořava–Lifshitz gravity (2.143), or by modified spacetime structures in canonical descriptions based on Equation (2.279). In addition to the dynamics, quantum spacetime effects could also change the form of the inflaton ground state, and therefore alter the amplitude (1.91) which in the standard treatment appears directly in the power spectrum (1.98).

The trans-Planckian problem is based on the fact that approximation methods used to study quantum field theory in inflation models cannot be applied with a fixed scale. (This feature is conceptually similar to infrared renormalization.) The rapidly expanding background drags along modes of perturbative inhomogeneity. When the wave length of these modes gets very large, the mode simply follows the expansion of space and stops evolving on its own, as described by the Mukhanov–Sasaki Equation (1.88), until much later when the expansion is no longer inflationary. While such an evolution can well be described semiclassically, a full quantum treatment with varying numbers of evolving degrees of freedom is challenging (Weiss 1985). It would, however, be required if full quantum gravity becomes relevant for some of the modes. As seen in the context of infrared renormalization, minisuperspace-based models are therefore insufficient once Planck-scale physics enters the equations.

6.3.3 Speculative Proposals

Potential properties of various approaches to quantum gravity have been used in myriad ways to suggest new ingredients for the standard Big Bang model. The main targets of such attempts are a non-singular description of the Big Bang singularity of general relativity, and alternatives or modifications of the inflation scenario of structure formation.

Another dichotomy exists between the origin of proposals from a background-independent or background-dependent theory, respectively. From a technical view-point, background-dependent approaches are more suitable for cosmological phenomenology because one can directly apply quantum effects to background spacetimes relevant for cosmology. The derivation of such effects for non-static backgrounds may not always be as direct as for static backgrounds, but in phenomenological investigations it is often possible to test possible non-static generalizations of known effects in static backgrounds. Some examples, mainly from string theory, will be given below. For a comprehensive exposition, see Calcagni (2017).

Background-independent Approaches
In background-independent approaches, by contrast, one must first show that it is even possible to have any states that in some approximation may be considered continuum background spacetimes. This problem is often challenging, making it all the more difficult to extend quantum effects to cosmologically relevant spacetimes. As described in Section 5.2.3, the only non-perturbative approach in which suitable cosmological backgrounds have been shown to exist is given by causal dynamical triangulations. But the task of working out cosmological scenarios remains challenging even in this approach.

Loop quantum gravity is probably unique among non-perturbative approaches in that a large number of cosmological applications has been published in this field, mainly in the context of loop quantum cosmology. Initially, loop quantum cosmology drew interest after it was shown in minisuperspace models that the Big Bang singularity may be resolved because wave functions can be extended uniquely across zero scale factor (Bojowald 2001). Since then, however, it has been difficult to extend derivations of cosmological implications beyond the minisuperspace level. Many studies have worked with combinations of minisuperspace models and perturbative inhomogeneity that were only later found to violate covariance; see Section 6.2.4. The application of minisuperspace models helps one to avoid the main difficulties of background-independent approaches because the assumption of spatial homogeneity implies a rather strong background structure, even if it does not completely determine the metric. The over-reliance on minisuperspace models in cosmological applications of loop quantum gravity therefore means that the background-independent nature that characterizes this approach has not yet been tested.

An application of strict minisuperspace models or perturbations also makes it impossible to implement infrared renormalization and to address the trans-Planckian problem, as already discussed in this section. In fact, most claims made in loop quantum cosmology are based on models that fail to implement basic properties of effective field theory (Bojowald 2019, 2020), such as the necessity of scale-dependent coupling parameters (or quantum corrections) because different energy scales usually activate different numbers of relevant degrees of freedom. Most derivations in loop quantum cosmology assume that a single effective theory without running parameters can (or even has to) be used at late times and at very early times (or well outside and deep inside black holes). Because of challenges in deriving such parameters from a full background-independent theory, such an assumption is completely unsupported, but it easily affects potential outcomes. While many very specific claims have been made in loop quantum cosmology, they are hardly reliable. Only qualitative features can at best be taken as a possible indication as to how high curvature might be approached in this framework (Bojowald 2015, 2020). The sensitivity to quantization ambiguities in this approach has already been discussed in Section 4.3.4.

There has been some interest in minisuperspace models in loop quantum cosmology because proposed modifications, such as the application of holonomy terms in Equation (4.57), can be rather drastic compared with the equations of general relativity. Such strong effects are necessary if one attempts to evade the conclusion, based on considerations of units, that quantum-gravity effects in observational cosmology should be suppressed by tiny factors given by the ratios of relevant scales: the Planck length divided by the typical length scale that appears in the physics of the cosmic microwave background. (Such dimensional expectations are also borne out in detail in canonical quantum gravity without loop modifications; Pinho & Pinto-Neto 2007; Falciano & Pinto-Neto 2009; Kiefer & Kraemer 2012.) However, the same effects that could imply significant implications in power spectra also have unexpected consequences in spacetime structures consistent with the underlying modifications, as shown in Section 6.2. In this context, we see a crucial implication of a background-independent approach: Since such an approach does not assume what spacetime structure it might realize, derivations of spacetime structures can potentially lead to counter-intuitive effects. These effects still remain to be understood in loop quantum gravity.

Background-dependent Approaches

Compared with general relativity coupled to standard matter ingredients, string theory provides many additional structures related to extra dimensions with non-trivial topology and supersymmetry that can be used to construct various new ingredients for cosmological phenomenology. In most cases, the same fields and interactions could be postulated as exotic matter terms in general relativity, but additional motivation from string theory can often provide additional support.

As usual with string theory, one of the main difficulties is to find suitable ingredients in the hay stack of degrees of freedom provided by compactified extra dimensions and supersymmetry. Once suitable ingredients have been identified, one must then show that they are stable with respect to cosmological dynamics. For instance, winding modes around compact extra dimensions are usually massive and, just based on the available energy, could therefore decay into lighter excitations such as momentum modes. Demonstrating stability or at least metastability requires an understanding of string interactions in a non-static background, which is often hard to achieve with some degree of rigor.

In this phenomenological spirit, string theory has made contributions to the two main branches of early-universe cosmology: identifying ingredients that can support inflation, and suggesting alternative scenarios that do not rely on inflation but can nevertheless explain observed properties of the nearly scale-free density power spectrum.

Inflation in String Theory: For reasons related to the presence of supersymmetry, string theory does not provide an obvious way to include a positive cosmological constant, or a field with a slightly varying positive potential that could play the role of an inflaton. Nevertheless, as shown in Kachru et al. (2003), it may be possible to do so with flux compactifications, given by higher-dimensional objects made out of interacting strings and winding around compact extra dimensions. In this context, one can address some of the stability questions to make it conceivable that this kind of inflation can last sufficiently long.

Another example, so-called Dirac–Born–Infeld theories, can be suggested for certain scalar fields in string theory (Silverstein & Tong 2004; Alishahiha et al. 2004). Any theory of this form is characterized by an action with a non-linear dependence on the usual kinetic energy of a scalar field, $X = -\frac{1}{2}\sum_{\alpha=0}^{3}\sum_{\beta=0}^{3} g^{\alpha\beta}(\partial\phi/\partial x^{\alpha})(\partial\phi/\partial x^{\beta})$, the same term (2.124) used in the definition of Horndeski theories (2.123) in Section 2.2.5. The Dirac–Born–Infeld version suggested by string theory is indeed of the general form (2.123), but with a more specific

$$H_1(\phi, X) = T(\phi)\left(1 - \sqrt{1 - 2\frac{X}{T(\phi)}}\right) - V(\phi) \tag{6.84}$$

as the only non-zero contribution, with free functions $T(\phi)$ and $V(\phi)$ depending only on the scalar field but not on its derivatives. For small $X/T(\phi)$, $H_1(\phi, X) \sim X - V(\phi)$ equals the standard Lagrangian of a scalar field, but the non-linear dependence on X for larger $X/T(\phi)$ can support inflation with weaker conditions on the potential than would be required for slow-roll inflation with an H_1 linear in X.

Alternatives to Inflation: Various ingredients from string theory have been used to suggest alternatives to inflation that could explain observational features such as a nearly scale-free power spectrum of the cosmic microwave background. Ekpyrotic and cyclic models (Khoury et al. 2001; Steinhardt & Turok 2002; Khoury et al. 2004) are based on the observation that the requirement of having an inflaton fluid with negative P such that $P \approx -\rho$, used in inflation, could be replaced by a stiff effective matter ingredient characterized by positive and large pressure, $P \gg \rho$. Certain perturbation modes with nearly scale-free power spectra can then be obtained. However, they are usually produced in a pre-expansion phase of cosmic collapse, and therefore require some well-defined replacement of the Big Bang singularity. This phase, in a fully dynamical context, remains hard to control in string theory.

The proposal of string-gas cosmology (Brandenberger & Vafa 1989; Kripfganz & Perlt 1988) replaces the dynamical era that is often used to generate a scale-free power spectrum, be it exponential expansion or an ekpyrotic transition, with a stationary equilibrium phase. In the presence of compact extra dimensions in a background spacetime of an ensemble of strings, one can expect a phase transition to take place at high temperature when winding modes become energetically accessible. If the temperature is sufficiently high to excite winding modes, an increased supply of energy no longer leads to enhanced thermal motion of strings described by momentum modes. The reduced thermal motion indicates that the temperature remains constant for some time even as more energy is supplied, just as it happens at well-known phase transitions such as evaporating water.

String-gas cosmology postulates that inflationary expansion can be replaced with a string-gas phase transition, starting with a high-temperature state at which many winding modes are excited. As these modes decay slowly, density perturbations are generated which, unlike in inflation where they originate from quantum fluctuations, are of thermal origin. Equilibrium thermodynamics can then be used to compute the expected power spectrum, which can indeed by nearly scale-free (Nayeri et al. 2006; Brandenberger et al. 2007). A largely open question in this context is how to combine string-gas cosmology, or any other scenario based on winding modes or T-duality such as pre-Big Bang cosmology (Veneziano 1991, 1999; Gasperini & Veneziano 1993, 2003), with manifestly T-dual spacetime structures described by double field theory where the metric $g_{\alpha\beta}$ is replaced by a larger matrix (5.131), in particular if non-associative spacetime features are realized as suggested by Equation (5.142); see Section 5.2.5.

Early-universe Cosmology without Planck-scale Physics

Given the difficulties encountered when trying to achieve rigorous control on Planck or string-scale physics in any approach to quantum gravity, it is of some interest to see whether non-singular scenarios of the early universe can be devised by completely avoiding the Planck scale. An example based on classical physics is the emergent universe (Ellis & Maartens 2004; Ellis et al. 2004), which starts with an initial configuration given by Einstein's static universe (Einstein 1917). A static isotropic solution with vanishing Hubble parameter and acceleration can be obtained in general relativity by balancing the cosmological constant, spatial curvature, and the matter density. Depending on the value of the cosmological constant at early times, the matter density can be well below the Planck density. Moreover, since Einstein's universe turns out to be unstable with respect to anisotropic perturbations, any sub-Planckian quantum fluctuation implies that it will eventually leave the static phase and could turn into an expanding universe.

The Hartle–Hawking proposal (Hartle & Hawking 1983) for an initial state of the universe closed off in Euclidean space is another example of an early-universe scenario that could avoid detailed Planck-scale physics; see Section 4.4.2. Here again, it is the cosmological constant that determines the curvature or density scale of the initial state and may be chosen to be sufficiently sub-Planckian. However, as discussed in more detail in Section 4.4.3, this proposal is unstable with respect to inhomogeneous perturbations (Feldbrugge et al. 2017, 2018; Di Tucci & Lehners 2018). Current proposals to solve the instability problem include quantum-gravity effects closer to the Planck scale, such as dynamical signature change (Bojowald & Brahma 2018), or special initial conditions with non-zero momenta (Vilenkin & Yamada 2018, 2019; Di Tucci & Lehners 2019) that may require new physics.

Finally, matter properties or higher-curvature terms in an effective action of gravity could avoid the Big Bang singularity by a bouncing solution that stays away from Planckian curvature (Wands 1999, 2009; Finelli & Brandenberger 2002; Battefeld & Peter 2015). It may also be possible to achieve such a bounce using quantum fluctuations without requiring detailed Planck-scale physics (Peter & Pinto-Neto 2008). In such scenarios, it is possible to obtain nearly scale-free power spectra from fluctuations seeded in the collapse phase. Open problems are related to the genericness of achieving bouncing behavior, and whether growing anisotropic shear in the collapse phase, (1.46), can be sufficiently controlled so as to ensure scale-free power spectra for a large set of initial values.

References

Agulló, I., Ashtekar, A., & Nelson, W. 2013, PhRvD, 87, 043507

Alishahiha, M., Silverstein, E., & Tong, D. 2004, PhRvD, 70, 123505

Alty, L. J., & Fewster, C. J. 1996, CQGra, 13, 1129

Ambjørn, J., Coumbe, D. N., Gizbert-Studnicki, J., & Jurkiewicz, J. 2015, JHEP, 08, 033

Aruga, D., Ben Achour, J., & Noui, K. 2019, arXiv:1912.02459

Ashtekar, A., Kaminski, W., & Lewandowski, J. 2009, PhRvD, 79, 064030

Ashtekar, A., Olmedo, J., & Singh, P. 2018, PhRvL, 121, 241301

Barrau, A., Bojowald, M., Calcagni, G., Grain, J., & Kagan, M. 2015, JCAP, 05, 051

Battefeld, D., & Peter, P. 2015, PhR, 571, 1

Bedroya, A., Brandenberger, R. H., Loverde, M., & Vafa, C. 2020, PhRvD, 101, 103502

Bedroya, A., & Vafa, C. 2019, arXiv:1909.11063

Ben Achour, J., Lamy, F., Liu, H., & Noui, K. 2018a, JCAP, 05, 072

Ben Achour, J., Lamy, F., Liu, H., & Noui, K. 2018b, EL, 123, 20006

Bojowald, M. 2001, PhRvL, 86, 5227

Bojowald, M. 2010, Canonical Gravity and Applications: Cosmology, Black Holes, and Quantum Gravity (Cambridge: Cambridge University Press)

Bojowald, M. 2015, FrP, 3, 33

Bojowald, M. 2015, RPPh, 78, 023901

Bojowald, M. 2019, Univ, 5, 44

Bojowald, M. 2019, JCAP, 01, 026

Bojowald, M. 2020, Univ, 6, 36

Bojowald, M. 2020, PhRvD, 102, 023532

Bojowald, M., & Brahma, S. 2018, PhRvL, 121, 201301

Bojowald, M., Brahma, S., Ding, D., & Ronco, M. 2020, PhRvD, 101, 026001

Bojowald, M., Brahma, S., & Reyes, J. D. 2015, PhRvD, 92, 045043

Bojowald, M., Brahma, S., & Yeom, D.-H. 2018, PhRvD, 98, 046015

Bojowald, M., Hossain, G., Kagan, M., & Shankaranarayanan, S. 2008, PhRvD, 78, 063547

Bojowald, M., & Mielczarek, J. 2015, JCAP, 08, 052

Bojowald, M., & Paily, G. M. 2012, PhRvD, 86, 104018

Bouhmadi-López, M., Brahma, S., Chen, C.-Y., Chen, P., & Yeom, D.-H. 2019, arXiv:1911.02129

Brandenberger, R. H., & Martin, J. 2001, MPLA, 16, 999

Brandenberger, R. H., Nayeri, A., Patil, S. P., & Vafa, C. 2007, IJMPA, 22, 3621

Brandenberger, R. H., & Vafa, C. 1989, NuPhB, 316, 391

Cailleteau, T., Linsefors, L., & Barrau, A. 2014, CQGra, 31, 125011

Cailleteau, T., Mielczarek, J., Barrau, A., & Grain, J. 2012, CQGra, 29, 095010

Calcagni, G. 2017, Classical and Quantum Cosmology (Berlin: Springer)

Chaney, A., Lu, L., & Stern, A. 2015, PhRvD, 92, 064021

Chaney, A., & Stern, A. 2017, PhRvD, 95, 046001

Cuttell, R. 2019, PhD thesis, King's College London

Cuttell, R., & Sakellariadou, M. 2014, PhRvD, 90, 104026

Dapor, A., Lewandowski, J., & Puchta, J. 2013, PhRvD, 87, 104038

Di Tucci, A., & Lehners, J.-L. 2018, PhRvD, 98, 103506

Di Tucci, A., & Lehners, J.-L. 2019, PhRvL, 122, 201302

Dray, T., Ellis, G. F. R., Hellaby, C., & Manogue, C. A. 1997, GReGr, 29, 591

Dray, T., Manogue, C. A., & Tucker, R. W. 1995, CQGra, 12, 2767

Einstein, A. 1917, Kosmologische Betrachtungen zur allgemeinen Relativitätstheorie (Berlin: Sitzungsberichte) 142

Ellis, G. F. R., & Maartens, R. 2004, CQGra, 21, 223

Ellis, G. F. R., Murugan, J., & Tsagas, C. G. 2004, CQGra, 21, 233

Ellis, G. F. R., Sumeruk, A., Coule, D., & Hellaby, C. 1992, CQGra, 9, 1535

Falciano, F. T., & Pinto-Neto, N. 2009, PhRvD, 79, 023507

Feldbrugge, J., Lehners, J.-L., & Turok, N. 2017, PhRvL, 119, 171301

Feldbrugge, J., Lehners, J.-L., & Turok, N. 2018, PhRvD, 97, 023509

Finelli, F., & Brandenberger, R. H. 2002, PhRvD, 65, 103522

Gambini, R., & Pullin, J. 2013, PhRvL, 110, 211301

Gasperini, M., & Veneziano, G. 1993, APh, 1, 317

Gasperini, M., & Veneziano, G. 2003, PhR, 373, 1

Hartle, J. B., & Hawking, S. W. 1983, PhRvD, 28, 2960

Hayward, S. A. 1993, arXiv:gr-qc/9303034

Horowitz, G. T., & Maldacena, J. M. 2004, JHEP, 0402, 008

Kachru, S., Kallosh, R., Linde, A., & Trivedi, S. P. 2003, PhRvD, 68, 046005

Kantowski, R., & Sachs, R. K. 1966, JMaPh, 7, 443

Khoury, J., Ovrut, B. A., Steinhardt, P. J., & Turok, N. 2001, PhRvD, 64, 123522

Khoury, J., Steinhardt, P. J., & Turok, N. 2004, PhRvL, 92, 031302

Kiefer, C., & Kraemer, M. 2012, PhRvL, 108, 021301

Kripfganz, J., & Perlt, H. 1988, CQGra, 5, 453

Kuchař, K. V. 1976a, JMaPh, 17, 777

Kuchař, K. V. 1976b, JMaPh, 17, 792

Kuchař, K. V. 1976c, JMaPh, 17, 801

Martin, J. 1994, PhRvD, 49, 5086

Martin, J., & Brandenberger, R. H. 2001, PhRvD, 63, 123501

Martín-Benito, M., Garay, L. J., & Mena Marugán, G. A. 2008, PhRvD, 78, 083516

Mielczarek, J. 2014, Springer Proceedings in Physics, Vol. 157 (Berlin: Springer), 555

Mielczarek, J. 2017, EL, 119, 60003

Nayeri, A., Brandenberger, R. H., & Vafa, C. 2006, PhRvL, 97, 021302

Niemeyer, J. C. 2001, PhRvD, 63, 123502

Perry, M. J., & Teo, E. 1993, PhRvL, 70, 2669

Peter, P., & Pinto-Neto, N. 2008, PhRvD, 78, 063506

Pinho, E. J. C., & Pinto-Neto, N. 2007, PhRvD, 76, 023506

Ripley, J. L., & Pretorius, F. 2019, CQGra, 36, 134001

Ripley, J. L., & Pretorius, F. 2020, CQGra, 37, 155003

Silverstein, E., & Tong, D. 2004, PhRvD, 70, 103505

Steinacker, H. 2018, JHEP, 02, 033

Steinhardt, P. J., & Turok, N. 2002, PhRvD, 65, 126003

Stern, A., & Xu, C. 2018, PhRvD, 98, 086015

Thiemann, T. 1998, CQGra, 15, 1281

Tricomi, F. G. 1968, Repertorium der Theorie der Differentialgleichungen (Berlin: Springer)

Vafa, C. 2005, arXiv:hep-th/0509212

Veneziano, G. 1991, PhLB, 265, 287

Veneziano, G. 2000, in The Primordial Universe—L'univers primordial ed. P. Binétruy, R. Schaeffer, J. Silk, & F. David Les Houches—Ecole d'Ete de Physique Theorique, Vol. 71 (Berlin-Heidelberg: Springer)

Vilenkin, A., & Yamada, M. 2018, PhRvD, 98, 066003

Vilenkin, A., & Yamada, M. 2019, PhRvD, 99, 066010

Wands, D. 1999, PhRvD, 60, 023507

Wands, D. 2009, ASL, 2, 194

Weiss, N. 1985, PhRvD, 32, 3228

www.ingramcontent.com/pod-product-compliance
Lightning Source LLC
Chambersburg PA
CBHW082136210326
41599CB00031B/5999